Leading cases in food law: EU and the Netherlands

European Institute for Food Law

Series editor

Bernd van der Meulen

VOLUME 17

The titles published in this series are listed at brill.com/eifl

Leading cases in food law: EU and the Netherlands

Cases selected by the Dutch Food Law Association

Edited by

prof. dr. Oswald Jansen
and
prof. dr. Bernd van der Meulen

BRILL | WAGENINGEN ACADEMIC

The Library of Congress Cataloging-in-Publication Data is available online at https://catalog.loc.gov
LC record available at https://lccn.loc.gov/

Typeface for the Latin, Greek, and Cyrillic scripts: "Brill". See and download: brill.com/brill-typeface.

ISSN 1871-3483
ISBN 978-90-04-74355-7 (hardback)
ISBN 978-90-04-76109-4 (e-book)
DOI 10.1163/9789004761094

Library of Congress Control Number: 2026933493

This book is printed on acid-free paper and produced in a sustainable manner.

Contents

Preface

EU food law is formed and informed by its sources. Among the sources of food law legislation and case law (jurisprudence) stand out, both at the EU level and at the level of Member States.

This book sets out the role case law plays in the shaping of food law. It presents landmark rulings from the Court of Justice of the European Union along with applications in one Member State, the Netherlands. The application of EU food law by courts and dispute settlement bodies in the Netherlands is not more or less important than applications in any other Member State. What matters is the interplay between the common EU level and the level of the Member States. For this, the Netherlands are as good an example as any. Readers will be able to connect the experience in their own home country to the experiences in the Netherlands as set out in this book.

The choice to address the Member States' level from the perspective of the Netherlands stems from the history of this book. Originally, the initiative to its creation was taken by the Dutch Food Law Association (Nederlandse Vereniging voor Levensmiddelenrecht – NVLR). For the celebration of its 20th anniversary, NVLR prepared a book in Dutch discussing the 20 most important rulings from the CJEU of that period, the 20 most important rulings from national courts and the 20 most important decisions from the self-regulatory advertisement board. This book was received so well that it was decided it should not be kept for only Dutch readers. Therefore we now present the current book in English. It is not a one-on-one translation of the Dutch book, but rather an adaptation and update for an international audience. By disclosing these food law sources and experiences to an international public of English readers we not only hope to encourage the exchange of national experiences with food law but also to contribute to the development of comparative food law.

Court rulings are provided online and can easily be found with the help of their ECLIs. With the available technology, language no longer needs to be a barrier to understanding what they are about. Emphasis in this book, therefore, is not on providing the texts of the rulings but rather on assessing their relevance for the wider audience in food law. Although we kept the original format of three times 20 chapters, most chapters address more than one ruling so the scope of the book is much wider than 60 cases.

This book is dedicated to Kenneth Defares. His last publication is included here as chapter 12. Kenneth was an attorney specialised in competition law and regulated markets, including – of course – the food sector. His office hosted the founding ceremony of NVLR on World Food Day 2003. He remained a member of the board of NVLR responsible for the secretariat until he passed unexpectedly on the 9th of

February 2024. He was a culinary magician, inspired consumer of excellent food, a feared adversary, respected colleague and dear friend. We miss him dearly.

Oswald Jansen and Bernd van der Meulen

Abbreviations used in this book

μg	microgram (equal to one millionth (1×10^{-6}) of a gram)
ABRvS	Administrative Jurisdiction Division of the Council of State
ACM	Netherlands Authority for Consumers and Markets
AG	Advocate General
Awb	Dutch General Administrative Law Act
BSE	Bovine Spongiform Encephalopathy (aka Mad Cow's Disease)
BuRO	Bureau for Risk Assessment & Research
CBb	Dutch Trade and Industry Appeals Tribunal
CBG	Dutch Medicines Agency
CBS	Central Bureau of Statistics
CvB	Board of Appeal (of the Advertising Code Committee)
CCMO	Central Committee on Research Involving Human Subjects
CJEC	Court of Justice of the European Communities (prior to Treaty of Lisbon)
CJEU	Court of Justice of the European Union (formerly CJEC)
CRvB	Central Appeals Council
CVZ	Code of Confectionery
DNA	deoxyribonucleic acid (the molecule that carries genetic instructions for all living organisms and some viruses)
DSB	Dispute Settlement Body (WTO)
ECOSOC	Economic and Social Council
EFSA	European Food Safety Authority
ETO	Ethylene oxide
FEBA	European Food Banks Federation
FIC	Food information to consumers regulation
GFL	General Food Law
GMOs	Genetically modified organisms
Gnw	Dutch Medicines Act
GPSD	General Product Safety Directive (Directive 2001/95/EC)
IER	*Intellectuele Eigendom en Reclamerecht* (intellectual property and advertising law) journal
IGJ	Health and Youth Care Inspectorate
IGZ	Health Care Inspectorate
ICESCR	International Covenant on Economic, Social and Cultural Rights
UNCRC	United Nations Convention on the Rights of the Child
MDMA	3,4-Methylenedioxymethamphetamine or XTC (in tablet form)
MOE	Margin of exposure
MPA	Medroxyprogesterone acetate (synthetic form of hormone progesterone)

NGOs	Non-governmental organisations
NVWA	Dutch Food and Consumer Product Safety Authority
Ppb	Parts per billion
Rb	District court
RCC	Dutch Advertising Code Committee
RIVM	National Institute for Public Health and the Environment
SCoFCAH	Former Standing Committee on the Food Chain and Animal Health
SCoPAFF	Standing Committee on Plants, Animals, Food and Feed
SCP	Social Cultural Planning Office
SPS	WTO Agreement on the Application of Sanitary and Phytosanitary Measures
TFEU	Treaty on the Functioning of the European Union
TVC	TV advertisement
VWS	Dutch Ministry of Public Health, Welfare and Sport
WD	Animal Act
Wm	Environmental Management Act
Wmo	Social Support Act
Wob	Former Freedom of Information Act (now Woo)
Woo	Open Government Act

PART I

European case-law

CHAPTER 1

Malagutti – No liability for RASFF damage

Annotated by Blanka Vitéz

Court of First Instance of 10 March 2004, ECLI:EU:T:2004:72

RASFF, compensation, precautionary principle.

Main considerations:

> If any doubt remains in that regard, it should be pointed out that under the precautionary principle prevailing in the matter of the protection of public health the competent authority may be obliged to take appropriate measures to prevent certain potential risks for public health without having to wait until the existence and seriousness of those risks has been fully demonstrated (see, to this effect, case T13/99 *Pfizer Animal Health* v *Council* [2002] ECR II-3305, paragraph 139 and the case-law cited, and Case T-392/02 *Solvay Pharmaceuticals* v *Council* [2003] ECR II-0000, paragraphs 121 and 122). If it was necessary to wait until all the research was completed before adopting such measures the precautionary principle would be rendered devoid of purpose (*Pfizer Animal Health* v *Council*, cited above, paragraphs 142, 386 and 387). That reasoning also applies in the case of a rapid information procedure such as the one introduced by the directive. The applicant, which is a victim of that alert system introduced in order to protect human health, must accept its adverse economic consequences, since the protection of public health must take precedence over economic considerations (*Solvay Pharmaceuticals* v *Council*, paragraph 121, and *Pfizer Animal Health* v *Council*, paragraph 456).

1.1 The case

The Rapid Alert System for Food and Feed (RASFF) has been ensuring exchanges of alerts on food since 1979 and feed since 2002. RASFF facilitates efficient correspondence between Member States[1] in cases where public health in the European Union (EU) may be at stake.

Malagutti-Vezinhet SA (Malagutti), a French company that exported fresh fruit[2], believes it has been adversely affected by this system.

1 Every Member State appoints a national body as contact point: in the Netherlands, this is the *Nederlandse Voedsel- en Warenautoriteit* (NVWA) or Netherlands Food and Consumer Product Safety Authority.

2 Malagutti has since ceased trading.

 | HTTPS://DOI.ORG/10.1163/9789004751699-002

On 4 September 2001, Iceland (part of the European Economic Area) withdrew a batch of apples from the market. These apples contained 0.8 mg/kg of dicofol residues – substantially more than the maximum permitted level of 0.02 mg/kg. Iceland notified the European Commission (Commission) on 6 September of its decision to do so. The Commission forwarded these, and subsequent, notifications to all RASFF members.

Notices distributed through RASFF indicated that the apples originated from France and had reached Iceland via the Netherlands. One message from the Netherlands identified Malagutti as supplier of the apples.

Following these alerts, two British agencies warned of the danger of dicofol and advised against importing Malagutti products. As a result, all trade between Malagutti and its customers in Britain was halted.

On 19 September 2001, the French authorities took samples of the same category of apples that had been rejected in Iceland. The results – no dicofol residues detected – were transmitted to the Commission. The Commission informed RASFF members accordingly. Around the same time, Malagutti disputed the validity of the messages exchanged with the Commission, stating that it had never exported apples to Iceland. Although the origin of the apples was not established before the Court, several accounts showed that Malagutti had sold apples in August 2001 to one of the parties named as the exporter of the withdrawn apples.

A month later, Malagutti unsuccessfully sought damages from the Commission. Malagutti appealed to the General Court against the Commission's rejection decision.

1.2 Before the General Court

European Union non contractual liability is subject to the following conditions:

1. unlawful conduct attributable to the institution,
2. actual damage, and
3. a causal link between the alleged conduct and the damage.

Malagutti presented several arguments in support of its claim that the Commission acted unlawfully. The Commission failed to consult Malagutti before disseminating the notice naming Malagutti as a supplier. According to Malagutti, the Commission was obliged to do so under Article 7(2) of Directive 92/59/EEC on general product safety (Directive 92/59).[3] The first paragraph of Article 7 requires Member States to notify the Commission when products are withdrawn or otherwise restricted; the

3 Directive 92/59/EEC of the Council of 29 June 1992 on general product safety, Official Journal L 228, 11.8.1992, p. 24–32

second paragraph states that the Commission shall consult with interested parties as soon as possible after such notification.

Malagutti also said that the Commission should have considered whether the measure taken by Iceland – withdrawing the apples from the market – complied with the principle of proportionality.

In addition, Malagutti stressed that there was no evidence of the exact origin of the apples inspected. In this regard, Malagutti pointed out that another French company was initially named in Iceland's messages.

1.3 Judgment of the General Court

The General Court held that there was no unlawful conduct by the Commission.

In this context, the Court pointed out that while Article 7(2) of Directive 92/59 requires the Commission to consult with the parties concerned, this was not relevant in this case as it did not apply to the RASFF.

Instead, Directive 92/59 related to two separate systems:

1. a system covering the actions of the Member States so that they could take the necessary measures to ensure that products placed on the market were safe, and
2. RASFF, which dealt with urgent situations at community level in cases of serious and immediate threat to the health and safety of consumers.

The form Iceland had used showed that it had notified the Commission within the context of the RASFF system. Any liability, the General Court ruled, could therefore only relate to that which the Commission had to assume under the RASFF framework.

The Court observed that under the RASFF framework, the Commission's obligations were limited to verifying whether the information provided was within scope. RASFF applies exclusively in situations where a product presents a serious and immediate risk to consumer health and safety and the effects of that risk (may) extend beyond the territory of the Member State concerned. Responsibility for the accuracy of this information lies with national authorities, not with the Commission whose role it is to disseminate the notifications. The Court therefore found no illegality on the part of the Commission.

Case closed, one might conclude: no unlawful conduct on the part of the Commission. Interestingly, however, the Court invoked the precautionary principle in an apparent attempt to also preclude the possibility of any national liability. In doing so it stated: '[a]pplicant, which is a victim of that alert system introduced in order to protect human health, must accept its adverse economic consequences, since the protection of public health must take precedence over economic considerations.'

1.4 RASFF notification and possible legal actions

Directive 92/59 is no longer in force. The EU has adopted general laws on product safety[4] and separate regulations for, among others, RASFF: Regulation 178/2002 (General Food Law or GFL)[5] and Regulation 2019/1715 (Information Management System for Official Controls or IMSOC).[6]

These new regulations provide rules on product traceability. For instance, Article 18 GFL obliges 'food business operators' to ensure that their products are traceable at all stages of production, processing and distribution. Scope for uncertainty about the exact origins of food products, such as apples, is thus reduced.

Article 24 of the IMSOC Regulation sets out the Commission's obligations within the framework of the RASFF system with regards to alert notices. In addition to checking whether notifications fall within the scope of the RASFF system, they must also be verified against several criteria. However, none of these criteria relate to the accuracy of the content of the notifications.

A RASFF notification therefore remains a potential nightmare for companies as they have no influence on the content. Designated authorities – in the Netherlands the Food and Consumer Product Safety Authority (*NVWA*) – could therefore inadvertently transmit incorrect or incomplete information.

When a company faces a RASFF notification concerning one or more of its products, it is important to consider what steps it can take.

Revoke and/or amend RASFF message

The first option is to engage directly with the Netherlands Food and Consumer Product Safety Authority (*NVWA*). Once the NVWA announces its intention to issue a RASFF notification or has already done so, a company may make its concerns known to the NVWA. The company could ask the NVWA not to post a notification, or to withdraw or amend a posted notification. Such requests must of course be

supported by compelling evidence: either demonstrating that a RASFF notification is not necessary, or that amendments are required to accurately reflect the situation.

4 Regulation (EU) 2023/988 of the European Parliament and of the Council of 10 May 2023 on general product safety, Official Journal L 135, 23.5.2023, p. 1-51.

5 Regulation (EC) No 178/2002 of the European Parliament and the Council of 28 January 2002 laying down the general principles and requirements of food law, establishing the European Food Safety Authority and laying down procedures in matters of food safety, Official Journal L 31, 1.2.2002, p. 1–24. As a result of the food crises, this regulation also added feedstuffs to the RASFF system. The rapid alert system as it existed in the context of Directive 92/59 did not concern feedstuffs.

6 Commission Implementing Regulation (EU) 2019/1715 of 30 September 2019 laying down rules for the functioning of the information management system for official controls and its system components (the IMSOC Regulation), Official Journal L 261, 14.10.2019, p. 37–96.

However, the NVWA is free to disregard a company's requests and may very well do so in the interest of speed: the NVWA is obliged under Article 50 GFL to immediately notify the Commission in cases of serious direct and indirect risks to human health. It is therefore important to submit conclusive evidence to the NVWA that no such risk exists. After all, risks are not certainties, as acknowledged by the definition of risk contained in the GFL: 'a function of the probability of an adverse health effect and the severity of that effect, consequential to a hazard'. This definition allows the NVWA a fair degree of discretion in its risk assessments.

If liaising with the NVWA does not lead to desired results, a company may consider going to court.

Case-law shows that appealing to the administrative court against an NVWA RASFF notification generally offers little relief. It is established case-law in the Netherlands that RASFF notifications are not intended to produce legal effects.[7] For this reason, they are not a decision within the meaning of the Dutch General Administrative Law Act and the administrative court does not have jurisdiction to adjudicate. This is different in, for example, the (rare) case where the NVWA imposes a 'recall' by decree.

A company is, however, welcome to act against a RASFF notification in the civil courts. Bringing summary proceedings is the obvious way to do this, due to the relative speed of these proceedings. Nevertheless, having a (forthcoming) RASFF notification withdrawn or amended is also a difficult ask in the civil court. Health and safety are thorny issues.

To date, therefore, there are no known published Dutch civil judgments in which a company has brought a successful action to have a RASFF notification amended or withdrawn.[8]

Damages

On top of this, there is a high probability that the damage has often already been suffered before a court judgment. It is therefore more interesting to assess whether the NVWA or the European Commission can be sued for damages.

As follows from Malagutti, the Commission is not responsible for the (accuracy

---

7 See, for example, the Court of Rotterdam, judgment of 21 July 2016, ECLI:NL:RBROT:2016:5622, Trade and Industry Appeals Tribunal (CBb), judgment of 23 August 2012, ECLI:NL:CBB:2012:BX6798 and CBb, judgment of 3 July 2018, ECLI:NL:CBB:2018:319.

8 For a discussion from a Spanish perspective of the inherent options for acquiring legal protection concerning food safety alerts and RASFF notifications, see; Vicente Rodrigues Fuentes, Food alters and RASFF, Chapter 9 in Altinay Urazbaeve e.a. *The functional field of food law. Reconciling the market to human rights*, Wageningen Academic Publishers 2019, pp. 277–294. For an Italian and Dutch perspective on this matter, among others, see: A.A. Corini e.a., *Enforcement of EU food law*, Chapter 8 in: M. Scholten and M. Luchtman, Law Enforcement by EU Authorities. Implications for Political and Judicial Accountability, EE Elgar, 2017, pp. 195–220.

of the) content of RASFF notifications. Only if it is clear from the content of the notification that it is not suitable for RASFF – and thus dissemination is not necessary – is it worth a go to request compensation from the Commission. Article 24 of the IMSOC Regulation – which did not exist at the time of Malagutti – can be of assistance here. For a successful request, the conditions mentioned earlier must be met: the Commission must have acted unlawfully; the company must have suffered actual damage; and there must be a causal link between the alleged conduct and the damage suffered.

For damages suffered due to the content of a RASFF notification (inaccuracy, incompleteness, and so on), a company would do better to investigate whether the NVWA can be held liable. A suitable basis for this is Section 6:162 of the Dutch Civil Code, tort. For this, a company must prove:

(i) a tortious act,
(ii) damage,
(iii) attributability of the tort to the Netherlands Food and Consumer Product Safety Authority (*NVWA*),
(iv) a causal link between the tort and the damage suffered.[9]

In this context, the relativity requirement of Article 6:163 of the Dutch Civil Code is also of relevance. This provision sets out that where a tort consists of a norm violation, there is no obligation to pay for damages if the violated norm was not meant to protect against the type of harm suffered.

Accordingly, a successful action against or in response to a RASFF notification is far from guaranteed – unless one can demonstrate that the NVWA acted with obvious carelessness.

Blanka Vitéz
Legal Adviser, Competition and Markets Authority, UK[10]

9 For info on the subject of administrative compensation/ damage compensation for unlawful government action, see Chapter 6 of this book.

10 Views expressed are solely my own.

CHAPTER 2

Kellogg's – Food enrichment should only be restricted on scientific grounds

Annotated by Imkje Tiesinga

Court of Justice of the European Union (CJEU) 2 December 2004, ECLI:EU:C:2004:762 (*Commission* v *Netherlands*)

Foods to which vitamins or minerals have been added, national regulation, measures having equivalent effect to an import restriction.

Main considerations:

> By adopting an administrative practice according to which ordinary foodstuffs enriched with vitamin A (in the form of retinoids), vitamin D, folic acid, selenium, copper or zinc, which are lawfully manufactured or marketed in other Member States, may be marketed in the Netherlands, when they are not substitution products or restored food or drink products within the meaning of Article 1(1)(c) and (d) of the *Warenwetbesluit Toevoeging micro-voedingsstoffen aan levensmiddelen* (Decree implementing the Commodities Law in relation to the addition of micronutrients to foodstuffs) of 24 May 1996, may be placed on the market only if that enrichment meets a nutritional need in the Dutch population, without verifying in addition that those foodstuffs are not intended as a substitute for foodstuffs already on the market for which the addition of those same nutrients is mandatory, the Kingdom of the Netherlands has failed to fulfil its obligations under Article 30 of the EC Treaty (now Article 28 EC).

2.1 Summary

Kellogg's and Inkosport are seeking permission in the Netherlands to market breakfast cereals and energy bars, respectively, with added vitamins and minerals. Dutch legislation (*Warenwetbesluit Toevoeging microstoffen aan levensmiddelen*) prohibited common foodstuffs enriched with vitamin A (in the form of retinoids), vitamin D, folic acid, selenium, copper or zinc, unless they are substitution products or restored food and drink products within the meaning of this national regulation.

Kellogg's and Inkosport already lawfully produced these vitaminised products in other Member States and also lawfully marketed them in countries other than the Netherlands. The Netherlands allowed the marketing of such products only if the fortification met a nutritional need of the Dutch population. The question

 | HTTPS://DOI.ORG/10.1163/9789004751699-003

was whether the Dutch ban violates Articles 30 and 36 of the EC Treaty (now, after amendment, Articles 28 and 30 TFEU).

As there is no harmonisation at the European level of permitted amounts (minima and maxima) of added vitamins and minerals, Member States are free to pursue their own policies. However, the principle of proportionality must be taken into account. In doing so, national authorities must demonstrate in each specific case, in the light of the dietary habits of their population and the results of international scientific research, that their rules are necessary for the protection of public health and, in particular, that the placing on the market of the products concerned poses a real risk to it. The Netherlands has not verified that Kellogg's breakfast cereals are not intended as a substitute for foods already on the market, for which the addition of the same nutrients is permitted or required.

The Court of Justice concludes that the Dutch administrative practice leads to a systematic prohibition of the placing on the market of enriched foodstuffs. That is not compatible with the requirement under Community law, relating to the identification and evaluation of a genuine risk to public health, of a thorough assessment, on a case-by-case basis, of the possible effects of the addition of vitamins and minerals to the foodstuff in question.

The Netherlands is acting in violation of Article 30 (now Article 28) of the EC Treaty because the administrative practice can be seen as a measure of equivalent effect to quantitative restriction.

This judgment is in line with both 'standard judgments' Dassonville and (to a lesser extent) Cassis de Dijon. A similar line was also followed in the later Noria and Solgar cases (from 2017 and 2010 respectively) (see Chapter 13 of this book). The general rule is that when Member States take measures that impede interstate trade, they require a strong justification such as the interests of public health. The necessity of the measures must be demonstrated by scientific risk analysis.

2.2 Follow-up

The requirement to notify the Minister prior to marketing a substitute product and/or restored edible or drink product already expired with effect from 24 September 2004.

In response to this ruling, the notice 'Addition of micronutrients to foodstuffs' (VGP/VL 2566186 of 8 March 2005) was published on 9 March 2005 (Official Gazette 2005, no. 48). This clarifies the grounds for exemption under Section 16(2) of the Dutch Commodities Act and how an exemption can be applied for.

Meanwhile, Dutch legislation has been further adapted, in particular to the European Regulation (EC) 1925/2006 on the addition of vitamins and minerals and certain other substances to foodstuffs.

Imkje Tiesinga
Consultant on (EU) food law

CHAPTER 3

Raverco, Coxon & Chatterton – Zero tolerance | Imports from third countries

Annotated by Sanne Heestermans

CJEU 28 September 2006, cases C-129/05 and C-130/05, ECLI:EU:C:2006:613 (*NV Raverco, Coxon & Chatterton Ltd. v Minister of Agriculture, Nature and Food Quality*)

Community rules, import conditions, imports, third countries, pharmacologically active substances, antibiotics, return of consignments, destruction of consignments, zero tolerance.

Main considerations:

7. In accordance with the checks prescribed by the Directive, sample extracts were taken from those consignments and analysed. Residues of chloramphenicol (1.4 ppb) and furazolidone (49 ppb) were detected in the duck breast consignments and residues of furazolidone (2.7 ppb) in the rabbit meat consignments.
14. It is clear from both the wording and the purpose of Article 17(2) of the Directive that the obstacles to the redispatch of products lie in the failure to meet Community requirements
15. Under provision, where the checks referred to in the Directive show the competent authority that the product does not satisfy the import conditions, redispatch may be precluded under certain conditions, Article 2(2) of the Directive defining 'import conditions' as veterinary requirements for products to be imported, as laid down in Community legislation.
16. As the Advocate General observed at point 22 of his Opinion, the Directive provides for a system of veterinary inspections to establish if products fulfil the Community import conditions. Moreover, it is abundantly clear from the wording of Article 8(2) and (4), Article 10(2)(a) and (f) and Article 12(2) to (6) of the Directive, in particular, that such inspections are based exclusively on Community legislation.
17. In the light of these considerations, the answer to the first question must be that Article 17(2)(a) of the Directive is to be interpreted as meaning that objection to the redispatch of a consignment that does not satisfy the import conditions must be based on the failure to meet Community requirements.
[...] Chloramphenicol and furazolidone, which were detected in the meat consignments at issue in the main proceedings in the course of veterinary inspections, are listed in Annex IV to the Regulation as pharmacologically active substances for which

 | HTTPS://DOI.ORG/10.1163/9789004751699-004

> no limits can be established. It is therefore accepted that a consignment of meat containing chloramphenicol and furazolidone must be regarded as constituting a hazard to the health of the consumer and as being prohibited under Article 5 of the Regulation.
> [...] Moreover, whereas Article 17(2) of the Directive provides for a general assessment to determine whether there are obstacles to the redispatch of a consignment that does not comply with Community legislation, the wording of Article 22(2) of that directive leaves no margin of discretion as to whether redispatch is possible once such a consignment has been found likely to constitute a danger to animal or human health. In such a case, the latter provision imperatively requires the seizure and destruction of the consignment in question.

3.1 The case

NV Raverco and Coxon & Chatterton Ltd offered frozen duck breast and rabbit meat from China for import into the European Union. These consignments of meat bore Chinese health and export certificates. When imported into the European Union, sample extracts were taken during the inspection of these consignments of meat and then analysed, with a view to determining the quality of the meat. Residues of chloramphenicol and furazolidone (both an antibiotic) were found in the consignments of duck breast, while only residues of furazolidone were found in the consignments of rabbit meat. Residue levels were in the order of parts per billion (ppb). Based on these results, the veterinarian involved in the inspection refused to import the consignments of meat and also ordered their destruction.

Both companies argued – in brief – that the consignments of meat should not have been destroyed but should have been sent back to China. The reasons given were as follows:

(a) Return did not pose a risk to human or animal health in the European Union;
(b) The Directive does not aim to protect users outside the European Union;
(c) Application of Article 5 of the Regulation is ambiguous; and
(d) Several return decisions were taken.

The present judgment concerns two joined cases in which four identical preliminary questions were raised concerning the interpretation of the provisions of Directive 97/78/EC and Regulation (EEC) No 2377/90.

The four intricately worded preliminary questions to be answered were as follows:

(a) Is Article 17(2)(a) of Directive to be interpreted as meaning that the objection to the redispatch of a consignment that does not meet the import conditions lies in the non-compliance with the Community requirement for importation or in the conditions applicable at the place agreed with the person interested in the shipment outside the territories listed in the Annex?

(b) Is Article 17(2)(a) of the Directive, read in conjunction with Article 22(2) of the Directive and Article 5 of the Regulation, to be interpreted as imperatively requiring the destruction of the consignments of animal products concerned in all cases in which, following one of the checks provided for by the Directive, it appears that a consignment of products may constitute a danger to human or animal health?

(c) Must Article 22 of the Directive, in conjunction with Article 5 of the Regulation, be interpreted as meaning that the mere fact that a residue of a substance listed in the Annex to the Regulation has been found in a consignment means that the consignment in question may constitute such a danger to human or animal health that return is precluded?

(d) If Question 2 is answered in the negative, is Article 17(2) of the Directive to be interpreted as meaning that it also seeks to protect the interests of the third country to which the consignment will be imported after its return, even if that interest does not also imply the protection of an interest that can be located in the Member States of the European Union?

3.2 Judgment of the Court

In assessing the first question referred for a preliminary ruling, the Court states that it is clear from both the wording and the purpose of Article 17 of Directive 97/78/EC that non-compliance with Community rules constitutes a problem for the return of products such as the duck and rabbit meat at issue in this case. Where checks have established that the product does not meet import conditions, Article 17 of the Directive allows returns to be excluded under certain conditions. Here, Article 2(2) of the Directive specifies that the import conditions are the same as veterinary requirements laid down in Community legislation. This means that all products to be imported within the Community must comply with these veterinary requirements. The answer to the first preliminary question is therefore in the affirmative.

As regards the second and third questions, the Court observes, first, that it follows from the wording of Article 5 of the Regulation that the presence in animal products of the pharmacologically active substances referred to in the present case, regardless of their maximum amount, constitutes, by definition, a danger to the health of the user. In view of this, the administration of these substances to food-producing animals is prohibited throughout the Community, without leaving room for the Member States' own assessment. The chloramphenicol and furazolidone found are pharmacologically active substances, for which no maximum level is set. For that reason, it is established that a consignment of meat in which these substances are found is considered to present a health risk to users.

3.3 Discussion

This judgment deserves a place in the selection of leading food law cases for two reasons. First, it touches on the relationship between law and science when law uses zero tolerances. Second, it touches on the relationship between the food law protection regime within the European Union and the interests of consumers outside the Union.

3.4 Zero tolerance

What does it actually mean when legislators express themselves in absolute terms such as 'absent' or 'zero'? Detection methods are rapidly becoming more sophisticated, allowing smaller and smaller quantities to be detected.

In this case, infections were found in ppb. That means in parts per billion, or in other words one molecule per 1,000,000,000 (1 billion) molecules. For comparison, Feyenoord's stadium, the Rotterdam Kuip, has a capacity of about one and a half billion litres. So if the Kuip were filled to the brim with water, we are talking about the presence of a few soft drinks' bottles, not even a bathtub in the Kuip. Is that a form of presence or a form of absence? Is it zero or is it something?

Within the scientific community, this ruling was noted with dismay.[11] For the Court to classify a contamination at ppb level as a presence and then also one that requires destruction of food is indigestible for some scientists. The Global Harmonization Initiative (GHI) were up in arms as a result of this ruling.[12] GHI is a global association of concerned scientists. At the time of writing, GHI has ambassadors in over 65 countries around the world. The concern is about food waste in a world that is food-deficient, when the waste is caused by irrational (differences in) food safety laws and regulations. Zero tolerance is perhaps the most telling example of irrational legislation. This is very much the case with antibiotics as discussed in this case. These antibiotics occur naturally in the environment. Added to this are the residues that enter the environment through the faeces of medicine users.

Meanwhile, the European legislator has partly addressed the criticism. For pesticides, zero tolerance has been replaced by a default maximum residue level (MRL), i.e. a very low maximum, but one expressed as a positive number. Thus, unlike zero tolerance, such a standard does not become stricter as detection methods become more sensitive. However, zero tolerances still exist for residues of certain unauthorised veterinary medicines – including those in this case – and unauthorised foodstuffs subject to authorisation, such as GMOs. Unlike the court in this case, the

11 See: https://www.globalharmonization.net/sites/default/files/pdf/articles/2007_TrendsFdSciTech_LelieveldKeener.pdf.

12 See: https://www.globalharmonization.net.

Rotterdam court[13] held that when enforcing the abovementioned zero tolerances, authorities enjoy a margin of appriciation.

3.5 Extraterritorial protection

The judgment also shows that Article 17(2) of the Directive no longer applies once it is established that a product is dangerous within the meaning of Article 22(2) of the Directive. Indeed, Article 17(2) of the Directive is generally applicable and covers all situations where the checks show that the product does not meet the import conditions.

However, Article 22(2) of the Directive refers to specific checks showing that a consignment of products may pose a risk to human and animal health. Article 17(2) of the Directive provides for a general assessment to determine whether or not reasons exist that prevent the return of a consignment, such as conflict with Community regulations. Article 22 of the Directive leaves no room for assessment as to whether a consignment can be returned if it may constitute a danger to human or animal health. This provision in fact forces the seizure and destruction of the product concerned.

In short, if a product contains substances that pose a danger to human health, the competent authorities must seize and destroy those consignments under Article 22(2) of the Directive in conjunction with Article 5 of the Regulation. According to the Court, the fact that the Directive is not intended to protect the health of persons outside the Community does not alter this. Such an interpretation would be contrary to the wording and purpose of the Directive, which is to lay down strict rules to protect human and animal health.

3.6 Relevance to practice

A landmark ruling by the European Court of Justice. The European Union is unique in the sense that it operates a system of internal law that has supranational and direct effect within the autonomous legal order of the Member States: the so-called Community legal order. However, the present judgment shows that the internal law of the European Union can also have effects on third countries, which are not part of the Community legal order. Even though internal law does not cover third countries, the European Court of Justice ruled in this case that Community law is so far-reaching that users from third countries must also be protected under European law against substances in food products that are dangerous to them. The effect of

13 See, in this book, Chapter 24, Rotterdam District Court 23 June 2008, ECLI:NL:RBROT:2008:BD5791, Greenpeace v VWS.

Community law is thus far-reaching and may even affect residents of third countries, who are in principle not bound by European law. Parties from third countries wishing to import food products within the European Union should therefore bear in mind that if these products do not comply with certain European standards, the competent European authorities will be forced to confiscate and destroy the product.

Sanne Heestermans
Head of Legal & Government relations YoungOnes

CHAPTER 4

Lidl Italia & Ute Reindl – Punitive strict liability in the food chain

Annotated by Bernd van der Meulen

Court of Justice of the European Communities, 23 November 2006, ECLI:EU:C:2006:736 (*Lidl Italia*)
Court of Justice of the European Union, 13 November 2014, ECLI:EU:C:2014:2370 (*Ute Reindl* v *Bezirkshauptmannschaft Innsbruck*)

Both with respect to defective labelling and with respect to food safety, retailers may be penalised for infringements in branded products caused by the producer.

Main considerations (*Lidl Italia*):

> 35 By its [...] questions, [...] the national court asks in essence whether Articles 2, 3 and 12 of [Labelling] Directive 2000/13 are to be interpreted as precluding a Member State's legislation [...] which makes it possible for an operator established in that Member State, which distributes a pre-packaged alcoholic beverage [...] produced by an operator established in another Member State, to be held liable for an infringement established by a public authority of that legislation resulting from the inaccuracy of the alcoholic strength by volume stated by the producer on the product label and, consequently, to be penalised by the imposition of an administrative fine where, as the mere distributor, it simply markets the product as delivered to it by the producer.
> 49 National legislation [...] which, where there is infringement of an obligation with regard to labelling [...], provides for liability not only on the part of the producers, but also of the distributors, is not such as to compromise achievement of the result prescribed by that directive.
> 50 On the contrary, such legislation, inasmuch as it gives a wide definition of the circle of operators which may be held liable for infringements of the obligations with regard to labelling [...], is manifestly such as to contribute to the achievement of the aim pursued by that directive of informing and protecting the ultimate consumer of foodstuffs.
> The Court rules:
> Articles 2, 3 and 12 [... [of the Labelling Directive 2000/13/EC] ...] are to be interpreted as not precluding legislation of a Member State [...], which makes it possible for an operator, established in that Member State, which distributes a pre-packaged alcoholic beverage to be delivered as such [...], produced by an operator established in another Member State, to be held liable for an infringement of that provision, estab-

 | HTTPS://DOI.ORG/10.1163/9789004751699-005

lished by a public authority, resulting from the producer's inaccurate statement on the product label of the alcoholic strength by volume of the product and, consequently, to be penalised by an administrative fine, even where, as the mere distributor, it simply markets the product as delivered to it by the producer.
Main considerations (*Ute Reindl*):
18 On 29 March 2012, during an on-the-spot check at that branch [of MPREIS], a food safety body took a sample of vacuum-packed fresh turkey breast produced and packed by another undertaking. As regards that sample, MPREIS was involved only at the distribution stage.
19 The sample underwent microbiological examination by the Austrian Food Safety Agency Innsbruck. In its expert report, that agency concluded that contamination by Salmonella Typhimurium was detectable and that, therefore, the sample was both 'unfit for human consumption' within the meaning of Paragraph 5(5)(2) of the LMSVG and 'unsafe' within the meaning of Article 14(2)(b) of Regulation No 178/2002. The criterion of food safety adopted by that agency in its expert report is that laid down in Annex I, Chapter 1, Row 1.28, to Regulation No 2073/2005.
20 On the basis of that report, the Bezirkshauptmannschaft Innsbruck brought proceedings against Ms Reindl for failing to comply with Paragraph 5(5)(2) of the LMSVG. The Bezirkshauptmannschaft Innsbruck held that Ms Reindl was liable for failing to comply with the limit value for Salmonella Typhimurium in Annex I, Chapter 1, Row 1.28, to Regulation No 2073/2005, and ordered her to pay a fine [...].
21 On appeal against the decision of the Bezirkshauptmannschaft Innsbruck, the *Unabhängiger Verwaltungssenat* in Tirol is unsure as to the extent of the liability of food business operators which are active only at the distribution stage.
31) By its [...] questions, [...], the referring court asks essentially whether EU law and, in particular, Regulations No 178/2002 and No 2073/2005, must be interpreted as precluding national legislation, [...], which penalises a food business operator active only at the food distribution stage for the placing on the market of a food stuff on account for the failure to comply with the microbiological criterion mentioned in [...] Regulation No 2073/2005.
36 It follows that EU law and, in particular, Regulations No 178/2002 and No 2073/2005 must be interpreted as meaning that, in principle, they do not preclude national legislation, [...], which penalises food business operators active only at the distribution stage for placing on the market foodstuffs which fail to comply with the microbiological criteria mentioned in [...] Regulation No 2073/2005.
42 Even if the system of penalties in the case in the main proceedings is a system of strict liability, it must be recalled that, according to the case-law of the Court, such a system is not, in itself, disproportionate to the objectives pursued, if that system is such as to encourage the persons concerned to comply with the provisions of a regulation and where the objective pursued is a matter of public interest which may justify the introduction of such a system (see judgment in Urbán, EU:C:2012:64, paragraph 48 and the case-law cited).
44 Having regard to all the foregoing, the answer to the [...] questions is that EU law, in particular Regulations No 178/2002 and 2073/2005, must be interpreted as meaning

> that, in principle, it does not preclude national law, such as that at issue in the main proceedings, which imposes a penalty on a food business operator active only at the distribution stage for placing a foodstuff on the market, on account of the failure to comply with the microbiological criterion laid down in Annex I [...] to Regulation No 2073/2005. It is for the national court to determine whether the penalty at issue in the main proceedings observes the principle of proportionality referred to in Article 17(2) of Regulation No 178/2002.

4.1 Punitive chain liability for labelling errors

Can a retailer be penalised for errors in the labelling of a branded product?

During an inspection at a Lidl shop, inspectors took away an 'amaro alle erbe' for laboratory testing. Amaro alle erbe, unlike its name suggests, is a drink produced in Germany. On examination, the drink was found to contain less alcohol than the 35% stated on the label. This was met with a fine under Italian food law. The product at issue was a branded product labelled by the producer. Lidl was not involved in the production and labelling and did not have its name attached to the product. The fine on Lidl therefore involved strict liability for mistakes made earlier in the chain. Lidl felt that it was wrongly blamed. The hearing judge inquired at the Court in Luxembourg whether the Italian system was compatible with European Union law. According to the court, it is. All players in the chain bear responsibility for the food business under their control. Holding the retailer liable benefits the objective of EU food law to provide a high level of protection.

4.2 Legislator's response in Food Information Regulation

The judgment was delivered under the operation of the old Labelling Directive 2000/13. The recodification of labelling law in the Food Information Regulation 1169/2011 provided an opportunity for the legislator to reconsider the issue. Apparently, the European Commission considered the Lidl case-law to be a bit over the top, as it proposed a provision that is now included as the first paragraph in Article 8 of the regulation:

'The food business operator responsible for the food information shall be the operator under whose name or business name the food is marketed or, if that operator is not established in the Union, the importer into the Union market.'

Reportedly, this mitigation of the Lidl case-law was not to the liking of France. The current fifth paragraph was allegedly inserted under pressure from that country. This provision seems to codify the Lidl case-law rather than correcting it:

'Without prejudice to paragraphs 2 to 4, food business operators, within the businesses under their control, shall ensure compliance with the requirements of food information law and relevant national provisions which are relevant to their activities and shall verify that such requirements are met.'

In one article, the Food Information Regulation expresses two opposing thoughts. The idea that responsibility for labelling only rests with the person whose name is connected to the product, and the idea that responsibility rests with all chain partners. Thus Lidl has been called into question, but has not been overturned. New case-law will have to restore legal certainty.

4.3 Punitive chain liability for food safety defects

After Lidl Italia, the question remained open as to whether the compatibility with EU law of punitive chain risk liability applies only to labelling or also to other food law requirements such as those relating to food safety in particular. This question was answered in the Ute Reindl judgment. Austrian inspectors had checked pre-packaged turkey fillet in an MPREIS supermarket that, upon laboratory testing, was found to be contaminated with Salmonella. Again, the question was whether it was compatible with EU law to penalise the retailer for mistakes made earlier in the chain concerning a product not marketed in the retailer's name. Again, the Court agreed with essentially the same reasoning that this benefits a high level of protection.[14]

4.4 Due diligence and private standards

In the UK, a similar view on chain liability combined with the possibility of escaping it through due diligence gave rise to the birth of private standards and certification systems in food law.[15] Under the due-diligence doctrine, the person who has demonstrably done everything possible to prevent the occurrence of the undesirable result escapes liability (both punitive and civil) when that result does nevertheless occur. The business that imposes clear food safety requirements on their supplier (through a standard) and monitors compliance (through audits and certification) can successfully escape liability under that system.

4.5 Food business operator

In the Ute Reindl case, a second aspect is striking. This is in the Court's designation of the parties. In the English language Reindl is designated as: 'Ute Reindl, represen-

14 The idea that the responsibility of a retailer might be lower than that of the producer is not far-fetched. With regard to recalls, it is expressed in Article 19(2) of Regulation 178/2002.

15 For more details see, Bernd van der Meulen (ed.) Private food law (2010) https://brill.com/edcoll-book-oa/title/68768 (open access).

tative of MPREIS Warenvertriebs GmbH, with liability'. The meaning of the last two words ('with liability') is not self-evident. The language of the case, however, was German. In that language the designation is as follows: '*Ute Reindl, strafrechtlich verantwortliches Organ der MPREIS Warenvertriebs GmbH*'. This explicitly means that she is responsible under criminal law. This phrasing is also found in other languages such as French – the Court's working language – and Dutch.

Ms. Reindl is a branch manager. Under recital 17 of the judgment, the Court says: 'As a result, she is responsible for compliance by that branch with all the applicable rules in the food sector.' The Court here uses the formula from the definition of food business operator. The definition in Regulation (EC) 178/2002 has not really been successful. The function of the definition is to indicate who is responsible for compliance with food law. The inclusion of this responsibility in the definition creates a circular argument in a practical sense. This judgment makes it clear that a branch manager can also be considered the responsible food business operator, even in terms of punitive strict liability in person. As far as I am concerned, this is rather extreme.

4.6 Relevance to practice

These rulings show that the Court leaves a lot of room for Member States to adopt systems of strict liability, even when applying punitive sanctions, even when imposed on managers in person. This may have far-reaching consequences in light of the Official Controls Regulation (625/2017). This regulation stipulates in the second paragraph of Article 139:

'Member States shall ensure that financial penalties for violations of this Regulation and of the rules referred to in Article 1(2), perpetrated through fraudulent or deceptive practices, reflect, in accordance with national law, at least either the economic advantage for the operator or, as appropriate, a percentage of the operator's turnover.'

It seems obvious to me that the legislator had the owner of the business in mind here, not a manager. The Union legislator's choice to focus not on the legal person but on the operator will no doubt give rise to many more headaches.

Perhaps needless to say, the Court does not rule that the retailer *should* be held liable under EU law, but only that EU law does not preclude it when the national legislator chooses to do so.

Bernd van der Meulen
At the time of writing this note, Professor of Comparative Food Law Renmin University of China School of Law

CHAPTER 5

Garlic – Distinction between food and medicine

Annotated by Irene Verheijen

Court of Justice EU 15 November 2007, case C-319/05 (*Commission v Germany*), ECLI:EU:C:2007:678

Medicine or foodstuff?, application criterion and administration criterion, priority of medicinal product legislation over food legislation, free movement of goods.

Main considerations:

> 44 In that context, a product is 'presented for treating or preventing disease' within the meaning of Directive 2001/83 when it is expressly 'indicated' or 'recommended' as such, possibly by means of labels, leaflets or oral representation [...].
> 46 A product is also 'presented for treating or preventing disease' whenever any averagely well-informed consumer gains the impression, which, provided it is definite, may even result from implication, that the product in question should, having regard to its presentation, have the properties in question [...].
> 47 In that regard, account must be taken of the attitude of an averagely well-informed consumer, in whom the form given to a product may inspire particular confidence similar to that normally inspired in him by proprietary medicinal products, having regard to the safeguards normally associated with their manufacture and marketing. Although the external form given to the product may serve as strong evidence of its classification as a medicinal product by presentation, the 'form' must be taken to mean not only the form of the product itself but also that of its packaging, which may, for reasons of marketing policy, tend to make it resemble a medicinal product [...].
> 64 [...] not sufficient that product has properties beneficial to health in general, but it must strictly speaking have the function of treating or preventing disease.
> 65 That statement is even more relevant in the case of products which, in addition to being food supplements, are recognised as having beneficial effects on health. As the Advocate General observed, in point 60 of her Opinion, there are many products generally recognised as foodstuffs which may also serve therapeutic purposes. That fact is not sufficient however to confer on them the status of medicinal product within the meaning of Directive 2001/83.
> 68 In those circumstances, it must be held that the product concerned, whose effect on physiological functions is no more than the effects of a foodstuff consumed in a reasonable quantity may have on those functions, does not have a significant effect on the metabolism and cannot, therefore, be classified as a products capable of restoring,

 | HTTPS://DOI.ORG/10.1163/9789004751699-006

correcting or modifying physiological functions within the meaning of the second subparagraph of Article 1(2) of Directive 2001/83.
97 It follows from all the foregoing considerations that, by classifying as a medicinal product a garlic preparation in capsule form not satisfying the definition of a medicinal product within the meaning of Article 1(2) of Directive 2001/83, the Federal Republic of Germany has failed to fulfil its obligations under Article 28 EC and Article 30 EC.

5.1 The case

This ruling follows a complaint to the European Commission from a company seeking to import a garlic preparation in capsule form into Germany. The German authorities classify the preparation as a medicinal product and do not authorise its importation and marketing in Germany. This while the product is marketed in other European Member States as a food supplement. The Commission considers this contrary to the free movement of goods within the EU.

The preparation is in no way indicated or recommended on the label, in the information on the packaging or otherwise, as a product with therapeutic or prophylactic properties. The product, which is manufactured from dried garlic powder, is presented in capsule form.

Germany takes the view that the product meets both the conditions for qualification as a foodstuff (food supplement) and those for qualification as a medicinal product. In that case, the specific provisions for medicinal products take precedence over those for foodstuffs or food supplements. It considers the garlic preparation in question to be a medicinal product by function, because it has pharmacological properties and therapeutic effects that have a preventive effect against arteriosclerosis and are therefore of decisive importance. In so far as there has been an infringement of the free movement of goods, it considers that this is justified by the protection of a mandatory requirement in the public interest: the protection of public health.

5.2 Medicinal product; administration and presentation criterion

The Court holds that, until there is complete harmonisation of the measures necessary to ensure health protection, differences will persist between the Member States in the classification of products as medicinal products or as foodstuffs. The fact that a product qualifies as a foodstuff in one Member State does not automatically preclude its qualification as a medicinal product in another Member State, where the product is imported.[16] In case of doubt, the (stricter) medicine regula-

16 CJEU 9 June 2005 joined cases C-211/03, C-299/03, C-316/03 and C-318/03 (HLH Warenvertrieb and Orthica v Germany), ECLI:EU:C:2005:370, rec. 56

tions take precedence over the (less strict) food regulations.[17] Member States have (broad) discretion in this respect.[18]

Briefly, a product qualifies as a medicinal product if:

a. the product has prophylactic or therapeutic properties, which is determined on the basis of, among other things, the composition, pharmacological effects and risks of the product: the administration criterion;[19]
b. the product is presented as having prophylactic or therapeutic properties, such as when properties are attributed to the product in advertisements that pertain to curing or preventing a disease: the presentation criterion.[20]

5.3 Review

Not a medicinal product by presentation

According to established case law, the criterion of presentation must be interpreted broadly. When relying on the presentation criterion, the Medicinal Products Directive seeks to bring within its scope not only medicinal products that have an actual therapeutic effect, but also products that are not sufficiently effective or that do not have the effect that the consumer is entitled to expect in view of their presentation method. The Directive thus seeks to protect consumers not only from medicinal products that are harmful or toxic as such, but also from various products used in place of adequate remedies (quackery).[21]

The decisive factor in that assessment is the way in which the product is presented or recommended (by the manufacturer or distributor), whether on the label, in the package information leaflet or by oral presentation.[22] A product is also presented as having therapeutic or prophylactic properties if, implicitly but no less emphatically, the way in which it is presented gives *an averagely well-informed consumer* the impression that it has such properties.[23] Here, not only the explicit commendations are

17 CJEU 20 October 1992, case C-219/91 (Ter Voort), ECLI:EU:C:1992:414, rec. 19; included in Article 2 of Directive 2001/83/EC (the Medicinal Products Directive).
18 CJEU 21 January 1999, case C-120/97 (Upjohn II), ECLI:EU:C:1999:14, rec. 34 et seq.; ruling HLH and Orthica, rec. 74 et seq.
19 Article 1(2)b of the Medicinal Products Directive.
20 Article 1(2)a of the Medicinal Products Directive.
21 CJEU 30 November 1983, case C-227/82 (Van Bennekom), ECLI:EU:C:1983:354, rec. 17.
22 CJEU 20 October 1992, case C-219/91 (Ter Voort), ECLI:EU:C:1992:414, rec. 26.
23 The perception of this consumer is also described as: '*the presumed expectations of an average consumer who is reasonably well-informed and reasonably observant and circumspect*'. See CJEU 16 July 1998, case C-210/96 (*Gut Springenheide*), ECLI:EU:C:1998:369, rec. 31 and 37; CJEU 13 January 2000, case C-220/98 (Estée Lauder/Lancaster), ECLI:EU:C:2000:8, rec. 27; G.A. van der Wal and F. van Schaik, *De afbakening tussen geneesmiddelen en levensmiddelen: een onderscheid van elastiek?*, JGRplus, 2007/2.

relevant, but also the nature of the product, the sales channel, the market to which the consumer will attribute the product and the external form of the product. Although the external form of the product may be a serious indication of its qualification as a medicinal product by presentation, it cannot be decisive for this purpose. Otherwise, certain food products that have traditionally been presented in a form similar to that of medicinal products (such as food supplements) would also be covered.[24]

The Court concludes that there is no question of a medicinal product by presentation, since no aspect on the packaging seeks to make the product in question resemble a medicinal product. The fact that the product is presented in capsule form is not sufficient to give rise, in the averagely well-informed consumer, to a level of confidence such as that which medicinal products generally inspire. The capsule form is not reserved for medicinal products but is also widely used for food supplements, for example.[25]

Not a medicinal product by administration

The competent national authorities, under the supervision of the courts, must determine on a case-by-case basis whether a product is a medicinal product by function.[26] In doing so, account must be taken of all the characteristics of the product, in particular its composition, its pharmacological properties as ascertainable in the present state of scientific knowledge, its methods of use, its volume and distribution, the familiarity of consumers with it and the risks that its use may entail.[27]

The Court considers that the function of the product is determined by its pharmacological properties and whether it can be administered to human beings in order to make a medical diagnosis or to restore, improve or modify physiological functions in human beings. However, this should not result in substances that act on the human body, but do not actually affect metabolism and therefore do not

24 Ruling Van Bennekom, rec. 18-19; CJEU 21 March 1991, case C-60/89 (Monteil and Samanni), ECLI:EU:C:1991:138, rec. 23-24; CJEU 21 March 1991, case C-369/88 (Delattre), ECLI:EU:C:1991:137, rec. 38 et seq.; HR 8 March 2002, ECLI:NL:PHR:2002:AD8178, JGR 2002/17. See also B.M.J. van der Meulen, *Medicinal Claims. Prohibition, Enforcement and Delineation: Food in Fact but Medicine in Law?* European Food and Feed Law Review 2017, pp. 392-405

25 Article 2(a) of Directive 2002/46 explicitly names the presentation in capsule form in the criteria used for the definition of the concept of 'food supplement'.

26 HR 22 September 2006, ECLI:NL:HR:2006:AX1576, JGR 2006/54, annotated by Vollebregt; J. Lisman, *De toelating van geneesmiddelen. Hoe effectief is het systeem?, Preadvies voor de Vereniging voor Gezondheidszorg,* 2006; ruling HLH and Orthica, rec. 74 and 94; CJEU 22 June 2000, case C-318/98 (Fornasar e.a.), ECLI:EU:C:2000:337, rec. 31-32; CJEU 16 April 1991, case C-112/89 (Upjohn/Farzoo), ECLI:EU:C:1991:147, rec. 23.

27 Ruling HLH and Orthica, rec. 51 et seq.; ruling Upjohn/Farzoo, rec. 24, CJEU 29 April 2004, case C-387/99 (Commission/Germany), ECLI:EU:C:2004:235, rec. 57; CJEU 29 April 2004, case C-150/00 (Commission/Austria), ECLI:EU:C:2004:237, rec. 64, rec. 29, CJEU 30 April 2009, case C-27/08 (BIOS Naturprodukte), ECLI:EU:C:2009:278, rec. 18.

actually alter the conditions under which it functions, being classified as medicinal products by function.[28]

The physiological effect is not specifically reserved for medicinal products, but is also part of the criteria used to define a food supplement.[29] It is therefore not sufficient that a product possesses beneficial properties for health in general, but it must actually have a preventive or curative effect. There are, after all, many products that are foodstuffs in the common sense of the term and that may be used objectively for therapeutic purposes; however, that circumstance cannot be sufficient to confer on them the status of medicinal products.[30]

The Court finds that the product in question, a powder of garlic extract, does not contain, apart from an excipient (adjuvant), any substance that does not itself occur in garlic in its natural state. The physiological effects that can be obtained with the ingestion of garlic powder capsules can similarly be obtained with the ingestion of garlic in its natural state. The product also has no other concomitant effects. Any risks, associated with the intake of the preparation, are no different from those associated with the intake of garlic in general. In any case, Germany has not sufficiently substantiated this.[31] The Court concludes that the product as such does not fall within the definition of medicinal product.

Infringement of free movement of goods

Germany's requirement of a marketing authorisation as a medicinal product constitutes a prohibited measure of equivalent effect as a quantitative restriction on imports.[32] Such a restriction on the free movement of goods may be permissible if it is justified, for example, by the protection of public health. In the absence of full harmonisation, it is for the Member States to determine the extent to which they wish to protect the health and life of individuals. However, the requirements of the free movement of goods within the EU must be taken into account.[33]

The Court does not consider the measure taken to be proportionate to the restriction that it places on the free movement of goods. In its view, the risks outlined

28 Case Upjohn/Farzoo, rec. 22: See also: CJEU 15 January 2009, case C-140/07 (Hecht/Pharma), ECLI:EU:C:2009:5, rec. 41

29 In Article 2 of Directive 2002/46 'physiological effect' is named as one of the criteria used to define the concept of a 'food supplement'.

30 See Opinion A-G Trstenjak, rec. 60.

31 See annotation by Van Wissen (JGR 2008/1).

32 Article 28 EU Treaty; now Article 34 TFEU. See: CJEU 11 July 1974, case C-8/74 (Dassonville), ECLI:EU:C:1974:82, rec. 5; CJEU 29 April 2004, case C-150/00 (Commission v Austria), ECLI:EU:C:2004:237, rec. 81.

33 CJEU 4 July 1993, case C-174/82 (Sandoz), ECLI:EU:C:1983:213, rec. 16; ruling Van Bennekom, rec. 37, CJEU 14 September 2006, added cases C-158/04 and C-159/04 (Alfa Vita Vassilopoulos and Carrefour-Marinopoulos), ECLI:EU:C:2006:562, rec. 21; CJEU 2 December 2004, case C-41/02 (Commission v Netherlands), ECLI:EU:C:2004:762, rec. 42.

do not relate specifically to the effects of the product, but to the consumption of garlic in general as a foodstuff, and those risks arise only in very special circumstances. It was therefore not necessary to use the onerous remedy of prior assessment as a medicinal product. Furthermore, Germany could have mitigated the risks by, for example, prescribing appropriate labelling warning consumers of potential risks associated with consumption of the product.

5.4 Conclusion

The present case makes it clear that a foodstuff with positive health effects should not be classified as a medicinal product too lightly.[34] Not every medicinal power present justifies classification as a medicinal product. When applying the administration criterion, it is important to establish, based on scientific evidence, that the product is actually suitable for making a medical diagnosis or for restoring, improving or modifying physiological functions.

5.5 Postscript

For further development of the distinction between medicinal products and other product categories (foodstuffs, medical devices), see contributions 18, 19, 25 and 39 elsewhere in this book.

I.E.M. (Irene) Verheijen
Lawyer-partner at Legaltree/at the time of writing President of the European Food Law Association (EFLA)

34 K.J. Defares and J. Langer, *Levensmiddel of geneesmiddel: een lastig onderscheid?*, NTER 11, November 2007, p. 221 et seq.

CHAPTER 6

Banana war – EU not liable for violation of WTO law

Annotated by Victor van Ahee

Court of Justice of the European Community, 9 September 2009, ECLI:EU: 2008:476 (*FIAMM and Fedon v Council and Commission*)

WTO Treaty, directly effective provision, EU/EC liability, import tariffs.

Main considerations:

> Indeed, the Court found that Community law as it stands does not provide for a rule under which the Community can be held liable for its normative action in a situation where the possible incompatibility of such action with the WTO agreements cannot be invoked before the Community courts. In particular, the appellants' claims for damages sought to hold the Community liable for such action. In those circumstances, the Court did have to dismiss those claims, irrespective of the appellants' arguments in support of them.

6.1 Introduction

These (joined) complex judgments involving claims for damages from producers of lead-acid batteries and spectacle cases do not seem to belong directly in a canon of food law jurisprudence. Nevertheless, this ruling has its origins in foodstuffs and its lessons are relevant to the European food industry.

The judgments still provide valid direction regarding the damages regime in Union law. It is also noteworthy that the Court of Justice of the European Community (the Court of Justice) indicates in this judgment that there is no general principle of law in Union law providing a basis for liability for a lawful act of government.

6.2 Background

This statement has its roots in the 'banana war' in the 1990s between the United States (US) and the European Community, the forerunner of the European Union (EC). At the time, the EC (in the eyes of the US) was treating some former EC colonies and underdeveloped economies in the Caribbean preferentially. At the urging of some major banana producers (such as Chiquita and Dole – based in the US),

 | HTTPS://DOI.ORG/10.1163/9789004751699-007

the US complained to the World Trade Organization, which oversees the General Agreement on Trade and Tariffs (WTO/GATT).

Under the WTO dispute settlement system, the Dispute Settlement Body (DSB) can settle trade disputes. The DSB ruled on the US complaint that the EC was in breach of the WTO agreements and had to comply with those agreements within 15 months. The EC did not do so in a timely and sufficient manner according to the US and the DSB. Then, under the WTO agreements, a retaliatory duty (at the option of the injured party) may be imposed. The US introduced retaliatory import tariffs in the amount of $520 million (the damage caused by the preferential treatment by the EC of bananas from the specific regions). The US imposed these retaliatory tariffs on various products imported to the US from the EC, such as Scottish wool, French cheese, as well as products such as spectacle cases and lead-acid batteries.

6.3 The case

The two plaintiffs in these proceedings are an Italian manufacturer of lead-acid batteries (FIAMM) and an Italian manufacturer of spectacle cases (Fedon), among other things. They suffered damages as a result of the US imposing (retaliatory) import tariffs following the Community's failure to comply with the DSB ruling. FIAMM and Fedon are now claiming the damages from the EC because in their view the EC can be held noncontractually liable for damages caused by its legislative institutions (such as the Council of the European Community and its European Commission). These claims can be brought before the Court of Justice under (current) Article 340 of the Treaty on the Functioning of the European Union (TFEU). Such claims are initially heard by the (then) Court of First Instance of the European Community (the General Court).

FIAMM and Fedor invoke the provisions of the WTO agreements and the DSB ruling declaring the 'banana regulation' incompatible with those provisions. Should the Court not follow them in this, FIAMM and Fedor also argue that they have been disproportionately disadvantaged by an admittedly lawful act of the EC institutions.

6.4 Treatment by the General Court

In the first instance, FIAMM and Fedor ended up before the General Court, where they pleaded breach of WTO rules and violations of general principles of law such as the principles of the protection of legitimate expectations and legal certainty, the right to property and freedom of economic initiative and the principles of good administration. FIAMM and Fedor argue on those grounds that the Community is liable for unlawful conduct by its institutions.

Among other things, FIAMM and Fedor invoke the Community's actions in violation of WTO agreements. The question then naturally arises as to whether a private citizen (like FIAMM or Fedor) is entitled to directly invoke the WTO agreements. According to the General Court's settled case-law, the WTO agreements are not among the norms against which the Community courts test the acts of Community institutions. As the General Court explains, this would lead to inequality between the Community and its trading partners. The review against WTO agreements is only permissible, according to the General Court, if the Community 'intended to give effect to a particular obligation assumed within the framework of the WTO or if the Community act expressly refers to specific provisions of the WTO agreements' (para 114).

This is not the case, according to the General Court, because although the DSB ruling contained the Community's instruction to comply with the WTO agreements, this does not mean that the parties cannot otherwise reach a solution (outside of the ruling). If a Community court could then review DSB rulings (as a special obligation of the WTO), this would limit the Community's freedom of action to reach such other solutions. Thus, even if a DSB ruling imposes an obligation on the Community, this does not mean that the General Court may review WTO agreements or DSB rulings at the request of an EC citizen. Thus, no unlawful act of the Community's institutions can be established either.

FIAMM and Fedor's reliance on liability for a lawful act of the Community goes a long way, but also fails. While the General Court finds that there can be such a basis for liability under the Dorsch Consult judgment, there must be damage actually suffered, showing a causal link with the Community's conduct and the damage must be abnormal and special in nature.[35] The Court held that while the damage is real and causally related to the conduct, the damage is not abnormal as it is a risk faced by 'Community entrepreneurs'.

On those grounds, the General Court dismissed the actions brought by FIAMM and Fedor by judgments of 14 December 2005. FIAMM and Fedor appealed against these judgments in the Court of Justice.

6.5 Treatment by the Court of Justice

Before the Court of Justice, FIAMM and Fedor bring a number of appeals against the judgments of the General Court. First, FIAMM and Fedor find that the General Court erred in failing to create a third exception under which review of WTO agreements in relation to private citizens should be possible. According to FIAMM and Fedor, this exception should apply in cases where the EC would not promptly

35 CJEU 15 June 2000, case C-237/98 P Van Dorsch v Council of the European Union and Commission of the European Communities, ECLI:EU:C:2000:321, rec. 19.

follow up on a condemnatory DSB ruling. The Court of Justice, however, rejects this line of reasoning by considering that it is not possible to separate the review of EC acts under WTO agreements from such a review of a DSB ruling based on those WTO agreements. This appeal therefore fails.

The two sides then address the position on the principle of liability for a lawful act. The Court of Justice stresses that Community law has a 'firmly established' framework in case-law and regulation regarding extra-contractual liability for a wrongful act. According to the Court of Justice, the same cannot be said of liability based on a lawful act of a Community institution. This form of liability is only at an early stage and, to the extent that it could be assumed, only after the conditions set out by the Court of Justice. As the Court of Justice explains in para. 169:

> As the Court of Justice noted in particular [...], the Court has on the contrary hitherto limited itself, as set out in settled case-law, to specifying some of the conditions under which such liability could be incurred in the event of the principle of Community liability for a lawful act being recognised in Community law [...]. It was solely on that basis that the Court note [...] that if the principle of such liability came to be recognised, at the very least three conditions, comprising the fact of damage, the existence of a causal link between it and the act concerned and the unusual and special nature of the damage, would all have to be satisfied in order for liability to be incurred.

Following this consideration, the Court of Justice holds that the policy freedom of institutions to adopt standards binding on individuals must not be unduly constrained or affected by the risk of liability of the Community. According to the Court of Justice, such liability should be limited to cases where the institution has manifestly and gravely exceeded the limits of its powers.

According to the Court of Justice, a principle of law generally applicable in the EC therefore does not exist where liability for a lawful act of government is concerned. For this reason, the Court concludes, in para. 176:

'In the light of all the foregoing considerations, it must be concluded that, as Community law currently stands, no liability regime exists under which the Community can incur liability for conduct falling within the sphere of its legislative competence in a situation where any failure of such conduct to comply with the WTO agreements cannot be relied upon before the Community courts.'

Thus, where the General Court concluded that a generally applicable legal principle of liability for lawful acts of public authority does exist in EU law, the Court of Justice held that this was not the case. This is a step too far for the Court of Justice.

As a result, while the Court of Justice annuls the judgments of the General Court, it leaves the outcomes intact – which it can do under case-law.

6.6 Discussion

There is, of course, much to be said for the position of FIAMM and Fedor. As producers of batteries and spectacle cases, they fell victim to an EC-US dispute over bananas. With the imposition of a hard deadline, the EC is ordered to amend its legislation but fails to do so, leaving FIAMM and Fedor the victims of retaliatory measures.

First, FIAMM and Fedor's appeal fails on the grounds that the WTO agreements (and the DSB ruling) do not have horizontal direct effect. Citizens like FIAMM and Fedor cannot rely on it. Thus, no illegality can be constructed and there will be no unlawful act either.

FIAMM and Fedor must then try to construct liability through lawful acts. The General Court found that there is a ground for this in Community law, but was reversed by the Court of Justice in this. The Community is therefore not liable in this case, regardless of whether its act was unlawful or lawful.

The rejection of Community liability is not surprising, as acceptance rarely happens. Community liability is limited to very special cases. For instance, there must be conduct attributable to a Community institution that violates a norm invoking rights for citizens. Next, the breach must be manifest and serious and the conduct must be causally linked to the damage. Only then does the Community incur tort liability.

For the time being, there is no place for liability for lawful acts, according to the Court of Justice. This does not mean that this rules out such liability. As this judgment shows, there may be a discussion as to whether the principle of liability for lawful acts is a general principle of law (the General Court found that it is). The Advocate General considered in his opinion that it is not required that the majority of Member States know a certain legal principle before it is a principle of Union law, citing as an example the '*Vertrauensschutsprinzip*', which was only applied in German law, but has become a principle of Union law.

In this case, the Court of Justice finds (in dissent from the General Court) that no such supported principle exists in the Union. However, at the moment more Member States recognise such liability, this may change. In the Netherlands, the loss compensation scheme in the Awb[36] can serve as an example of this. That is an obligation to compensate damages for a lawful government act. Although this principle has already been determined in case-law, such developments, namely codification, are circumstances that contribute to a more supported Union law principle of liability for lawful government acts. Perhaps by the next anniversary of the NVLR, the world will be a different place.

Victor van Ahee
Lawyer at Van Doorne law firm, Amsterdam

36 The Dutch General Administrative Law Act (Algemene wet bestuursrecht).

CHAPTER 7

Processing aids – No positive list requirement without risk analysis

Annotated by Victor van Ahee

Court of Justice of the European Union, 28 January 2010, ECLI:EU:2010:44 (*European Commission v French Republic*)

Processing aids, infringement procedure, TFEU, mutual recognition.

Main considerations:

> A marketing ban, otherwise the most serious obstacle to trade in products lawfully manufactured and marketed in other Member States, may be issued only if the alleged real risk to human health appears to be sufficiently plausible on the basis of the latest scientific data available at the date of issue of that ban. In that context, the risk assessment to be carried out by the Member State concerns the degree of likelihood of adverse effects on human health of the use of processing aids in the preparation of food and the severity of those potential effects.

7.1 Introduction

France has put in place a prior authorisation system for the placing on the market and use of processing aids (and foodstuffs manufactured with processing aids). In an infringement procedure, the Commission considers that France has thereby introduced a measure with equivalent effect to a quantitative restriction on imports, in breach of EU law. The measure would in the view of the Commission not be justified or proportionate.

In this ruling, the Court indicates where the (lower) limits are for establishing an authorisation requirement in non-harmonised areas of food law and how such a system can be justified.

7.2 The French regulation

This ruling concerns the regulation of processing aids, which are – in short – substances that are added to food during production and are not present in the final

 | HTTPS://DOI.ORG/10.1163/9789004751699-008

product or are no longer present with a function in the food.[37] In France, at the time of the ruling, a decree from 1912 was in force that included a system under which a so-called positive list for (among others) processing aids was established (in several decrees). A positive list contains an overview of substances that are allowed on the market (and under which conditions).[38] This list was subsequently extended by separate decrees (per individual processing aid), but only a few of those decrees contained a provision on the mutual recognition of substances from other Member States.

In 2001, the French government adopted a new decree that would repeal the 1912 decree. The 2001 decree provided a basis for establishing a (new) positive list of processing aids. The 2001 decree also contained a provision allowing mutual recognition for processing aids lawfully marketed in another Member State, but which would not comply with the positive list in France. If an equivalent level of safety were to be guaranteed for that processing aid (as in the French system), it would be allowed on the French market.

However, the 2001 decree was only due to come into force once a new positive list had been adopted. That did not happen until 2006, so until then the 1912 decree was still applicable and the mutual recognition regime of the 2001 decree had not yet entered into force.

The Commission, which had become aware of the French situation through the notification requirement of Article 8 of Directive 98/34/EC,[39] considered that France was acting in violation of Article 28 of the EC Treaty by introducing a prior authorisation system for processing aids through the delayed entry into force of the 2001 decree. According to the Commission this was not allowed, and it therefore launched an infringement procedure.

7.3 Infringement procedure

Under Article 258 of the Treaty on the Functioning of the European Union (TFEU), the Commission can initiate what is known as an infringement procedure before the Court of Justice of the European Union (CJEU) if it considers that a Member State is in breach of an obligation under the EU treaties. The TFEU prescribes a procedure that starts with the administrative or pre-litigation phase in which the Commission, via a letter of formal notice with a reasoned opinion, recommends that the Member State still comply with its obligations. If the Member State fails

37 At the time of the regulation, a definition was absent. It has since been included in Article 3 of the Additives Regulation (EC) 1333/2008.

38 In contrast to a negative list which contains an overview of those substances that are not allowed on the market.

39 Now Directive (EU) 2015/1535.

to comply (and does not convince the Commission), the Commission can start an infringement procedure against it.

While the Commission may act at the request of another Member State, it may also act on its own initiative to initiate the pre-litigation phase. In this case, the Commission acted on its own, following the French government's notification of the introduction of the 2006 decree (which would have brought the 2001 decree into force). A number of views were exchanged during the pre-litigation phase, but neither the Commission nor France became convinced that the other was right. Therefore, the Commission decided to apply to CJEU to rule that France was in breach of the TFEU.

7.4 Free movement of goods

The Commission believes that France is violating the free movement of goods by introducing a prior authorisation system. The principle of free movement of goods within the EU is set out in Article 34 TFEU (at the time of the ruling, Article 28 of the EC Treaty) and requires that no quantitative import restrictions or measures of equivalent effect may exist in the EU between Member States. Quantitative import restrictions include, for example, import quotas, but these are not common anymore. Particularly relevant are measures of equivalent effect. Since the Dassonville ruling,[40] all trade measures that directly or indirectly, actually or potentially restrict intra-EU trade qualify as such measures.

Measures of equivalent effect that restrict trade include labelling requirements, packaging requirements, preparation requirements, etc. Such measures can restrict intra-community trade, as a 'foreign' producer has to adapt its product to be able to sell it across the border. Under Article 36 TFEU, however, there are measures that are allowed, but for this to be the case, these measures must meet a certain (non-economic) justification, such as public policy or public health. The principle of proportionality must also be met, in the sense that the measure actually serves the purpose and that no less intrusive measures are possible.

7.5 The grounds

The Commission considers that France violates Article 28 of the EC Treaty[41] with its system of prior authorisation for the marketing of processing aids (or foodstuffs prepared with them) on three grounds. Firstly, the Commission considers that the

40 CJEU 11 July 1974, Procureur du Roi v Benoît and Gustave Dassonville, case C-8/74, ECLI:EU:C:1974:82.

41 Now Article 34 TFEU.

measure is not based on a sufficient justification, secondly that the French system creates legal uncertainty and thirdly that France does not provide an adequate opportunity to obtain an 'authorisation'.

Regarding the first ground, the French government indicates that it considers an authorisation (inclusion on the positive list) necessary because of the health risks of processing aids. The French government substantiates this by referring to a scientific note from the French research institute AFSSA, which substantiates that the risks of processing aids are not sufficiently clear if hazards are not assessed in advance (as part of an authorisation procedure). Based on the precautionary principle the French government has therefore introduced a prior authorisation system (the positive list) in order to always carry out a risk analysis. Thus, while the system is a measure with an equivalent effect to a quantitative restriction on imports, it is based on a proper justification (public health).

The CJEU ruled in paragraph 89 the circumstances under which a trading ban (without a licence) is permissible:

'A decision to prohibit marketing, which indeed constitutes the most restrictive obstacle to trade in products lawfully manufactured and marketed in other Member States, can be adopted only if the real risk alleged for public health appears sufficiently established on the basis of the latest scientific data available at the date of the adoption of such decision.'

However, in cases where such a real risk exists, the precautionary principle may be invoked by a Member State, subject to the principle of proportionality. According to the CJEU, this principle means that a Member State may introduce objective and non-discriminatory measures if the existence or extent of a risk cannot be established with certainty but real harm to public health is nevertheless likely. According to the CJEU (in paragraph 92), the conditions for applying this precautionary principle are as follows:

'A correct application of the precautionary principle presupposes, first, identification of the potentially negative consequences for health of the proposed use of processing aids, and, secondly, a comprehensive assessment of the risk to health based on the most reliable scientific data available and the most recent results of international research.'

The French government's argument is that given the broad application and relatively unlimited quantities of processing aids, it is impossible to determine the risk to health, so it is entitled to draw up a general ban on the basis of the precautionary principle. However, the CJEU does not follow this argument, considering that the French regulation violates the principle of proportionality. In view of Articles 6 and 7 General Food Law (GFL),[42] the French government's approach is contrary to the system of risk assessment and the precautionary principle as it applies in European

42 Regulation (EC) No 178/2002 laying down the general principles and requirements of food law.

food law. According to CJEU, the French system is therefore contrary to Article 28 of the EC Treaty (Article 34 TFEU).

The second and third grounds were then also accepted by the CJEU. The regulations that applied in France were very complex, as the 2001 decree had not actually entered into force yet, while a directive from the French government seemed to assume this. The procedure that would have to be followed to get a processing aid on the positive list in France was also unclear and did not have a sufficiently reasonable deadline by which an application would have been processed.

7.6 Practical application

For the French government, this ruling is a 'slap on the wrist', as by the time of the CJEU ruling the regulations had already been amended again in line with what was required, with the inclusion of a proper mutual recognition regime. Nevertheless, this ruling is not without value for practice.

Infringements in the area of free movement of goods occur mainly on aspects of legislation that are not harmonised. Deviations from harmonised aspects (such as under the FIC Regulation[43] or the Nutrition and Health Claims Regulation[44]) are not allowed at all. For non-harmonised aspects, each Member State may still legislate those elements (to some extent), provided it does not violate Article 34 TFEU. Of practical relevance here, for example, are regulations on the limits of food supplements.

The principle of mutual recognition is not always well applied in the Member States. Regulators prefer to wait and see rather than rely on the regulations in other Member States. This CJEU ruling indicates that a Member State may only ban in advance a product lawfully marketed in another Member State if sufficient research shows that a real risk exists, or – if scientific research is not complete or possible – that real harm to public health is likely. Only after a sufficient risk analysis may a Member State invoke the precautionary principle when introducing barriers to trade in foodstuffs.

By definition, authorisation requirements with positive lists cover entire categories of products. As the French government points out, the rationale behind such systems is that there is a lack of knowledge regarding new substances that producers may wish to apply. Why are food additives, genetically modified organisms and novel foods subject to an authorisation requirement in the EU? Not because there is knowledge that they pose a risk, but because in practice there is a lack of knowledge as to whether they are safe. Therefore, by definition, the burden of proof imposed

43 Regulation (EU) No 1169/2011 on the provision of food information to consumers.

44 Regulation (EC) No 1924/2006 on nutrition and health claims made on foods.

by the Court to justify an authorisation requirement cannot be met. The literature[45] has further pointed out that this judgment may turn against the EU in the WTO context. Also under the WTO SPS agreement,[46] trade barriers must be justified by prior risk assessment. The EU believes that on the basis of the precautionary principle, the burden of proof may be reversed. It is then not the government that has to prove that there is a risk, but the industry that has to prove in an authorisation procedure that the product is safe. Surely this argument loses some persuasiveness now that the EU Court of Justice itself apparently does not find it sustainable.

Victor van Ahee
Lawyer at Van Doorne law firm, Amsterdam

45 See Anna Szajkowska and Bernd van der Meulen, Science Based Governance? EU Food Regulation Submitted to Risk Analysis, in Mark Fenwick et al. *Networked Governance, Transnational Business and the Law*, Springer Heidelberg 2014, pp. 57–81, and B.M.J. van der Meulen, Prior authorisation schemes: trade barriers in need of scientific justification, case C-333/08 Commission v. French Republic 'processing aids'; European Journal of Risk Regulation, 2010 vol.1, no.4, pp. 465–471.

46 The agreement on the application of sanitary and phytosanitary measures.

CHAPTER 8

Bread and paper towels – HACCP: the regulator should not sit in the operator's chair

Annotated by Bernd van der Meulen

Court of Justice of the European Union 6 October 2011 Case C 382/10, ECLI:EU: C:2011:639 (*Erich Albrecht*)
Court of Justice of the European Union 6 October 2011 Case C-381/10, ECLI:EU: C:2011:638 (*Astrid Preis*)

Food safety inspectors should not rely on hypothetical risks but should substantially assess the Hazard Analysis and Critical Control Points (HACCP) system chosen by the food business.

Main considerations (*Erich Albrecht*):

> 22 It follows that, in a situation such as that in the main proceedings, where the competent authorities do not appear to have concluded that there was actual contamination, it cannot be concluded that food business operators have infringed the abovementioned paragraph 3 on the basis only of the finding that a potential purchaser could conceivably have touched the foodstuffs by hand or sneezed on them, without considering the measures taken by those operators under Article 5 [HACCP]of the regulation [852/2004] in order to prevent, eliminate or reduce to acceptable levels the hazard which the contamination referred to in paragraph 3 of Chapter IX of Annex II to the regulation may present and without determining that the measures taken in that regard were insufficient in the light of all the available relevant data.
> 23 In that last regard, in particular, it cannot be concluded that those measures were insufficient without duly taking into consideration the possible expert evidence submitted by those operators to show that such containers, used for self-service retail, do not pose hygiene problems.
> 24 Consequently, the answer to the questions referred is that paragraph 3 of Chapter IX of Annex II to the regulation [852/2004] must be interpreted as meaning that, in circumstances such as those at issue in the main proceedings, with regard to containers used for self-service retail of bread and bakery products, the fact that a potential purchaser could conceivably have touched the foodstuffs offered for sale by hand or sneezed on them does not make it possible, on that basis alone, to hold that those foodstuffs were not protected against any contamination likely to render them unfit for human consumption, injurious to health or contaminated in such a way that it would be unreasonable to expect them to be consumed in that state.

 | HTTPS://DOI.ORG/10.1163/9789004751699-009

8.1 Background and question

On the same day, the Court handed down two judgments relating to means-oriented requirements imposed by Austrian food safety inspectors to implement open standards in Hygiene Regulation 852/2004. In the other case,[47] a company had been ordered to install a sink in the staff toilet equipped with hot and cold running water, a soap dispenser and a paper towel dispenser for hygienically drying hands. It also prescribed that the taps in the washroom should be operable without touching by hand. The outcome was similar to that in the present case.

In the present case, the Court states in consideration 9:

> The referring court has before it a number of actions brought by franchised traders offering bread and bakery products for sale. The competent authorities instructed those traders to construct containers for self-service retail of those products in such a way that the products in question can be removed only by technical means, such as tongs or a sliding mechanism, and items already removed from the container cannot be replaced.

In supermarkets in several EU countries, it is quite common for consumers to be able to pick up bread and pastries themselves and put them in a plastic bag – or put them back if they so desire.

Food business operators are responsible for compliance with food law under Article 17 of Regulation (EC) 178/2002. What does this mean in the relationship to inspection when food law uses open standards whose implementation is mandated to the businesses, as in the case of HACCP?

In this 'bread case', the court had the opportunity to rule on this question.

8.2 Responsibility

Pursuant to Article 17 of Regulation (EC) 178/2002 the food business operator is responsible for compliance with food safety standards. The parliamentary history of this provision shows that the European legislator intended this provision to leave to the operator the freedom to determine the means of achieving the objective. The European Commission writes[48] among other things:

---

47 CJEU 6 October 2011, C-381/10, ECLI:EU:C:2011:638 (*Astrid Preis sl*).

48 Para 1.7 of the Proposal for a Regulation of the European Parliament and of the Council laying down the general principles and requirements of food law, establishing the European Food Authority, and laying down procedures in matters of food (COM(2000)716 def), PB C 96E of 27.3.2001, pp. 247–268, https://eur-lex.europa.eu/legal-content/EN/TXT/?uri=uriserv%3AOJ.CE.2001.096.01.0247.01.ENG. N.b. the pdf-version contains only the proposal, the html version also contains the explanatory notes.

'In some areas of European food law, notably in hygiene legislation, the primary responsibility for ensuring compliance with food law, and in particular the safety of the food, rests with the food business. To complement and support this principle there must be adequate and effective controls organised by the competent authorities of the Member States. In other areas of food law this principle is not so widely applicable. This proposal will extend this principle to all food law, and lead to a general review of food law to establish if this principle is respected or whether there are rules where Community legislation has unnecessarily taken responsibility away from the feed or food business by prescribing how a given objective has to be achieved instead of fixing the objective' The preamble to Regulation (EC) 178/2002 takes this line of thought even further in recital 30:

'A food business operator is best placed to devise a safe system for supplying food and ensuring that the food it supplies is safe; thus, it should have primary legal responsibility for ensuring food safety. Although this principle exists in some Member States and areas of food law, in other areas this is either not explicit or else responsibility is assumed by the competent authorities of the Member State through the control activities they carry out.'

8.3 Court considerations

The Court explicitly affirms the primacy of business. To this end, the Court discusses in detail the relationship between food business operator and regulator when it comes to safety assessment.

The Austrian administrative court had asked preliminary questions in a case in which administrative measures had been imposed. The measures had been imposed on shops that sold bread and bakery products that consumers could take out themselves from plastic containers or baskets and put in a plastic bag. This method of selling is also common in supermarkets in other countries.

> 9 (…) The competent authorities instructed those traders to construct containers for self-service retail of those products in such a way that the products in question can be removed only by technical means, such as tongs or a sliding mechanism, and items already removed from the container cannot be replaced.
>
> 10 Those requirements were imposed after checks by the authorities which established that containers had been installed in the food shops in question in the main proceedings for the purpose of self-service retail of bread and bakery products. The facts as established indicate that the covers of those containers have handles, making it possible to lift the covers with one hand while removing the product using tongs made available to the customer with the other. The customer should then replace the tongs and close the cover again.
>
> 11 According to Vienna's Landeshauptmann, the risk of those self-service retail con-

> tainers is that the customer can remove and touch the foodstuffs by hand, or cough or sneeze on them. In addition, that authority pointed out that the mechanism in place does not prevent the customer from returning a product to the container. According to that authority, the exposure of those food products to sneezing by customers can cause germs or viruses to be deposited on the products. Equally, removing the food by hand can contribute to the spreading of germs.

Among other things, the Court notes in consideration 20:

> the European Union legislature's objective of allocating primary responsibility for food safety to food business operators.' Against this background, the court takes the view in recital 22 that:
> '[...] in a situation such as that in the main proceedings, where the competent authorities do not appear to have concluded that there was actual contamination, it cannot be concluded that food business operators have infringed the abovementioned paragraph 3 on the basis only of the finding that a potential purchaser could conceivably have touched the foodstuffs by hand or sneezed on them, without considering the measures taken by those operators under Article 5 of the regulation in order to prevent, eliminate or reduce to acceptable levels the hazard which the contamination referred to in paragraph 3 of Chapter IX of Annex II to the regulation may present and without determining that the measures taken in that regard were insufficient in the light of all the available relevant data.

8.4 Importance for practice

This judgment shows the following. The operator of a food business is responsible for food safety and must establish an HACCP system for that purpose. HACCP is a form of imposed self-regulation. The food business must decide for itself what rules apply to its business processes. These rules, which are created by the company itself, must also be taken as a starting point by the regulator. The regulator may impose measures when concrete food safety breaches have been found. Except for that case, the regulator must materially assess the adequacy of the food safety system taking account of its substantiation (among other things). The regulator may not impose measures on the basis of hypothetical problems and the regulator may certainly not substitute its own preferred tools for the system chosen by the operator. In other words, the regulator should not sit in the operator's chair.

Only the legislator may set target requirements and – possibly – means-oriented requirements.

Bernd van der Meulen
Food legal consultant (*www.BerndvanderMeulen.eu*)

CHAPTER 9

Weintor – Absolutely forbidden claims

Annotated by Bernd van der Meulen

CJEU 6 September 2012, Case C-544/10 (*Deutsches Weintor eG v Land Rheinland-Pfalz*) ECLI:EU:C:2012:526

The suggestion that discomfort occurring on consumption is reduced is a health claim. Any health claim regarding alcoholic beverages is per se misleading. A complete ban on claims for alcoholic beverages is a permissible restriction of fundamental rights.

Main considerations:

15 The parties are therefore in dispute as to whether the fact that a wine is described as 'easily digestible', in conjunction with a reference to low acidity levels, constitutes a 'health claim' within the meaning of the first subparagraph of Article 4(3) of Regulation No 1924/2006, which is normally prohibited in relation to alcoholic beverages.
35 Thus, the concept of a 'health claim' must cover not only a relationship implying an improvement in health as a result of the consumption of a food, but also any relationship which implies the absence or reduction of effects that are adverse or harmful to health and which would otherwise accompany or follow such consumption, and, therefore, the mere preservation of a good state of health despite that potentially harmful consumption.
36 Moreover, the concept of a 'health claim' is deemed to refer not only to the effects of the consumption – in a specific instance – of a precise quantity of a food which is likely, normally, to have only temporary or fleeting effects, but also to those of the repeated, regular, even frequent consumption of such a food, the effects of which are, by contrast, not necessarily only temporary and fleeting.
46 In those circumstances, the compatibility of the prohibition, without exception, of a claim of the kind at issue in the main proceedings must be assessed in the light not only of the freedom to choose an occupation and the freedom to conduct a business, but also of the protection of health.
47 It follows from this that such an assessment must be carried out in accordance with the need to reconcile the requirements of the protection of those various fundamental rights protected by the Union legal order, and striking a fair balance between them […]
51 However, in a case such as that in the main proceedings, even if the claim at issue can be regarded as being substantively inherently correct in that it indicates reduced acidity levels, the fact remains that it is incomplete. The claim highlights a certain quality that facilitates digestion, but is silent as to the fact that, regardless of a sound

 | HTTPS://DOI.ORG/10.1163/9789004751699-010

> digestion, the dangers inherent in the consumption of alcoholic beverages are not in any way removed, or even limited.
>
> 52 Thus, the European Union legislature was fully entitled to take the view that claims such as that at issue in the main proceedings are ambiguous or even misleading where they relate to an alcoholic beverage. By highlighting only the easy digestion of the wine concerned, the claim at issue is likely to encourage its consumption and, ultimately, to increase the risks for consumers' health inherent in the immoderate consumption of any alcoholic beverage. Consequently, the prohibition of such claims is warranted in the light of the requirement to ensure a high level of health protection for consumers.

9.1 Background

The Weintor judgment is important for two reasons. It sheds some light on the scope of the concept of 'health' in the context of health claims, and it is one of the few examples where a food law provision is tested against the Charter of Fundamental Rights of the European Union.

Wine is among the most intensively regulated food products in the EU. Oenological practices – practices that may be applied to wine – are listed exhaustively in Regulations 1308/2013 and 2019/934. De-acidification is one of the permitted practices (de-alcoholisation is not[49]). Deutsches Weintor is a wine producer that applies deacidification. Consumers are notified of this, stating that the wine is '*bekömmlich*'. This German expression is difficult to translate. However, it certainly is less emphatic than the English translation of the ruling 'easily digestible'. Regulators in the German federal state of Rhineland-Palatinate took action against this statement because, in their opinion, it was an unauthorised health claim. With regard to beverages with an alcohol content of more than 1.2%, all health claims are prohibited under article 4(3) of the Nutrition and Health Claims Regulation (1924/2006).[50] The court asks preliminary questions about the scope of the term 'health claim' and the admissibility of a complete ban without exception.

49 The alcohol-free wines that are becoming increasingly popular are therefore strictly speaking in conflict with the law. In contrast to protected dairy and meat names of origin, to date no interested party or court seems bothered about this.

50 The provision states 'Beverages containing more than 1,2 % by volume of alcohol shall not bear health claims. As far as nutrition claims are concerned, only nutrition claims referring to low alcohol levels, or the reduction of the alcohol content, or the reduction of the energy content for beverages containing more than 1,2 % by volume of alcohol, shall be permitted.'

9.2 Health claim

First and foremost is the question of whether the predicate '*bekömmlich*' constitutes a health claim. Underlying that question is the matter of what we actually mean by 'health' in the context of health claims. The legislature has not defined this term. Does it encompass only a more structural condition of well-being or also fleeting moments of well-being? Does it refer only to positive effects of consumption or also comparatively less negative ones?

The lower acidity of the wine could contribute to more pleasant digestion and possibly less abdominal pain. According to the Court, this all falls within the range. Therefore, in respect of alcoholic beverages, statements related to digestion are prohibited.

9.3 Claims ban without exception

The follow-up question is whether it is not a bit much to categorically prohibit the provision of certain information even in cases where they are scientifically proven truths. The referring court expressly places the question in the context of the hypothesis that the assertion is true in a material sense. The referring court asks the question in the light of the Charter of Fundamental Rights of the European Union, more specifically the freedom to pursue a professional activity (Article 15 of the Charter) and the freedom to conduct a business (Article 16 of the Charter). As far as I am concerned, freedom of expression and information (Article 11 of the Charter), in particular, would also have deserved attention. However, it doesn't make much difference. All Charter human rights are subject to the same limitation clause laid down in Article 52 of the Charter.

> Article 52 Scope of guaranteed rights
> 1. Any limitation on the exercise of the rights and freedoms recognised by this Charter must be provided for by law and respect the essence of those rights and freedoms. Subject to the principle of proportionality, limitations may be made only if they are necessary and genuinely meet objectives of general interest recognised by the Union or the need to protect the rights and freedoms of others.

The Court focuses its review on a clash of fundamental rights. It evidently has difficulty accepting the hypothesis of correctness of the claim put forward in the question. On the one hand, it accepts this hypothesis verbally, but on the other hand it reasons this hypothesis off the table with an argument that favourable information about alcoholic beverages is by definition incomplete and (therefore) misleading. In that light, the Court considers proportionality and necessity satisfied.

9.4 Discussion

Per square centimetre, the food label is perhaps the most intensively regulated communication channel in the world. The legislator prescribes in detail what information must be provided, what information may be provided under what conditions and what information may not be provided.

When it comes to prohibited information, the legislator's main concern is to prevent consumers from being misled. The ban on claims on alcoholic beverages is a far-reaching example of censorship. The ban on medical claims (Article 7(3) of the Food Information to Consumers Regulation)[51] is another example. These bans are far-reaching because they also extend to the sharing of science-based facts.[52] The scope is considerable. It appears to involve not only commercial information from producers to consumers but also information to health professionals (CJEU 14 July 2016 case C-19/15 (Innova Vital) ECLI:EU:C:2016:563) and potentially also to information provided by bloggers, influencers and journalists (CJEU 2 April 2009 case C-421/07, Criminal proceedings against Frede Damgaard, ECLI:EU:C:2009:222) even when they are independent of the commercial interest in question.

In the CJEU case of 15 November 2007 (C-319/05, *Commission v Federal Republic of Germany* ('garlic') ECLI:EU:C:2007:678)[53], the Court itself, following the AG, expressed the view that many foodstuffs have therapeutic properties. However, communication about these properties is strictly prohibited.

In the Netherlands, the ban on medical claims is (was?) enforced through medicines legislation. Information about the possible therapeutic role of foods is classified as the presentation of unauthorised medicines.[54]

In the United States, the courts have taken a different route than the Court in this case. According to the US courts, expressing unauthorised health claims is not necessarily misleading – as the regulator FDA had argued – but only potentially misleading. If the misleading element can be removed with a disclaimer, a total ban

---

51 This provision states: 'Subject to derogations provided for by Union law applicable to natural mineral waters and foods for particular nutritional uses, food information shall not attribute to any food the property of preventing, treating or curing a human disease, nor refer to such properties.'

52 Court of Rotterdam 6 March 2014, ECLI:NL:RBROT:2014:1615, for example, concerns a fine that is partly based on a reference on a website to a scientific study which appears to show that food supplements containing magnesium limit the risk of stroke to a small but nevertheless significant extent. The study in question is S.C. Larsson, N. Orsini, A. Wolk (2012). *Dietary magnesium intake and risk of stroke: a meta-analysis of prospective studies, American Journal of Clinical Nutrition*, doi:10.3945/ajcn.111.022376.

53 Included in this book and discussed by Irene Verheijen.

54 See, for example, The Hague District Court 14 November 2012, ECLI:NL:RBSGR:2012:BY5486 discussed in this book by Irene Verheijen.

is too far-reaching a curtailment of constitutionally guaranteed freedoms. Our EU Court does exactly the opposite.[55]

9.5 Side effects

The distinction between foodstuffs and pharmaceuticals partly enshrined in food law has unforeseen side effects. Because the role of nutrition in preventing and treating disease cannot be communicated, industry does not invest in research in this area. Knowledge thus remains relatively limited. At the time of the Covid pandemic, certain – now reviled – doctors used nutritional supplementation in the absence of a drug. According to them, with some success. In particular, this concerned vitamins and minerals for which the European Food Safety Authority (EFSA) recognised claims regarding immune system function.[56] The Nutrition Centre in the Netherlands made the objection that healthy diet, sleep and exercise are sufficient for proper immune system function. According to RIVM (National Institute for Public Health and the Environment) figures, those who feed themselves healthily are in the minority. In terms of compliance with the advice to eat enough fruits and vegetables, that is not even 20%.[57] It therefore seems plausible that some part of the population in the Netherlands is qualitatively malnourished. This suspicion is confirmed by the prevalence of non-communicable diseases that are often nutrition-related such as cardiovascular disease (1.55 million people)[58], diabetes (1.2 million people)[59], obesity (14% of the population)[60] and more generally overweight (half of the population)[61]. Under these circumstances, it seems legitimate to ask whether it is possible to use nutritional supplementation to accelerate the elimination of accumulated nutritional quality deficits. An answer to this question has not been given to date. However, the Netherlands Food and Consumer Product Safety Authority (*NVWA*) did issue a public warning on 26 March 2020 entitled: 'nutritional supplements do not help against coronavirus'. It wrote, among other things: 'Taking more vitamins does not give you a higher resistance nor does it protect you better against pathogenic bacteria and viruses.' The problem with the Nutrition

55 For more details on this issue, see farewell lecture, Bernd van der Meulen, The Ultimate Truth. Demons, Dogma, Dominance, Definitions and Democracy - Farewell Address, European Institute for Food Law, working paper 2019/02 https://food-law.nl/Working-papers.

56 In particular vitamins A, C, D, B6, B12 and folic acid and the minerals zinc, iron, selenium and copper. See the EU register of approved claims: https://food.ec.europa.eu/food-safety/labelling-and-nutrition/nutrition-and-health-claims/eu-register-health-claims_en.

57 See https://www.wateetnederland.nl/resultaten/voedingsmiddelen/richtlijnen/alle_richtlijnen.

58 Heart foundation.

59 Diabetes foundation.

60 CBS https://www.cbs.nl/nl-nl/nieuws/2018/27/100-duizend-volwassenen-hebben-morbide-obesitas.

61 VWS https://www.cbs.nl/nl-nl/nieuws/2024/10/obesitas-afgelopen-40-jaar-verdrievoudigd.

Centre's advice and the NVWA's warning is that it takes the healthy person as its starting point and ignores the large groups with diet-related health deficiencies. Part of the explanation is undoubtedly that within food law, disease cannot be talked about.[62]

If the Covid pandemic teaches us one thing in food law, it is that we need to rethink the relationship between nutrition and health, including disease prevention and cure. To do so, we must move away from the distinction between food and pharma. To that end, we must move away from communication bans that discourage scientific research and make it harder for consumers to make informed choices.

Even before the pandemic, the European Commission's Joint Research Centre (JRC)[63] envisaged a possible future in which 'phood' plays a central role. 'Phood' is the symbiosis between pharma and food. One of its elements is personalised nutrition. This involves nutrition directly related to personal health status. A future with 'phood' must get rid of communication bans within the field of nutrition and health. It is therefore very much to be hoped that on the next occasion the Court will rule differently on article 7(3) of Food Information Regulation than it did in this case on article 4(3) of the Claims Regulation.

Bernd van der Meulen
Director European Institute for Food Law (www.food-law.nl)

62 With the exception of disease risk reduction and medical nutrition.

63 JRC, *Delivering on EU Food Safety and Nutrition in 2050 - Future challenges and policy preparedness* (2016) https://publications.jrc.ec.europa.eu/repository/handle/JRC101971.

CHAPTER 10

Berger – Naming & shaming | Unfit for consumption is by definition unsafe

Annotated by Barbara Mutsaers

Court of Justice of the European Union 11 April 2013, case C-636/11, ECLI:EU:C:2013:227 (*Berger v Freistaat Bayern*)

General food law, unfit for human consumption, safety warning, enforcement, national authorities.

Main considerations:

> *The Court (Fourth Chamber) rules:*
> Article 10 of Regulation (EC) No 178/2002 of the European Parliament and of the Council of 28 January 2002 laying down the general principles and requirements of food law, establishing the European Food Safety Authority and laying down procedures in matters of food safety, must be interpreted as not precluding national legislation allowing information to be issued to the public mentioning the name of a food and the name or trade name of the food manufacturer, processor or distributor, in a case where that food, though not injurious to health, is unfit for human consumption. The second subparagraph of Article 17(2) of that regulation must be interpreted as allowing, in circumstances such as those of the case in the main proceedings, national authorities to issue such information to the public in accordance with the requirements of Article 7 of Regulation (EC) No 882/2004 of the European Parliament and of the Council of 29 April 2004 on official controls performed to ensure the verification of compliance with feed and food law, animal health and animal welfare rules.

10.1 The case

Article 14 of Regulation 178/2002[64] – which regulation is also known as the General Food Law (GFL) – states that food may not be placed on the market if it is unsafe. Food is deemed unsafe if it is considered to be:

a. injurious to health;
b. unfit for human consumption.

64 Regulation (EC) No 178/2002 of the European Parliament and of the Council of 28 January 2002 laying down the general principles and requirements of food law, establishing the European Food Safety Authority and laying down procedures in matters of food safety.

 | HTTPS://DOI.ORG/10.1163/9789004751699-011

In assessing whether a food is unfit for human consumption, consideration shall be given to whether a food is unacceptable for human consumption, having regard to its intended use, as a result of contamination by extraneous matter or otherwise, or through putrefaction, deterioration or decay.

Member States enforce food law and should verify food business operators' compliance with the applicable requirements of food law at all stages of production, processing and distribution. To this end, Member States should maintain a system of official inspections and other situation-appropriate activities, including communication with the public on the safety requirement and risks of food.[65] On 16 and 18 January 2006, the veterinary service in Passau, Germany, carried out official inspections at a number of establishments of the Berger Wild GmbH group (hereinafter Berger Wild'), which is engaged in the processing and sale of game. As the authorities found poor hygienic conditions during the control, samples of that meat were taken and sent to the *Bayerische Landesamt für Gesundheit und Lebensmittelsicherheit* (Bavarian Health and Food Safety Authority (hereinafter 'LGL') for analysis. Those analyses showed that the foodstuffs in question were unfit for human consumption and therefore unsafe under Regulation 178/2002.

The *Bayerische Staatsministerium für Umwelt, Gesundheit und Verbraucherschutz* (Bavarian Ministry for the Environment, Health and Consumer Protection) subsequently informed Berger Wild on 23 January 2006 that – in accordance with §40 of the *Lebensmittel-, Bedarfsgegenstände und Futtermittelgesetzbuch* (hereinafter 'LFGB') – it intended to inform the public that the food in which peculiarities had been found during the above-mentioned checks were unfit for human consumption.[66] Berger Wild opposed this intention. They considered it disproportionate and proposed to publish a 'product warning' inviting its customers to come and exchange the products mentioned in the warning at the usual points of sale, since sensory abnormalities could occur with these products without posing a health risk.

In three press releases dated 24, 25 and 26 January 2006, the Minister for Consumer Protection of the Free State of Bavaria (hereinafter the 'Minister') an-

65 Article 17(2) Regulation (EC) No 178/2002.

66 Article 40 of the Lebensmittel-, Bedarfsgegenstände und Futtermittelgesetzbuch (Federal Republic of Germany Code covering foods, consumer goods and feedstuffs) of 1 September 2005 (BGBl. 2005 I, p. 2618), amended on 18 October 2005 (BGBl. 2005 I, p. 3007), in the applicable version of it from 17 September 2005 up to and including 24 April 2006 (hereinafter: 'LFGB'), states, among other things: 'Within the framework of Article 10 of Regulation (EC) No 178/2002, the competent authority may inform the public of the name of the food or animal feed and the name or trade name of the food or animal feed manufacturer, processor or distributor; where doing so is better able to prevent risks, it may also release the name of the operator responsible for placing on the market. A public information measure, within the meaning of and in accordance with the above rules, may also be taken in the following cases (...)4. where food, which is not injurious to health but is unfit for human consumption, in particular because it is nauseating, is or has been distributed in significant quantities or where, because of its specificity, it has been distributed only in small quantities but over a relatively lengthy period of time.'

nounced the recall of game-based products marketed by Berger Wild. The press release stated that investigations had shown that meat samples from the said batches smelt rancid, musty, spoiled or sour and that in certain meat samples examined, the putrefaction process had already begun. The press release also stated that inspections at Berger Wild establishments had revealed appalling hygiene conditions and that the competent authorities had immediately imposed a temporary ban on marketing the products produced or processed by Berger Wild at those establishments.

In a speech to the Bavarian parliament on 31 January 2006, the Minister stated, among other things, that Berger Wild could no longer market its products, had filed for bankruptcy the same day, and that consequently a health risk from newly marketed products could be ruled out.

Since Berger Wild/Karl Berger[67] believed that they had suffered significant damage as a result of press releases issued by the authorities of the Freistaat Bayern, they brought an action for damages against the Freistaat Bayern before the Landgericht München I. This is a matter of state liability.

Berger Wild/Karl Berger argued that Article 10 of Regulation 178/2002 allows the public to be informed only when there is an actual danger to health, and not when it is only a case of food being unfit for human consumption. The Freistaat Bayern, on the other hand, argued that under the aforementioned Article 10, the competent national authorities may also issue a warning to the public where there is no actual danger to health.

The Landgericht München I, before which the case was pending, assumed that the foodstuffs covered by the warnings – under § 40 LFGB – could not be classified as harmful to human health. The Landgericht München I also assumed in its preliminary assessment that the warnings addressed to consumers under the LFGB were regular, but it questioned whether that arrangement complied with Regulation 178/2002. In that context, the Landgericht München I suspended the proceedings and asked the Court of Justice of the European Union (hereinafter the 'Court') for a preliminary ruling on the following questions:

> (1) Does Article 10 of [Regulation (EC) No 178/2002] preclude rules of national law allowing information to be issued to the public mentioning the name of a food or animal feed and the name or trade name of the food or animal feed manufacturer, processor or distributor, in the event that the food is not injurious to health but is unfit for human consumption, particularly food that is nauseating, is or has been distributed in significant quantities or, because of its specificity, has been distributed only in small quantities but over a relatively lengthy period of time?'

67 In some places in the Court ruling it says that Berger Wild brought an action before Landgericht München I for damages against the Freistaat Bayern. However, in other places in the ruling it says that there were proceedings between Karl Berger and the Freistaat Bayern.

(2) If the answer to the previous question is in the affirmative:
Would the answer to the first question be different if the situation at issue arose prior to 1 January 2007, but at a time at which national law had already been brought into line with the above regulation?

10.2 Judgment of the Court

Article 10 of Regulation 178/2002, to which the preliminary questions relate, imposes an information obligation on public authorities only where there are reasonable grounds to suspect that a food or feed presents a risk to human or animal health. The article reads as follows:

'Without prejudice to the applicable provisions of Community and national law on access to documents, where there are reasonable grounds to suspect that a food or feed may present a risk for human or animal health, then, depending on the nature, seriousness and extent of that risk, public authorities shall take appropriate steps to inform the general public of the nature of the risk to health, identifying to the fullest extent possible the food or feed, or type of food or feed, the risk that it may present, and the measures which are taken or about to be taken to prevent, reduce or eliminate that risk.' Thus, the aforementioned provision in itself does not prohibit the government from informing the public if a food is unfit for human consumption but not harmful to health.

The Court considers that, in order to provide a useful answer for the resolution of the dispute before the referring court, Article 17(2) of Regulation 178/2002 must also be interpreted. Although that provision is not expressly mentioned in the questions referred for a preliminary ruling, the Court considers itself competent to do so, referring to the judgment of the Court of 14 October 2010, in case C-243/09, ECLI:EU:C:2010:609, paragraph 39 and the case-law cited there. Indeed, it was held there that, in the context of the procedure of cooperation between national courts and the Court established by Article 267 of the Treaty on the Functioning of the European Union, the Court's task is to provide the national court with a useful answer for the resolution of the dispute before it. To that end, the Court must, where appropriate, reformulate the questions submitted to it. The Court's task is, after all, to interpret all the provisions of European Union law necessary for resolving disputes pending before the national courts, even where those provisions are not expressly mentioned in the questions referred by those courts.[68]

Article 17(2) of Regulation 178/2002 provides, inter alia, that Member States are

68 See, in that sense, the rulings of CJEU 8 March 2007, Campina, C45/06, ECLI:EU:C:2007:154, rec 30 and 31; CJEU 26 June 2008, Wiedemann and Funk, C329/06 and C343/06, ECLI:EU:C:2008:366, rec. 45, and CJEU 2 September 2010, Kirin Amgen, C66/09, ECLI:EU:C:2010:484, rec. 27 and the judgment of the Court of 14 October 2010, in C-243/09, ECLI:EU:C:2010:609.

to maintain a system of official controls and other activities as appropriate to the situation, including communication with the public concerning food safety and risks. The Court considers that that article must be interpreted as allowing the national authorities, in circumstances such as those in the main proceedings, to inform the public to that effect. Indeed, Article 14 of Regulation 178/2002 provides that a food unfit for human consumption must be considered 'unsafe'. A food that is unacceptable and therefore unfit for human consumption does not satisfy the food safety requirements set out in Article 14(5) of Regulation 178/2002 and is in any event likely to harm consumers' interests. This while Article 5 of Regulation 178/2002 explicitly states that protecting human life and health and the interests of consumers is precisely one of the objectives of food law.

In view of this, when food is not injurious to human health but does not comply with the said food safety requirements because it is unfit for human consumption, national authorities may inform consumers accordingly, as provided for in Article 17(2) of Regulation 178/2002. The Court ruled that this must, however, comply with the requirements of Article 7 of Regulation (EC) No 882/2004 of the European Parliament and of the Council of 29 April 2004 on official controls performed to ensure the verification of compliance with feed and food law, animal health and animal welfare rules. Article 7 of Regulation 882/2004 provided, inter alia, that the public in general shall have access to information on the control activities of competent authorities and their effectiveness and, secondly, that the competent authority shall take the necessary measures to ensure that its staff do not disclose information obtained in the course of official controls which, in duly justified cases, by its nature is covered by professional secrecy.

The Court therefore ruled that Article 10 of Regulation 178/2002 does not preclude national legislation, such as the German provision in question, which allows the public to be informed – with the name of the food and that of the company under whose name or business name the food was produced, processed or marketed – if such a food is not harmful to health but is unfit for human consumption. Before Berger, there was a difference of opinion in the literature as to whether by classifying unfit food as 'unsafe' the legislator indeed meant to subject certain demonstrably non-harmful foodstuffs to the full weight of the food safety regime as well, or whether, in the case of foodstuffs showing defects as mentioned in the fifth paragraph of Article 14 of Regulation (EC) 178/2002, it was much more about a (rebuttable) presumption of injuriousness.[69] In my view, the Court's judgment is logical. After all, the European legislator deliberately made the choice to qualify

69 For this discussion, see for example Bernd van der Meulen, 'The safe food principle. A critical reflection on the key concept of EU food safety law', European Institute for Food Law working paper 2018/08, and Daniela Polinski and Bernd van der Meulen, Unfit for Human Consumption. The elusive element in the EU food safety concept of Article 14 GFL, European Institute for Food Law working paper 2019/01 (www.food-law.nl).

unfit foodstuffs as unsafe and thus prohibited their marketing. It would therefore be illogical if the same regulation did prohibit authorities from informing the public about food placed on the market in violation of the law.

With regard to the second question – which, incidentally, the Court does not deal with explicitly but rather implicitly in the judgment – the following applies. Indeed, the Court repeatedly points out that, according to Article 65(2) of Regulation 178/2002, Articles 14 to 22 apply from 1 January 2005 and that Regulation 882/2004 – according to Article 67 of that Regulation – applies from 1 January 2006. It can therefore be argued that the Court implicitly determined that it does not matter if the facts took place before 1 January 2007 (the end of the implementation period). After all, the aforementioned regulations/articles were already in force at the time of the publications and statements and the regulations have direct effect.

10.3 Subsequent developments

The ruling dates back several years. Regulation 178/2002 is still in force and Articles 10, 14 and 17 have not been amended since then. In contrast, Regulation 882/2004 is no longer in force and was repealed by Regulation 2017/625 with effect from 15 December 2019.[70] However, Article 8 of Regulation 2017/625 contains a similar clause to Article 7 of Regulation 882/2004. For example, Article 8(1) of Regulation 2017/625 provides – in summary – that unless there is an overriding public interest in the disclosure of information covered by professional secrecy, the competent authorities shall ensure that information acquired by them in the course of carrying out their duties in the context of official controls and other official activities, which by its nature is covered by professional secrecy in accordance with Union or national law, is not disclosed to third parties. Member States should ensure that they impose appropriate confidentiality obligations on staff and other persons employed during official controls and other official activities for that purpose. In doing so, the aforementioned also applies to control authorities for organic products and authorised bodies to which natural persons to whom certain official control tasks have been delegated, and official laboratories. In addition, Article 8(5) of Regulation 2017/625

70 Article 146 of Regulation (EU) 2017/625 of the European Parliament and of the Council of 15 March 2017 on official controls and other official activities performed to ensure the application of food and feed law, rules on animal health and welfare, plant health and plant protection products, amending Regulations (EC) No 999/2001, (EC) No 396/2005, (EC) No 1069/2009, (EC) No 1107/2009, (EU) No 1151/2012, (EU) No 652/2014, (EU) 2016/429 and (EU) 2016/2031 of the European Parliament and of the Council, Council Regulations (EC) No 1/2005 and (EC) No 1099/2009 and Council Directives 98/58/EC, 1999/74/EC, 2007/43/EC, 2008/119/EC and 2008/120/EC, and repealing Regulations (EC) No 854/2004 and (EC) No 882/2004 of the European Parliament and of the Council, Council Directives 89/608/EEC, 89/662/EEC, 90/425/EEC, 91/496/EEC, 96/23/EC, 96/93/EC and 97/78/EC and Council Decision 92/438/EEC (Official Controls Regulation).

explicitly states once again that the confidentiality obligations in that Article do not prevent competent authorities from publishing or otherwise disclosing information on the outcome of official controls concerning individual operators, provided that, without prejudice to situations where disclosure is required by Union or national legislation, the following conditions are met:

> (a) the operator concerned is given the opportunity to comment on the information that the competent authority intends to publish or make otherwise available to the public, prior to its publication or release, taking into account the urgency of the situation; and
> (b) the information which is published or made otherwise available to the public takes into account the comments expressed by the operator concerned or is published or released together with such comments.

10.4 Netherlands

Since 2006, the Ministers of Economic Affairs (now Agriculture) and Health, Welfare and Sport – the names of these ministries have changed a number of times over the years – have already made control data from the Netherlands Food and Consumer Product Safety Authority (*NVWA*) public.[71] This was done by invoking Article 8 of the Public Access Act of administration (Wob)/now Article 3.3 of the Open Government Act (Woo).[72] So there is also a national regulation in the Netherlands, which gives the power to do this in certain cases.

The wording of Article 3.3 Woo makes it clear that administrative bodies in the Netherlands not only have the power, but in some cases even the obligation, to provide information. For example, that article stipulates that the administrative body that directly concerns it must provide information on its own initiative about the policy, including its preparation and implementation, as soon as this is in the interests of good and democratic administration. The administrative body shall ensure that the information is provided in a comprehensible form, in such a way that stakeholders and interested citizens are reached as far as possible and at such times that they can make their views known to the administrative body in good time. Publication of NVWA's inspection data generally serves the following purposes:

1. Offers transparency to parties within the chain so that they can base their choices on it;

---

71 *Kamerstukken* 2014-2015, 34.111, no. 3, *Memorie van Toelichting*.

72 Woo makes a distinction between two forms of information provision: in response to a request to this effect (passive information provision) or on one's own initiative (active information provision).

2. Contributes to consumer confidence in the hands of government, food and other products/services; and
3. Increases compliance.[73]

In addition, the disclosure of inspection data aims to promote transparency of the NVWA's supervisory activities.[74]

On the basis of Article 8 3.3 Woo, a decision can be made on active disclosure on a case-by-case basis, whereby the exceptions and limitations contained in Woo apply and must be taken into account when making a decision. In the Decree containing policy rules on disclosure of inspection data by NVWA (hereinafter 'Policy Rules'), the policy is elaborated further.[75] For example, Article 2 in conjunction with Article 5c of the Policy Rules stipulates that with regard to the control of the application of food safety rules – namely the rules and requirements arising from food legislation, as referred to in Article 3 of Regulation 178/2002, with which a food business must comply – the Minister of Agriculture, Nature and Food Quality decides with due observance of Articles 5.1 and 5.2 Woo to publish on the NVWA's website, the following information, among other things: the name and address of the company or location where the inspection took place, the rules and requirements on which the inspections took place, in which of one of the categories 'business suspended', 'enhanced supervision', 'points for improvement established' or 'complies' the business or premises was placed on the basis of the inspection findings, and, where relevant, product type and brand name of the product. Because of the present Court judgment, we know that this type of national regulation and the active informing

73 The Minister of Public Health, Welfare and Sport (VWS) and the Minister of Agriculture, Nature and Food Quality (LNV) communicated this principle to the Chair of the Lower House of the States-General in the letters of 23 June 2005 (*Kamerstukken* II 2004/2005, 26 991, no. 122), 23 February 2006 (*Kamerstukken* II 2005/2006, 26 991, no. 138) and 10 December 2009 (*Kamerstukken* II 2009/2010, 26 991, no. 276). It is also, for example, emphasised by the VWS Minister and the Secretary of State for Economic Affairs, Agriculture and Innovation in the letter dated 14 April 2011 (*Kamerstukken* II 2010/2011, 26 991, no. 314) and in the letter dated 7 August 2012 (*Kamerstukken* II 2012/2013, 33 289, no. 3). This is just a selection.

74 Framework for publishing NVWA control data of companies (at aggregated and individual level) d.d. 19 November 2015, https://www.nvwa.nl/over-de-nvwa/hoe-de-nvwa-werkt/namen-bekend-maken-van-gecontroleerde-bedrijven-en-producten/waarom-nvwa-namen-noemt-van-bedrijven-of-producten. https://www.openbare-inspectieresultaten.nvwa.nl/.

75 For the sake of completeness, it should be noted that there are other rules in the Netherlands, for example for publication of information that is obtained on the observance of other legislation such as the Medicines Act. In this way, for example, the Healthcare Act and Youth Act in the Netherlands have changed. This was done as a result of legislative proposal 34.111, whereby a specific principle and obligation is created for the active publication of regulatory data and sanction decisions for the Netherlands Food and Consumer Product Safety Authority (*NVWA*), the Healthcare Inspectorate (IGZ), the Youth Inspectorate (IJZ) and the Security and Justice Inspectorate (IVenJ). In addition, the Decision on disclosing regulatory and implementation data Healthcare Act and Youth Act has come into force.

of the public on that basis, indicating the name of the food and that of the business under the name or trade name of which the food was produced, processed or marketed, is permissible under European law even in cases where the food in question is merely unfit.

The Court – having regard inter alia to the question put by the referring court – confined itself in this case only to Regulation 178/2002. Article 16 of Directive 2001/95/EC of the European Parliament and of the Council of 3 December 2001 (GPSD) on general product safety also contains a provision entitling authorities of Member States to make available to the public, in certain cases, information on the risks posed by products to the health of consumers. For example, Article 16 of GPSD reads as follows:

> 1. Information available to the authorities of the Member States or the Commission relating to risks to consumer health and safety posed by products shall in general be available to the public, in accordance with the requirements of transparency and without prejudice to the restrictions required for monitoring and investigation activities. In particular the public shall have access to information on product identification, the nature of the risk and the measures taken.
> However, Member States and the Commission shall take the steps necessary to ensure that their officials and agents are required not to disclose information obtained for the purposes of this Directive which, by its nature, is covered by professional secrecy in duly justified cases, except for information relating to the safety properties of products which must be made public if circumstances so require, in order to protect the health and safety of consumers.
> 2. Protection of professional secrecy shall not prevent the dissemination to the competent authorities of information relevant for ensuring the effectiveness of market monitoring and surveillance activities. The authorities receiving information covered by professional secrecy shall ensure its protection.

At the time, it was considered that Article 16 GPSD did not need to be implemented separately in Dutch law, given Article 8 Wob. However, Article 8 Wob extends beyond the aforementioned Article 16 of Directive 2001/95/EC, as mentioned above. This is relevant, because Article 16 GPSD refers to 'risks to consumer health and safety posed by products', whereas in the case of unsuitability of a product, the product is not harmful to health. Although GPSD does not pronounce on its nature, it is generally considered that Member States are allowed to take more far-reaching measures when it comes to general product safety, in so far as they are in line with the provisions of the Union Treaty.[76] This is the reason why, in my view, it can also

---

76 P. Dejemeppe, 'La nouvelle loi sur la sécurité des consommateurs', Journal des tribunaux (Belgium) 1994, 413 and G. Howells, Consumer product safety, Aldershot, Dartmouth Publishing Company, 1998, p. 69. Howells mentions explicitly a minimum harmonisation.

be considered that GPSD does not preclude national legislation that allows the public to be informed – with an indication of the name of the food and that of the company under whose name or trade name the food was produced, processed or marketed – if such a food is not harmful to health but is unfit for human consumption.

10.5 Judgment of the Court

The judgment does show that such active disclosures by authorities can have major (financial) consequences for a company. Damages might have been partly avoided if Berger Wild itself had complied with its obligations. Article 19 of Regulation 178/2002 applies both in the case where there is a suspicion of an unfit foodstuff as well as in the case of a harmful foodstuff. Indeed, Article 19 of Regulation 178/2002 states: 'If a food business operator considers or has reason to believe that a food which it has imported, produced, processed, manufactured or distributed is not in compliance with the food safety requirements …'. And as the present Court judgment shows, Article 14 of Regulation 178/2002 contains food safety requirements, and a food that is unacceptable and therefore unfit for human consumption does not comply with the food safety requirements as set out in Article 14(5) of Regulation 178/2002.

Barbara Mutsaers
Food law lawyer at AKD Benelux Lawyers and member of the editorial board of Waar & Wet.

CHAPTER 11

Teekanne – Misleading labelling landmark

Annotated by Ebba Hoogenraad

Court of Justice of the European Union 4 June 2015, C-195/14, ECLI:EU:C:2015:361 (*Teekanne*)[77]

Foods, labelling, presentation, ingredient declaration, deception, average consumer.

Main considerations:

> 38 However, the fact that the list of ingredients is displayed on the packaging of the goods at issue in the main proceedings does not in itself exclude the possibility that the labelling of those goods and methods used for it may be such as to mislead the purchaser within the meaning of Article 2(1)(a)(i) of Directive 2000/13.
> 39 The labelling, as defined in Article 1(3)(a) of that directive, is composed of any words, particulars, trade marks, brand name, pictorial matter or symbol relating to a foodstuff and placed on its packaging. Some of those items may in practice be misleading, erroneous, ambiguous, contradictory or incomprehensible.
> 40 In that case, the list of ingredients, even though correct and comprehensive, may in some situations not be capable of correcting sufficiently the consumer's erroneous or misleading impression concerning the characteristics of a foodstuff that stems from the other items comprising its labelling.
> 41 Therefore, where the labelling of a foodstuff and methods used for the labelling, taken as a whole, give the impression that a particular ingredient is present in that foodstuff, even though that ingredient is not in fact present, such labelling is such as could mislead the purchaser as to the characteristics of the foodstuff.

11.1 Summary and discussion

When assessing whether a package is likely to create a misleading impression on the average consumer regarding the ingredients of the food, the entire package must be assessed. A correct ingredient declaration cannot always eliminate the deception created by other elements on the packaging. In particular, the court must take into account the wording and images used and the location, size, colour, font, language, sentence structure and punctuation used of the various elements on the packaging.

77 This CJEU decision is annotated in IER 2015/53 by E.H. Hoogenraad & B. Duivenvoorde.

 | HTTPS://DOI.ORG/10.1163/9789004751699-012

11.2 Prejudicial question and answer

German company Teekanne markets fruit tea under the name '*Felix Himbeer-Vanille Abenteuer*' (Felix raspberry-vanilla adventure). The packaging includes images of raspberries and vanilla blossoms with accompanying red-and-yellow colouring, Also on the packaging is a logo with the text '*nur natürliche Zutaten*' (only natural ingredients) and the product is described as '*Früchtentee mit natürlichen Aromen*' (fruit tea with natural aromas).

A German consumer organisation is filing a lawsuit against Teekanne. The complaint is that the packaging is misleading: it gives consumers the impression that the tea contains components of raspberry and vanilla, or at least natural raspberry or vanilla flavourings. This turns out not to be the case: the flavourings are not obtained from vanilla and raspberry, despite the fact that they do taste like vanilla and raspberry. The list of ingredients clearly communicates that Teekanne does not contain raspberry, vanilla or their flavourings.

The *Bundesgerichtshof* in Germany refers a preliminary question to the Court of Justice of the European Union (CJEU): is it permitted for food packaging to give the impression that the product contains a certain ingredient when this is not the case, and this latter is only apparent from the ingredient declaration?

The CJEU states first that consumers should be provided with correct and non-misleading food information. The guiding principle is the expected behaviour of the *Gut-Springenheide*[78] consumer (the normally informed and reasonably prudent and observant average consumer). The CJEU emphasises that this average consumer, whose purchase is determined by the composition of the product, first reads the ingredient declaration. But the fact that the list of ingredients is on the packaging does not preclude that the labelling or packaging can mislead the purchaser. The list of ingredients, even if correct and complete, may in certain situations be insufficient to adequately correct the consumer's erroneous or ambiguous impression about the characteristics of the food arising from other elements of the packaging/labelling. A label which, taken as a whole, gives a false impression that the food contains an ingredient which it does not in fact contain is misleading. The national court will have to examine, in the light of all the various elements of the labelling, whether the labelling can mislead the average consumer as to the presence of raspberry and vanilla blossom ingredients or flavourings obtained from those ingredients. In so doing, the national court must take particular account of the wording and illustrations used and the placement, size, colour, font, language, sentence structure and punctuation used of the various elements on the packaging.

---

78 CJEU 16 July 1998, case C-210/96 (*Gut Springenheide GmbH, R. Tusky v. Oberkreisdirektor des Kreises Steinfurt – Amt für Lebensmittelüberwachung, in the presence of Oberbundesanwalt beim Bundesverwaltungsgericht*) ECLI:EU:C:1998:369.

Answer to the preliminary question:
In the light of the foregoing considerations, the answer to the question referred is that Articles 2(1)(a)(i) and 3(1)(2) of Directive 2000/13 must be interpreted as precluding the labelling of a foodstuff and methods used for the labelling from giving the impression, by means of the appearance, description or pictorial representation of a particular ingredient, that that ingredient is present, even though it is not in fact present and this is apparent solely from the list of ingredients on the foodstuff's packaging.

The *Bundesgerichtshof*'s final verdict[79] is that the product presentation of the Teekanne tea is indeed misleading. The eye-catching information '*Himbeer-Vanille-Abenteuer*' and the images of the raspberries and vanilla blossom lead the average consumer to expect real raspberry and vanilla, while that is not what the consumer will get. The correct information in the ingredient declaration does not negate this deception.

11.3 Discussion

With its question to the CJEU, the *Bundesgerichtshof* challenges the so-called labelling doctrine. The content of this doctrine is as follows: the consumer who wants to know what is what, reads the entire packaging (including ingredient declaration) and is not – or more mildly phrased: not easily – misled by suggestive names, claims or (images on) packaging when the consumer can read exactly what is the actual content of the product based on the ingredient declaration.

The labelling doctrine of the Court of Justice of the European Union dates back to the *Cassis de Dijon* decision of 1979,[80] in which the CJEU ruled that stating the alcohol content on a bottle of *Cassis de Dijon* is sufficient to inform consumers about the composition of the product, and that for this purpose, legislating a specific alcohol content is unnecessary. The rationale: imposing mandatory rules is often unnecessary, as consumers can make an excellent choice based on information made available.

Until the Teekanne decision, the labelling doctrine was standard case-law in the Netherlands, including at the self-regulatory Advertising Code Committee (*Reclame Code Commissie: RCC*) and its Board of Appeal (*College van Beroep: CvB*). The idea was that those who care to know exactly how the product is composed, will do the research. Those who do not care, do not and are therefore not misled. That has now completely changed by the Teekanne decision.

The CJEU begins by stating that consumers must be provided with accurate and

79 Bundesgerichtshof, 02-12-2015 – I ZR 45/13 http://juris.bundesgerichtshof.de/cgi-bin/rechtsprechung/document.py?Gericht=bgh&Art=en&nr=74666&pos=0&anz=1.

80 CJEU 20 February 1979, C-120/78, ECLI:EU:C:1979:42 (*Cassis de Dijon*).

non-misleading food information. The guiding principle is the expected behaviour of the average consumer, who is reasonably well informed and reasonably observant and circumspect. The CJEU emphasises that this average consumer, whose purchase is determined by the composition of the product, first reads the ingredient declaration. So far, nothing new. But then the CJEU continues with a landmark ruling. The CJEU holds that correct information on the back of the label does not eliminate any misleading impression ('cannot sufficiently correct'). In assessing whether the average consumer might be misled, account should be taken of the packaging as a whole. The CJEU goes on to state that the national court should examine, with reference to 'all the various elements of the labelling', whether a normally informed and reasonably observant and circumspect average consumer may be misled.

The CJEU thereby sets out the criterion for assessment: in its examination, the national court must take into account in particular (but not exclusively) the wording and images used and the placement, size, colour, font, language, sentence structure, and punctuation used of the various elements on the packaging.

In the annotation of the Teekanne decision in the Dutch (legal) magazine Intellectual Property and Advertising law (IER), the annotators signalled that the entire packaging remains the starting point, including the list of ingredients. Some degree of exaggeration will remain possible; it is all about the expectation of the average consumer. It is also possible that the standard colour schemes (pink is strawberry flavour, orange refers to orange flavour) can remain in place without objection because the consumer understands that this does not imply the presence of actual fruit (extract). The annotators predicted:

> One thing is certain. The Teekanne case will have far-reaching consequences. Critical consumer groups will invoke Teekanne to substantiate their case. It will come down to a 'circumstances-of-the-case' assessment. An important responsibility lies with the courts and the ACC to strike a careful balance between deception and persuasion. And this also means that the consumer needs to be valued appropriately, without disregarding or patronising them. The consumer is not an absolute food expert, but they are neither ignorant or naive.

The ACC and BoA have indeed, in a long series of decisions, systematically applied the assessment framework that follows from the Teekanne decision (see, for example, the *Van Mook Muesli* ruling of the BoA, also published in this book in Chapter 53)[81]. The muesli in question actually contains hazelnuts, but so much less than the packaging suggests that it was still misleading. In the *Holy Soda* ruling[82] (see Chapter 48 of this book), shortly after the Teekanne decision, the ACC ruled that

81 CvB 12 October 2018, file 2017/00469 (*Van Mook Muesli*), also in Chapter 53 of this book.

82 RCC 15 September 2015, file 2015/00862 (*Holy Soda*).

the statement '100% natural' seen in the context of the entire label was not incorrect. The ACC and the BoA are on a case-by-case basis, striking to find the right balance.[83]

Foodwatch advocates a strict application of the Teekanne decision. With Teekanne in hand, the watchdog has already initiated several proceedings before the ACC and has sought extensive publicity in doing so.

The response of the Dutch food industry to the Teekanne decision can be found on the supermarket shelves. Since the decision, a consistent addition of the word 'flavour' appears on many packages. At supermarket Albert Heijn, for example, you no longer buy *Frisse Frambozen Vla* (fresh raspberry custard), but *Frisse Frambozen Smaak Vla* (fresh raspberry-flavoured custard). In this way, it is made clear to the consumer that the custard does not contain real raspberries (or raspberry extract), but flavouring that tastes like raspberries. The Dutch supervisory authority NVWA (Netherlands Food and Consumer Product Safety Authority) has also given detailed guidance in its Labelling Manual on the use of 'pictorials' on packaging, including in cases where only a flavouring is added to the foodstuff and a flavour disclaimer is required.[84]

The Teekanne decision is an instant classic and clearly set the tone in the field of misleading practices in food law; which is why it must certainly be included in this edition.

Ebba Hoogenraad
Partner, food law and advertising at Hoogenraad & Haak, Amsterdam

83 See also Chapters 41, 43, 55, 56 and 60 in this book.
84 https://www.nvwa.nl/documenten/consument/eten-drinken-roken/etikettering/publicaties/handboek-etikettering-van-levensmiddelen.

CHAPTER 12

De Tox Forte – A general residual category of novel foods

Annotated by Kenneth Defares

CJEU 9 November 2016, C-448/14, ECLI:EU:C:2016:839 (*Davitas*)

Food supplements, novel food.

Main considerations:

> 33 Only an interpretation of Article 1(2)(c) of Regulation No 258/97 as meaning that the expression 'new primary molecular structure' covers foods or food ingredients which were not used for human consumption in the EU before 15 May 1997 is consistent with the pursuit of those objectives. That interpretation makes it possible to ensure the effective protection of public health against potential risks produced by novel foods and food ingredients, in that a single safety assessment will be required whenever it is proposed to use for human consumption a substance which has not hitherto been consumed by man as a food.
> 34 Conversely, if the concept of foods or food ingredients with a 'new primary molecular structure' covered only substances which did not exist in nature with the same primary molecular structure before 15 May 1997, any substance existing on that date which has not yet been used for human consumption and does not fall within any of the specific categories in Article 1(2)(d) to (f) of Regulation No 258/97 would be automatically exempted from the safety assessment laid down by that regulation before being put on the market in the EU, without it being possible to evaluate the possibility of a risk to health.

12.1 Background and facts

In 2012, Davitas, a German manufacturer and vendor of dietary supplements, launched the nutritional product 'De Tox Forte'. This product consists of one ingredient: clinoptilolite, a mineral substance of volcanic origin. The city of Aschaffenburg classified 'De Tox Forte' as a 'novel food' within the meaning of Regulation 258/97[85] and prohibited Davitas from marketing the product without authorisation

85 Regulation (EC) No. 258/97 of the European Parliament and of the Council of 27 January 1997 concerning novel foods and novel food ingredients.

 | HTTPS://DOI.ORG/10.1163/9789004751699-013

on the basis of the Regulation. Davitas disagreed with this decision and took the position in court that the case was not a 'novel food' because clinoptilolite already existed in nature before Regulation 258/97 came into force on 15 May 1997. Against this background, the German administrative law judge referred the following preliminary questions to the Court of Justice of the European Union ('the Court'):

- 'Is the product "De Tox Forte" marketed by the appellant a food or food ingredient with a new molecular structure within the meaning of Article 1(2)(c) of Regulation No 258/97?
- In particular, does it suffice, in order to be able to answer this question in the affirmative, that that product, which contains the substance clinoptilolite in its particular primary molecular structure, was not yet being used as a food prior to 15 May 1997, or is it also necessary that that product is produced by means of a production process which results in a new or intentionally modified molecular structure, that is, it must be a substance which did not previously exist in nature in that form?'

The court addressed these two questions together.

12.2 Court ruling

Regulation 258/97 contains an authorisation requirement for the placing on the market of a 'novel food' or a 'novel food ingredient'. Article 1(2) of the Regulation imposes two conditions on the definition of 'novel': first, that before the Regulation came into force (15 May 1997), the food or food ingredient has not been used for human consumption to a significant degree in the EU; second, that the food falls into one of the categories listed in the article. The invoked category in this case is (c): 'foods and food ingredients with a new or intentionally modified primary molecular structure'.

That the first condition was met was not in doubt since clinoptilolite was not used for human food to any significant extent before 15 May 1997. The question the Court faced concerned the second condition: does the product belong to one of the categories of novel foods, more specifically, does the term 'new primary molecular structure' in the c-category refer to foods not used for human consumption before 15 May 1997 or to foods with a molecular structure that did not exist as such in nature before that date? The Court interpreted the provision to refer to foodstuffs that were not used for human consumption, regardless of whether they already existed in nature.

To this end, the Court considered as follows: that the term 'primary molecular structure' is not defined in the Regulation. Nor can it follow the usual meaning of the term, since it can refer to both a primary molecular structure newly used by man and a primary molecular structure newly created or modified by man. To interpret the term, it is therefore necessary to look at the context of the article and the purpose of the Regulation. With regard to the context, the Court notes that the Regulation is a general instrument covering all novel foods or ingredients. If the provision were interpreted to mean that substances that already existed in nature before 15 May 1997 were not covered, the Regulation would be deprived of its general character.

The Court further considered that the Regulation has a dual objective: to ensure the functioning of the internal market in novel foods and to protect public health from the hazards that these foods may pose. To this end, the Regulation established uniform safety controls for novel foods. An interpretation that would automatically exempt substances already occurring in nature from such safety control would be contrary to this objective. However, the interpretation that substances that have not been used for human food are subject to such safety controls is consistent with this objective.

The term 'new primary molecular structure' should therefore be interpreted as referring to foods or food ingredients that were not used for human consumption within the territory of the European Union before 15 May 1997.

12.3 Opinion of the Advocate General (AG)

The Court's ruling is broadly in line with the AG's opinion. Unlike the Court, the AG points out that interpreting the provision as relating to substances not used for human consumption before 15 May 1997 amounts to an overlap with the first condition in Article 1(2) of the Regulation. The AG explains this overlap by the fact that the Union legislature did not provide a specific category for novel foods consisting of substances of mineral origin. In the new Regulation 2015/2283, which replaces Regulation 258/97, this omission was remedied by the inclusion of a separate category for substances of mineral origin. Since the new Regulation does not apply to this case, the AG – like the Court – resolves the omission by interpreting the provision in the light of the purpose and general intent of the legislation in question. The AG also points out that category (c) contains a generic reference, while the other categories are limited. Given the general scope of the Regulation, category (c) should not be interpreted strictly. Furthermore, there is an established practice where clinoptilolite has been classified as a 'novel food' in previous attempts to market it in the Union. Finally, the AG refers to safety controls and the objective of ensuring a high level of human health. Based on those considerations, in the AG's view, category (c) includes a 'substance of mineral origin which exists in nature and to which no manufacturing process capable of modifying its molecular structure has been applied, where that structure was not an ingredient of a foodstuff consumed in the territory of the Union before 15 May 1997'.

12.4 Comment

The Court in this case was faced with an 'omission' in the legislation on novel foods. Whereas the Union legislator had included a separate category for substances derived from animals, plants, algae, fungi and microorganisms, the Regulation did not include such a specific category for substances of mineral origin. Why this category was missing is unclear, but it seems that the Union legislator simply forgot to include this group of substances. Indeed, in Regulation 2015/2283,[86] the new Article 2(2)(a)(iii) does include substances of mineral origin as a specific category. The category 'new primary molecular structure' also still exists as a separate category in the new Regulation (albeit with the word 'primary' deleted), which raises the question of why a separate category was needed for substances of mineral origin if, according to the Court, these substances already fall within the scope of category (c), the 'new primary molecular structure'.

In itself, it should not be surprising that there has been a debate as to whether or not the product falls into one of the categories. It is inherent in a demarcation based on an exhaustive list of closed categories that there are products to be distinguished that fall within and products that fall outside the categories. The necessary consequence of this is – in the words of the Court – that 'any substance which … does not fall within one of the specific categories referred to … is automatically exempted from the safety checks provided for by this regulation before being placed on the Union market'. Of course, they must be safe within the meaning of Article 14 of Regulation (EC) 178/2002. If the legislator wants to catch every conceivable innovation, it would not be logical to work with closed categories but rather with a general open formula (as in the United States: 'anything added to food') or at least an open residual category. With this ruling, the Court seems to have thus created a general open residual category.

Since the dispute occurred before the new regulation came into force, the new category for food of mineral origin did not provide relief. Instead, category (c) was followed. Both the Court and the AG followed the established case-law according to which, in the absence of an EU law definition, the interpretation of a term must be determined in accordance with its usual meaning in everyday language, having regard to the context in which it is used and the purposes pursued by the legislation of which it forms part. On the basis of that method, the AG and the Court also brought substances of mineral origin that already existed in nature within the concept of 'new primary molecular structure'.

The AG acknowledged that there were some snags in this interpretation. First,

86 Regulation (EU) 2015/2283 of the European Parliament and the Council of 25 November 2015 on novel foods, amending Regulation (EU) 1169/2011 of the European Parliament and the Council and repealing Regulation (EC) 258/97 of the European Parliament and of the Council, and Commission Regulation (EC) 1852/2001.

the genesis showed that the provision initially referred to substances that were not used as foodstuffs or ingredients. The fact that the latter phrase was not included in the final text was not, according to the AG, an indication that it was the intention of the Union legislator to limit the scope of the category. Incidentally, this phrase was included in the same category in Regulation 2015/2283. In addition, the AG acknowledges that the interpretation results in the two conditions overlapping. As a reminder, the first condition requires that the food has not been used in human food to a significant degree before 15 May 1997. Category (c) is now interpreted as referring to a substance with a molecular structure that has not been used for human consumption in the Union before 15 May 1997.

It is worth noting that neither the AG nor the Court (nor the parties apparently) pay attention to the fact that the legislature speaks of a 'primary' molecular structure. In most substances, primary and secondary structures are not distinguished. This is only the case with proteins. In fact, the structure of proteins is broken down into four layers: primary, secondary, tertiary and quaternary. The application of this category to mineral substances is therefore not obvious, to say the least.

Nevertheless, substances of mineral origin are classified under category (c). Both the Court and the AG give considerable weight to the general nature of the Regulation and the aim of ensuring a high level of human health protection. This ruling highlights the importance of context and purpose, in particular the aim of ensuring a high level of human protection, when interpreting concepts. Especially in a case like the present one, where the Union legislature had forgotten to include a separate category and the wording was open to more than one interpretation, this method of interpretation offered a solution.

After this judgment declaring the former (c) category (now the (i) category) a general residual category, is there still any added value in using categories in the Novel Foods Regulation? Part of that value could lie in the exceptions added to some categories. For example, when a new animal or plant variety is obtained from common farm animals or crops using traditional breeding techniques, the resulting food products do not count as novel food (see categories (iv) and (v)). The reference to the primary molecular structure in the former (c) category meant that minor modifications in food proteins (namely at the secondary molecular level or lower) escaped the Novel Food Regulation's authorisation regime. Because the word 'primary' has been deleted, we must take into account the possibility that this exception has lapsed.

Kenneth Defares
was lawyer-partner Defares law firm. This version of his contribution is published after his demise. He died February 9th 2024.

CHAPTER 13

Food supplements – No limits without risk analysis

Annotated by Sarah Arayess

Court of Justice of the European Union 27 April 2017, C-672/15, ECLI:EU:C:2017:310 (*Noria*)

Court of Justice of the European Union 29 April 2010, C-446/08, ECLI:EU:C:2010:233 (*Solgar*)

Free movement of goods, principle of mutual recognition, derogation, food supplements, maximum amounts, generally accepted scientific data.

Main considerations (Solgar):

4 Article 5 of Directive 2002/46 on the approximation of the laws of the Member States relating to food supplements must be interpreted as meaning that the fact that appropriate labelling might dissuade the group of consumers to which it is addressed from using a nutrient beneficial to them in small doses is not a relevant criterion for setting the maximum amounts of vitamins and minerals that may be used in the manufacture of food supplements. Taking account of the varying degrees of sensitivity of different consumer groups allows a Member State to apply a maximum amount appropriate for a specific group of consumers, such as children, to the whole population only if that measure is limited to what is necessary in order to protect the health of the persons belonging to that group and only if that measure is proportionate to the objective it pursues, and only if that objective cannot be attained by measures less restrictive of trade within the European Union, which is a matter to be ascertained by the national court.

In exercising their discretion relating to the protection of public health, the Member States must observe the principle of proportionality. Furthermore, it is for the national authorities to show in each case, in the light of national nutritional habits and in the light of the results of international scientific research, that their rules are necessary in order effectively to safeguard the interests referred to in Article 30 EC and, in particular, that the marketing of the products in question poses a real risk for public health. In that context, it is for the national court to examine, in particular, whether appropriate labelling informing consumers about the nature, the ingredients and the characteristics of the food supplements concerned constitutes a measure sufficient to ensure the protection of the health of those persons, in particular in order to avoid the harmful effects relating to an excessive consumption of the nutrients concerned.

 | HTTPS://DOI.ORG/10.1163/9789004751699-014

13.1 Case and background

Under discussion here is the free movement of goods, services, capital and people: the four freedoms of the European Union. Very briefly, one of the aims of the single market is to facilitate trade within the EU's borders. On paper, this is a smoothly functioning and understandable system, but practice proves more recalcitrant. So, too, for the food industry. The introduction of national rules creates ambiguity and confusion for companies that would prefer to market one product in several Member States without too many modifications. In practice, the label or sometimes even the composition of the product has to be changed. Well-known examples are national mandatory indications of origin or legally described limits of, for example, minerals or vitamins. A product may then be marketed in one Member State, for example, while a regulation in another prevents the same product from being offered. This practice hampers the functioning of the free movement of goods in the internal market. The European Court of Justice regularly rules on such cases. In this annotation, two impressive judgments, from 2010 and 2017, are discussed together.

13.2 Court's judgment in the Solgar case

In both cases, the French authorities refused market access to the products of Solgar and Noria, respectively, because they exceeded the maximum levels laid down in national regulations. The Solgar case[87] concerns the ingredient fluorine, which was not allowed to be used in food supplements in France. The French authorities note that it is impossible to determine the intake of fluorine from other food sources. For France, this justifies the zero tolerance policy for fluorine. Solgar finds that the way the maximum amounts were set by a national measure is contrary to Directive 2002/46/EC. The European Court of Justice is asked for an opinion. As long as the European Commission has not yet set maximum amounts for vitamins and minerals in food supplements, are Member States permitted to adopt legislation themselves? The Court answers this question in the affirmative. Member States may set their own national limits, but in doing so they must respect the requirements of Articles 28 and 30 EC Treaty (now: Articles 34 and 36 TFEU). In addition, Article 5(1) and (2) of the Directive must be followed. This stipulates, among other things, that a risk assessment should be based on generally accepted scientific data. Where it is impossible to calculate accurately the intake of a specific mineral from dietary sources and where there is a proven risk, a maximum level may be set at zero. Where a scientific risk assessment shows that there is still uncertainty about the risks, the precautionary principle may be applied. Previous Court of Jus-

87 CJEU 29 April 2010, case C-446/08, ECLI:EU:C:2010:233 (*Solgar France*).

tice judgments already show that, prior to setting maximum amounts, a risk assessment based on generally accepted scientific data is carried out. Purely hypothetical assumptions about health risks do not in any case justify the setting of maximum quantities.

The theoretical starting point is clear: goods should be able to be traded freely and without barriers in the (internal) market of European Member States. By invoking one of the justifications in Article 36 TFEU, free movement can be restricted by a Member State using a non-tariff measure. The limitative list of justifications consists of the following interests: public morality, public policy, public security, health and life of humans, animals or plants, historical/archaeological heritage and industrial or commercial property. Important interests, in other words.

13.3 The Court's judgment in the Noria case

French company Noria Distribution is also facing criminal charges domestically for offering dietary supplements with higher levels of vitamins and minerals than the standard prescribed by French law.[88] According to Noria, the decision serving as the basis for the criminal case is not compatible with EU law. Briefly, the European Court of Justice has been asked whether France, given the Directive and the principles of free movement of goods and mutual recognition, is permitted to introduce a national legal maximum without allowing companies to follow a procedure to obtain authorisation on the basis of mutual recognition. Furthermore, the European Court of Justice is considering how to set maximum levels for vitamins and minerals. Is a Member State permitted to set maximum levels solely on the basis of national scientific opinions when the conclusions of recent, international scientific opinions advocate higher levels under identical conditions of use?

According to the Court, such national schemes with a statutory cap are only permissible if a number of conditions are met. One of the conditions is that the scheme must provide for a derogation procedure. That procedure must allow manufacturers to place on the market food supplements lawfully manufactured or marketed in another Member State despite the fact that they exceed the maximum levels set by the national legislation. The procedure must also be easily accessible and possible to be completed within a reasonable time. A decision taken under the exemption procedure must be subject to appeal in a court. The French regulation lacked such a procedure, making the national regulation contrary to the free movement of goods.

The Court in this case also considers how maximum levels are set. As discussed in the context of the Solgar case, maximum levels may be set on a case-by-case basis backed up by a thorough scientific risk assessment based on generally accept-

88 CJEU 27 April 2017, case C-672/15, ECLI:EU:C:2017:310 (*Noria Distribution*).

ed scientific data, but not on general or hypothetical considerations. Furthermore, the Court holds that the Directive and the free movement of goods preclude the scientific risk assessment being carried out solely on the basis of national scientific opinions when, at the date of adoption of the measure in question, there are also recent and reliable international scientific opinions which conclude that higher maximum levels may be set.

13.4 Discussion

One of the pillars of the European community is the free movement of goods, services and people. In practice, this is proving to be a tricky issue. Increasingly, Member States are introducing national regulations. This creates a patchwork of rules and a lack of transparency for companies. Whereas in theory, barring special exceptions, it should be possible to sell a product throughout the European Union, in practice the risk of being confronted with a national law turns out to be high. The result: mandatory modifications to the product, which can lead to a barrier to crossing the border. Smaller companies in particular therefore are increasingly deciding to withhold from engaging in cross-border trade.

The free movement of goods is reflected in Article 34 TFEU, which prohibits quantitative restrictions on imports and measures having equivalent effect between Member States. A rule that limits the quantity of goods to be imported is a quantitative restriction on imports, for example, an import ban. A measure of equivalent effect does not aim to limit quantity but does produce that effect. All rules that directly or indirectly, actually or potentially hinder trade in the single market fall within the scope of this concept. A ban on marketing a product by definition restricts the free movement of goods and should thus be considered a measure having equivalent effect within the meaning of Article 34 TFEU.

However, the free movement of goods is not infinite and it can be lawfully restricted by a Member State in some cases. The bases for the restriction are given in Article 36 TFEU. Some serious grounds for restriction are: danger to public safety, health and life of persons, animals or plants. In the case of foodstuffs, the second basis, health and life of persons, will often come into play. The Member State then restricts the free movement of goods with the protection of persons as justification. In many cases, it is not clear whether, for example, exceeding maximum levels of vitamins or minerals will result in harm to public health. In such cases, the Member State may act on the basis of the precautionary principle. Applying the precautionary principle, a competent authority may take precautionary measures to prevent possible harmful effects of a product or process, for example in case of possible damage to health (health exception). The precautionary principle is backed by safeguards. An overall assessment of the risk to health must be made on the basis of the most reliable scientific data and the latest results of international research.[89] This is also subject to very strict requirements: '*Where it proves to be impossible to determine with certainty the existence or extent of the alleged risk because of the insufficiency, inconclusiveness or imprecision of the results of studies conducted, but the likelihood of real harm to public health persists should the risk materialise, the precautionary principle justifies the adoption of restrictive measures*'.[90] In practice, however, the strict requirements for Member States are not always closely followed. Sometimes this leads to proceedings in court. In this context, see a long-running procedure between the Ministry of Health, Welfare and Sport and Solgar over vitamin D in the Netherlands.[91] In many cases, this problem remains invisible. Companies are unaware of the applicable (European) rules or decide not to start proceedings in principle for other reasons.[92]

As in France before Noria, the Netherlands also had no legal option for companies to apply for a derogation. For example, the Netherlands also set a legal maximum level of vitamin D, while other countries had no such limit or a much higher maximum. The lack of such regulation made it impossible for companies to apply for a waiver. Meanwhile, this shortcoming for food supplements has been solved by the introduction of the Temporary policy rules on exemption of vitamin preparations in 2017. This allows those who market a vitamin preparation that does not meet the requirements set by the Commodities Act Regulation Exemption Vitamin Preparations, but which is lawfully marketed in another Member State of the European Union, to apply for an exemption as referred to in Section 16(2) of the Commodities Act from the Minister for Medical Care. It is an important step for manufacturers of food supplements who can now actively apply for an exemption.

Closely related to the free movement of goods is the principle of mutual recognition. Mutual recognition applies to products for which the rules are not harmon-

---

89 CJEU 2 December 2004, case C-41/02, ECLI:EU:C:2004:762, rec. 53 (*Commission v Netherlands*), CJEU 23 September 2003, C-192/01, ECLI:EU:C:2003:492, rec. 51 (*Commission v Denmark*), CJEU 29 April 2010, C-446/08, ECLI:EU:C:2010:233, rec. 69 (*Solgar France*), CJEU 28 January 2010, C-333/08, ECLI:EU:C:2010:44, rec. 92 (*Commission v Republic of France*).

90 CJEU 2 December 2004, C-41/02, ECLI:EU:C:2004:762, rec. 54 (*Commission v Netherlands*), CJEU 23 September 2003, C-192/01, ECLI:EU:C:2003:492, rec. 52 and 53 (*Commission v Denmark*), CJEU 29 April 2010, C-446/08, ECLI:EU:C:2010:233, rec. 70 (*Solgar France*) and CJEU 2 December 2004, C-41/02, ECLI:EU:C:2004:762, rec. 93 (*Commission v Netherlands*). See also the first NVLR lustrum edition: Kenneth Defares and Bernd van der Meulen, *Het voorzorgsbeginsel. Preadvies voor de Nederlandse Vereniging voor Levensmiddelenrecht*, Sdu 2009.

91 See among others: Trade and Industry Appeals Tribunal (*CBb*) 28 December 2016, ECLI:NL:CBB:2016:405 and District Court of Rotterdam, 11 May 2017, ECLI:NL:RBROT:2017:3571 (Solgar Vitamins). This case was handled by S. Arayess for Solgar.

92 Commission staff working document executive summary of the refit evaluation accompanying the document Proposal for a Regulation of the European Parliament and of the Council on the mutual recognition on goods lawfully marketed in another Member State, 19.12.2017 SWD(2017) 476 final.

ised, and this concept follows from the case-law of the Court of Justice. In 2008, the practical implementation of this was laid down in a Regulation. Problems associated with the implementation of mutual recognition and the free movement of goods also permeated the European Commission. In 2015/2016, the European Commission therefore launched an evaluation of the 'Single Market' and the Mutual Recognition Regulation. A number of conclusions quickly followed from this: the scope of the Mutual Recognition Regulation – which products are covered by it – is unclear. How can a company show that a product has been lawfully marketed in another Member State? National authorities still too often prioritise national rules. It is difficult and expensive to challenge decisions by national authorities and there is insufficient communication between all actors. An 'impact assessment' and a proposal for a new Regulation followed in 2017. In 19 March 2019, the new Mutual Recognition Regulation (Regulation (EU) 2019/515) came into force. This Regulation replaced Regulation (EC) 764/2008.

The new Mutual Recognition Regulation has four main new components. The most important is the introduction of the voluntary mutual recognition declaration. Furthermore, the installation of national product contact points and the 'single digital gateway'[93] should provide specific and reliable information on legislation in force in a specific Member State and how mutual recognition is applied. Also, SOLVIT, the European problem-solving mechanism, has received more attention and a more important role. SOLVIT itself is not new, but SOLVIT's role in a situation where mutual recognition is rejected by a Member State is described in more detail in the Mutual Recognition Regulation. The Mutual Recognition Regulation should also ensure better cooperation between competent authorities and product contact points.[94] Furthermore, the Mutual Recognition Regulation stipulates that the EU can release funds for, among other things, information campaigns, education and training exchanges of officials and best practices. The EU can also fund cooperation between product contact points and competent authorities and collect data on the functioning of the principle of mutual recognition and its impact on the single market.

Time will tell whether in practice the new Regulation will also have the desired effect: more awareness for the free movement of goods for a smoother functioning European Union.

Sarah Arayess
Lawyer/founder Say Legal

93 Regulation (EU) 2018/1724 of the European Parliament and of the Council of 2 October 2018 establishing a single digital gateway to provide access to information, to procedures and to assistance and problem-solving services and amending Regulation (EU) No 1024/2012.

94 Recital 43 in the new Mutual Recognition Regulation.

CHAPTER 14

Fidenato – No precaution without science

Annotated by Nathan Meijer

Court of Justice of the European Union 13 September 2017 (*Fidenato and Others*) ECLI:EU:C:2017:676
Court of Justice of the European Union 8 September 2011 (*Monsanto and Others*) ECLI:EU:C:2011:553

Agriculture, genetically modified food and feed, emergency measures, national measure seeking to prohibit the cultivation of genetically modified MON 810 maize, maintenance or renewal of the measure, Regulation (EC) No 1829/2003, Article 34, Regulation (EC) No 178/2002, Articles 53 and 54, conditions of application, precautionary principle.

Main considerations (*Fidenato*):

> Article 34 of Regulation No 1829/2003, read in conjunction with the precautionary principle as set out in Article 7 of Regulation No 178/2002, must be interpreted as meaning that it does not give Member States the option of adopting, in accordance with Article 54 of Regulation No 178/2002, interim emergency measures solely on the basis of that principle, without the conditions set out in Article 34 of Directive No 1829/2003 being satisfied.

14.1 Introduction

This case concerns the application of emergency measures, based on the precautionary principle, to genetically modified food and feed (GMOs). The general precautionary principle is set out in Article 7 of the General Food Law (GFL), Regulation (EC) No 178/2002:[95]

> In specific circumstances where, following an assessment of available information, the possibility of harmful effects on health is identified but scientific uncertainty persists, provisional risk management measures necessary to ensure the high level of health protection chosen in the Community may be adopted, pending further scientific information for a more comprehensive risk assessment.

95 Regulation (EC) No 178/2002. The Dutch Food Law Association (NVLR) celebrated its first anniversary with a book about this precautionary principle (Sdu, 2009).

Thus, any measures taken on the basis of the precautionary principle must be temporary in nature. The material threshold for these temporary measures is, that there is 'the possibility of harmful effects' about which there is 'scientific uncertainty'. In Fidenato and Others, the main issue is the degree of (un)certainty about adverse effects concerning which certain measures may or may not be taken.

14.2 The case

The substantive context of this case is as follows. In 1998, a genetically modified maize variety (MON 810)[96] was authorised by the European Commission (EC) for the European market. In 2013, the Italian government asked the EC to take emergency measures to ban the cultivation of that maize, referring to a report by two Italian research institutes. The EC disagreed with the urgency of these emergency measures, but asked the European Food Safety Authority (EFSA) to evaluate the evidence within four months (i.e. by the end of September 2013). However, the Italian government took emergency measures and the cultivation of this GMO maize was banned in the meantime (12 July 2013). On 24 September, EFSA issued its opinion; there was no new evidence for the emergency measures and the initial conclusions on the safety of this GMO maize remained applicable. Against this backdrop, Fidenato and Others were prosecuted for cultivating this variety of GMO maize in Italy, and four questions on this matter were referred to the Court of Justice (CJEU) by the Italian court.

14.3 Authorisation procedure for GMOs

The procedure that companies must follow before they can market GMOs in the EU is an expression of the precautionary principle.[97] Precisely because the Commission wants to ensure a 'high level of protection of human life and health', all GMOs are subject to a comprehensive safety assessment before they can be placed on the market.[98]

In its answer to the Italian court's preliminary questions in Fidenato and Others, the CJEU referred several times to the Monsanto and Others judgment.[99] It is therefore appropriate to briefly address that judgment here as well. That case con-

96 This seems to be a legally sensitive variety: it was also the subject of case C-36/11, Pioneer Hi Bred Italia (ECLI:EU:C:2012:534), C-442/09, Bablok e.a. (ECLI:EU:C:2011:541), and C-236/01, Monsanto Agricoltura Italia e.a. (ECLI:EU:C:2003:431).

97 Regulation (EC) No 1829/2003, recitals 2, 3, 5.

98 Ibid.

99 CJEU 8 September 2011, Monsanto e.a. (C-58/10–C-68/10, ECLI:EU:C:2011:553).

cerned the same maize variety MON 810: in that case the French government withdrew the licence. It is important to note that the authorisation to market the MON 810 maize was issued by the Commission in 1998 and then by the French authority: at that time, Directive 90/220/EEC[100] still applied to the GMO authorisation procedure. The main reasons for amending the legislation were that there was not yet an authorisation procedure for animal feed,[101] and because the Commission wanted to delegate the risk assessment of GMO products to EFSA.[102]

Monsanto had informed the Commission after the entry into force of the GMO Regulation that MON 810 maize was an 'existing product',[103] but did not submit a notification to the French authority under the environmental directive on GMOs.[104] When Monsanto subsequently applied for renewal of the authorisation with the French authority,[105] the application was rejected – a decision that was later declared an emergency measure under Regulation 1829/2003,[106] and again later under Directive 2001/18/EC.[107] CJEU did not go along with this: although the licence could in principle be revoked under the emergency measures of Regulation 1829/2003, Directive 2001/18/EC did not apply here.[108] Whether the procedural conditions for these emergency measures had been met was for the French court to assess[109] – but CJEU did add that it had to be established that there was

> in addition to urgency, the existence of a situation which is likely to constitute a clear and serious risk to human health, animal health or the environment.[110]

---

100 Directive 90/220/EEC of the Council of 23 April 1990 on the deliberate release into the environment of genetically modified organisms.

101 Regulation (EC) No 1829/2003, preamble recital 7

102 Regulation (EC) No 1829/2003, preamble recitals 9 and 31.

103 CJEU 8 September 2011 Monsanto e.a. (C-58/10–C-68/10, ECLI:EU:C:2011:553), rec. 27 and Article 20(1)(a) of Regulation (EC) No 1829/2003.

104 CJEU 8 September 2011 Monsanto e.a. (C-58/10–C-68/10, ECLI:EU:C:2011:553), rec. 28, Article 17(2) of Directive 2001/18/EC of the European Parliament and of the Council of 12 March 2001 on the deliberate release into the environment of genetically modified organisms and repealing Council Directive 90/220/EEC – Commission Declaration.

105 Under Article 20(4) of Regulation (EC) No 1829/2003 (ECJ 8 September 2011 Monsanto e.a., rec. 29).

106 Article 34 concerning the emergency measures, discussed below – CJEU 8 September 2011 Monsanto e.a., rec. 34.

107 CJEU 8 September 2011 Monsanto e.a., rec. 35.

108 CJEU 8 September 2011 Monsanto e.a., rec. 43–63.

109 CJEU 8 September 2011 Monsanto e.a., rec. 74.

110 CJEU 8 September 2011 Monsanto e.a., rec. 81.

14.4 Emergency measures

Material requirements

Article 53 of the GFL deals with emergency measures:

> Where it is evident that food or feed [...] is likely to constitute a serious risk to human health, animal health or the environment, and that such risk cannot be contained satisfactorily by means of measures taken by the Member State(s) concerned, the Commission, acting in accordance with the procedure provided for in Article 58(2) on its own initiative or at the request of a Member State, shall immediately adopt one or more of the following measures, depending on the gravity of the situation [...]

According to Article 54 GFL, if a Member State informs the EC of the need for emergency measures but the EC takes no action, Member States can adopt and enforce 'interim protective measures' until EU-wide measures are adopted. Finally, Article 34 of Regulation 1829/2003 on GMOs stipulates:

> Where it is evident that products authorised by or in accordance with this Regulation are likely to constitute a serious risk to human health, animal health or the environment, [...] the need to suspend or modify urgently an authorisation arises, measures shall be taken under the procedures provided for in Articles 53 and 54 of Regulation (EC) No 178/2002.

The Fidenato and Others case refers to two types of emergency measures. First, the aforementioned general emergency measures regulated by Articles 53 and 54 GFL. In addition, the GMO legislation also provides for emergency measures through Article 34, which refers to Articles 53 and 54 GFL for procedural implementation. The material threshold for emergency measures in both cases is that products are 'likely to constitute a serious risk to human health, animal health or the environment'. The Court first recalls that Article 7 defines the precautionary principle for all food law, while Article 34 of the GMO Regulation specifically defines the conditions for GMO products under which emergency measures can be taken.[111] Although the Court concedes, citing Monsanto and Others,[112] that provisional emergency measures[113] fall within the scope of the precautionary principle,[114] 'nevertheless, that principle cannot be interpreted as meaning that the provisions set out in Article 34 of Regulation No 1829/2003 may be disregarded or altered, in particular by relax-

111 CJEU 13 September 2017 Fidenato and Others, rec. 46.
112 CJEU 8 September 2011 Monsanto e.a.
113 Article 34 of Regulation (EC) No 1829/2003 and Article 53 and 54 GFL.
114 Article 7 GFL.

ing them'.[115] Moreover, the Court notes that the material threshold for application of the precautionary principle is merely that 'the possibility of harmful effects on health is identified but that scientific uncertainty persists'.[116] In contrast, Article 34 of the GMO Regulation allows the use of emergency measures only 'where products authorised under or pursuant to this Regulation are "likely" to present a "serious" risk to human, health, animal health or to the environment'.[117] The reason for this difference in the 'required level of risk' for the application of emergency measures is that GMO products had to undergo a risk assessment before they were authorised – while emergency measures based on the precautionary principle apply to 'the whole area of food law'.[118] Thus the Court concludes:

> Consequently, in order to prevent Article 7 of Regulation No 178/2002 failing to reduce the level of uncertainty required by the rule laid down in Article 34 of Regulation No 1829/2003 for the adoption of emergency measures, such an autonomous application of the precautionary principle, as laid down in Article 7 of Regulation No 178/2002, without the substantive conditions laid down by Article 34 of Regulation No 1829/2003 being satisfied in light of the adoption of the emergency measures set out in that article, cannot be accepted.[119]

Procedural conditions: rights and obligations of the Commission and Member States

In Fidenato and Others, in addition to the question concerning the substantive conditions for the imposition of emergency measures, there were also some questions concerning their procedural nature. First, the question of whether the Commission is required to take emergency measures if requested by a Member State ('although it is not clear that products authorised by Regulation No 1829/2003 or in accordance with that regulation are likely to constitute a serious risk').[120] In addition, if the Commission does not act on a request for emergency measures, may the Member State itself maintain such measures at national level?[121] The CJEU answered the first question by indicating that the Commission is not obliged to take measures despite the Member State's request, because the Commission also has a margin of discretion.[122] On the other question, the answer was that the Member State was in

115 CJEU 13 September 2017 Fidenato and Other, rec. 48.
116 Rec. 50 and Article 7 GFL.
117 Ibid.
118 Rec. 52.
119 Rec. 53.
120 Rec. 23.
121 Rec. 31.
122 Rec. 29 and 30.

principle allowed to take these measures 'so long as the Commission has not adopted [...] a decision'.[123]

14.5 In conclusion

To summarise: in principle, a Member State can take emergency measures based on the precautionary principle, even in the case of GMO products, if the EC does not act – but only until the EC takes a decision on the issue. The EC has discretion in this area and does not have to take emergency measures just because it is requested by a Member State. Finally, a Member State is not allowed to 'circumvent' the conditions for emergency measures set out in the GMO Regulation by invoking the general precautionary principle laid down in the GFL – which requires a much lower level of risk to take emergency action.

In a context where scientific risk assessment has led to an unambiguous outcome, (more) strict requirements apply before scientific uncertainty can be successfully invoked.

Nathan Meijer
Wageningen Food Safety Research (*WFSR*)

123 Rec. 42.

CHAPTER 15

Confédération paysanne – New breeding techniques and GMOs

Annotated by Bernd van der Meulen

Court of Justice of the European Union, 25 July 2018, ECLI:EU:C:2018:583 (*Confédération paysanne and Others v Premier ministre and ministre de l'Agriculture, de l'Agroalimentaire et de la Forêt*)
Opinion of A-G Bobek, 18 January 2018, ECLI:EU:C:2018:20
Court of Justice of the European Union, 7 February 2023, ECLI:EU:C:2023:75 (*Confédération paysanne II*)

Definition of GMO and new breeding techniques
The process criterion is decisive for the classification of a product as GMO. The exception for mutagenesis does not apply in respect of techniques that developed mainly after the adoption of the GMO Directive.

Main considerations (Paysanne I):

What is a GMO?

> 29 (...) as is apparent from the order for reference, certain of those techniques/methods involve the use of chemical or physical mutageneous agents, and others involve the use of genetic engineering, those techniques/methods alter the genetic material of an organism in a way that does not occur naturally, within the meaning of that provision.
> 30 It follows that organisms obtained by means of techniques/methods of mutagenesis must be considered to be GMOs within the meaning of Article 2(2) of Directive 2001/18.

What is the scope of the mutagenesis exception?

> 48 As the referring court states in essence, the risks linked to the use of those new techniques/methods of mutagenesis might prove to be similar to those which result from the production and release of a GMO through transgenesis. It thus follows from the material before the Court, first, that the direct modification of the genetic material of an organism through mutagenesis makes it possible to obtain the same effects as the introduction of a foreign gene into that organism and, secondly, that the development of those new techniques/methods makes it possible to produce genetically

 | HTTPS://DOI.ORG/10.1163/9789004751699-016

modified varieties at a rate and in quantities quite unlike those resulting from the application of conventional methods of random mutagenesis.
50 Furthermore, it has been emphasised, in recital 8 of that directive, that the precautionary principle was taken into account in the drafting of the directive and must also be taken into account in its implementation. Emphasis is also placed, in recital 55 of Directive 2001/18, on the need to follow closely the development and use of GMOs.
51 In those circumstances, Article 3(1) of Directive 2001/18, read in conjunction with point 1 of Annex I B to that directive, cannot be interpreted as excluding, from the scope of the directive, organisms obtained by means of new techniques/methods of mutagenesis which have appeared or have been mostly developed since Directive 2001/18 was adopted. Such an interpretation would fail to have regard to the intention of the EU legislature, reflected in recital 17 of the directive, to exclude from the scope of the directive only organisms obtained by means of techniques/methods which have conventionally been used in a number of applications and have a long safety record.

What are the implications for the common catalogue of varieties of agricultural crops?

67 It follows that genetically modified varieties obtained by means of techniques/methods of mutagenesis such as those at issue in the main proceedings, with the exception of varieties obtained by means of techniques/methods of mutagenesis which have conventionally been used in a number of applications and have a long safety record, come within the scope of Article 4(4) of Directive 2002/53 and the obligations with regard to the protection of health and the environment laid down in that provision for the purpose of acceptance for inclusion of the varieties in the common catalogue.

What regulatory powers do Member States have?

78 By contrast, organisms obtained by means of techniques/methods of mutagenesis which have conventionally been used in a number of applications and have a long safety record do not come within the scope of that directive, in accordance with Article 3(1) of that directive, read in conjunction with point 1 of Annex I B thereto.
79 Consequently, and to the extent to which the EU legislature has not regulated those organisms, Member States have the option of defining their legal regime by subjecting them, in compliance with EU law, in particular the rules on the free movement of goods set out in Articles 34 to 36 TFEU, to the obligations laid down by Directive 2001/18 or to other obligations.

Main considerations (Paysanne II):

> 51 In that regard, it must be stated that a general extension of the benefit of the exemption provided for in Article 3(1) of Directive 2001/18 to organisms obtained by the application of a technique/method of mutagenesis which is based on the same processes of modification, by the mutagenic agent, of the genetic material of the organism concerned as a technique/method of mutagenesis which has been conventionally used in a number of applications and which has a long safety record, but which combines those processes of modification with other characteristics, distinct from those of that second technique/method of mutagenesis, would not respect the intention of the EU legislature set out in paragraph 48 of the present judgment.
> 52 It cannot be ruled out that the application of a technique/method with such characteristics may lead to genetic modifications of the organism concerned which differ, by their nature or by the rate at which they occur, from those obtained by the application of that second technique/method of mutagenesis.
> 53 It follows that the limitation of the examination carried out for the purposes of applying the exemption provided for in Article 3(1) of Directive 2001/18, read in conjunction with point 1 of Annex I B thereto, solely to the processes of modification, by the mutagenic agent, of the genetic material of the organism concerned, would present the risk that, under cover of the application of a technique/method of mutagenesis conventionally used in a number of applications and with a long safety record, organisms may ultimately be obtained whose genetic material is different from those obtained by the application of that technique/method of mutagenesis, whereas it is precisely the experience gained as regards the latter organisms which enables establishing that the dual criterion resulting from that provision is satisfied.

15.1 Introduction

With its ruling in Paysanne I, the Court clarifies the EU law concept of a genetically modified organism (GMO). In so doing, the Court puts an end to the legal uncertainty that has existed for years concerning the legal status of crops resulting from new breeding techniques that do not involve the introduction of foreign genes into the plant.

15.2 The facts and the proceedings

The preliminary questions are referred by the Conseil d'État, the highest administrative court in France. The questions arose in proceedings brought by a French agricultural organisation and eight associations working to protect the environment and disseminate information about dangers associated with GMOs. They demand that the French government repeal a regulation in national legislation implementing Directive 2001/18, in particular in so far as mutagenesis is thereby exempted from GMO regulations. They want a ban on the cultivation and marketing of herbicide-resistant oilseed rape varieties, and a moratorium on herbicide-resistant plant varieties obtained by mutagenesis. The appellants take the view that herbicide-resistant crops pose environmental or health risks arising in particular from the spread of their genetic material resulting in weeds carrying the herbicide-resistant gene, from the associated need to use more herbicide and to diversify the types of herbicide used from the resulting environmental pollution, from unintended effects such as undesirable mutations or mutations which miss their target and affect other parts of the genome, and from the accumulation of carcinogenic molecules or endocrine disruptors in the crops grown for food or feed.

Third-party stakeholders, such as those engaged in genetic engineering, were not involved in the proceedings either by the French court or the CJEU.

15.3 Evolution, breeding and genetic modification

In sexual reproduction, plants and animals (and humans too) evolve by forming a new and unique DNA pattern (in the chromosomes) based on the DNA of the parents. Growers and breeders take advantage of this by crossing parent pairs that have certain desirable traits or lack certain less desirable traits. Those specimens in which more desirable or less undesirable traits are expressed are selected from the offspring. In this way, plant and animal varieties can be obtained that better meet people's needs.[124] This is called 'breeding'. Carrots, for example, produce carotene as a result of breeding (making them orange instead of their original white colour), cucumber and courgette have become edible because it was possible to prevent the formation of toxins (cucurbitacin).[125] The flower base of the plant we now know as 'cauliflower' mutated to the extent that we can now make a whole meal with it. Breeding is a labour-intensive process because large numbers of offspring must be observed to determine what traits they have in order to make further selections based on them. The development of new varieties involves several generations (of the species in question) in most cases. Depending on the time it takes for the species in question to become sexually mature, breeding can therefore take decades – think, for example, of fruit trees. To speed up the process, techniques are used to break the chromosomes of the parents into pieces of DNA prior to fusion into the chromosome of the offspring. In this way, the variety among the offspring is

124 See also, Niels P. Louwaars, Genetics, chapter 7 in Bart Wernaart and Bernd van der Meulen (eds.) Applied food science, Brill 2022, https://brill.com/edcollbook-oa/title/68695 (open access).

125 This toxin can return if plants are homegrown. See for example: https://www.telegraaf.nl/nieuws/544262/duitser-dood-na-eten-courgette.

increased and the likelihood of desirable combinations of traits is also increased. Breaking up the chromosome is called 'mutagenesis'. It takes place using chemicals or irradiation.

Through breeding, only traits that are present in the gene pool can be expressed in the offspring. The gene pool includes all individuals that can be crossed with each other. The gene pool includes the species and sometimes something more. In the last century, techniques have been developed that allow traits from outside the gene pool to be added. In the judgment, this is referred to as 'transgenesis'. This gave rise to the creation of a legal framework for genetically modified organisms: GMOs.

Meanwhile, more subtle techniques than transgenesis have emerged that generally take the gene pool as their starting point. This is known as cisgenesis. What is new is that instead of chopping up the chromosomes and waiting to see what new combinations arise spontaneously (the Court calls this 'random mutagenesis'), the desired gene or genes are cut from one chromosome and inserted into the recipient chromosome ('site-directed mutagenesis'). The best-known technique for site-directed mutagenesis is called CRISPR/Cas. Because the gene pool is taken as a starting point, offspring are created that in many cases could also have been created via sexual reproduction – although the process would have taken longer.

Site-directed mutagenesis is also known as 'gene editing'. In this expression, the genome (the entire genetic code as expressed in the chromosomes) is compared to a text. By removing and/or inserting the right piece of text in exactly the right place, the text is edited so that it tells the desired story, so to speak.

For years, there has been a great need for clarity as to whether crops obtained through new breeding techniques using site-directed mutagenesis fall within or outside the GMO legal framework. The great significance of this judgment is that this clarity has now been provided.

15.4 Interpretation of the GMO definition

The answer to the underlying question, 'what is a GMO in the legal sense?', must be found by interpreting the legal definition of the term. This definition is laid down in Article 2(2) of Directive 2001/18 on the deliberate release into the environment of genetically modified organisms. Directive 2002/53/EC on the common catalogue of varieties of agricultural plant species refers to this definition. Regulation (EC) 1829/2003 on genetically modified food and feed and Regulation (EC) 1830/2003 on traceability and labelling of genetically modified organisms and traceability of food and feed products produced from genetically modified organisms do the same (Articles 2(5) and 1(3), respectively). The definition from Directive 2001/18 (Article 2(2)) is thus the general EU definition.

It reads as follows:

> […] genetically modified organism (GMO) means an organism, with the exception of human beings, in which the genetic material has been altered in a way that does not occur naturally by mating and/or natural recombination.' According to this definition: '(a) genetic modification occurs at least through the use of the techniques listed in Annex I A, part 1;
> (b) the techniques listed in Annex I A, part 2, are not considered to result in genetic modification […]

Of particular note is the phrase: 'in which the genetic material has been altered in a way that does not occur naturally by mating and/or natural recombination'. In English and also almost all other EU languages[126], the expression 'altered in a way' is ambiguous. The expression can refer to both the result (i.e. the change concerns a result that is not possible by nature) and the way ('manner') by which this result has been achieved (i.e. the change has been brought about in a way that is not possible by nature). Prior to the emergence of the new breeding techniques, this ambiguity did not matter. In genetic modification in the sense of transgenesis, it was always both at once: a method that could not occur in nature and a result that could not occur in nature. However, in cases where the result could occur in nature even if advanced techniques were employed, the correct interpretation of this phrase becomes of paramount importance. Is it about the process or is it about the product?

In the Opinion of the Advocate General (AG) the product is put at the centre. Under point 56, he wrote: 'In my view, *provided that they meet the substantive conditions of Article 2(2) of the GMO Directive*, organisms obtained by mutagenesis are GMOs within the meaning of the GMO Directive […].' [emphasis added]. Apparently, he considers the substantive conditions and not the process to be decisive.

Instead, in recital 29, the CJEU focuses on the process whereby 'since […] certain of those techniques/methods involve the use of chemical or physical mutageneous agents, and others involve the use of genetic engineering, those techniques/methods alter the genetic material of an organism in a way that does not occur naturally'.

Whether the result could occur in nature plays no role (anymore) in the new interpretation of the GMO concept in European law.

15.5 Excluded GMOs

Pursuant to Article 3 j⁰ Annex IB of Directive 2001/18, GMOs obtained by certain techniques, including mutagenesis, are exempted from the operation of the Directive. The Court also addressed the question of the scope of this exemption for

126 On this and other the aspects important in this context, see: Gema Fernández Albújar and Bernd van der Meulen, 'The Legal GMO Concept Reassessment of the GMO definition in the light of new breeding techniques (NBTs)', European Institute for Food Law Working Paper 2017/03, available via: http://www.food-law.nl/Working-papers.

mutagenesis. In the view of the AG, the Directive uses a dynamic reference to mutagenesis in the sense that it would include forms of mutagenesis dating from after the Directive. On this point, too, the Court takes a different course from the AG. New forms are excluded from the exception. GMO law applies to them without question. In substantiating that view, in recital 48, the Court leans heavily on the referring court's factual (determination) assertion that 'the risks linked to the use of those new techniques/methods of mutagenesis might prove to be similar to those which result from the production and release of a GMO through transgenesis.'

15.6 Scope

Since the GMO definition applies to all Union regulations involving GMOs, the entire agri-food domain is affected. The judgment itself makes this explicit for plant variety rights (Directive 2002/53). However, food law in Regulations 1829/2003 and 1830/2003 is also affected. Foods (ingredients) derived from crops obtained by new breeding techniques are (currently) subject to the authorisation, traceability and labelling requirements that apply to GMOs.

15.7 Precaution

The Court relies in part on the precautionary principle. It should be noted that this is not the precautionary principle within the meaning of Article 7 of Regulation 178/2002 (but perhaps within the meaning of Article 191 of the Treaty on the Functioning of the European Union, TFEU). Article 7 concerns scientific uncertainty. According to the Court's own established case-law, this can only be invoked after prior risk assessment. However, this case is not about what to do when there is scientific uncertainty after risk assessment, but about which products should be subject to prior risk analysis.[127]

If the Court had ruled otherwise, it would presumably have made little difference as far as prior risk assessment concerning food safety is concerned, as the Novel Foods Regulation would probably apply (Regulation 2015/2283). It would, however, have made a difference concerning prior risk assessment for the environment. That would not be applicable.

15.8 Comment

The Court has brought to an end a crippling legal uncertainty. However, the legal certainty that has taken its place is not without its problems.

The European Union now has a legal concept of GMO that no longer coincides

127 See more on the different meanings of the precautionary principle in Chapter 20 in this book.

with transgenesis and thus differs from what is understood internationally in science as GMO, but also, for example, from the Convention on Biological Diversity and the Nagoya Protocol. Unlike transgenesis, products obtained through cisgenesis will often not be detectable. Enforcement of GMO law with regard to these products will therefore require robust administrative traceability systems, including in international trade. Already on 19 January 2019, alarmist reports appeared in the *Financieel Dagblad* and the *Telegraaf* under the headline 'Mutated food comes to the Netherlands via Canada'. The newspapers believe this is due not so much to the inherent properties of the products, but to the Ceta Treaty.

Genetic modification has been demonised in Europe. Unlike medicines and animal feed, there is no market for GMO-based foods. Not even when they are found to be safe. Consequently, there is also little room for scientific development in the European Union. When the European Food Safety Authority (EFSA) assesses a GMO as safe in a concrete case, this does not lead to an increase in confidence in the product in question, but to a decrease in confidence in EFSA's independence and integrity.

The scientists who have been engaged in developing and applying new breeding techniques and their legal advisers could be blamed for allowing the initiative to establish the legal status of their products to be taken out of their hands. They were not the ones who initiated legal proceedings. They were not the ones involved in the formulation of the preliminary questions. They were not the ones involved in determining the facts.[128] On all these points, NGOs have been too quick for them. Those NGOs therefore won a resounding victory and new breeding techniques seem doomed in the European Union.

15.9 Paysanne II

Of course, once Pandora's box was opened, there were bound to be problems. After the above judgment (Paysanne I) was delivered, the French Conseil d'État instructed the French authorities to draw up an exhaustive list of mutagenesis techniques or methods that have traditionally been used in a number of application cases and that have already been proved safe. Using this list, it would be possible to distinguish between mutagenesis techniques that existed before the entry into force of

128 If it would have been established that the risk from cisgenesis is comparable to the risk from transgenesis, then, if there had been scientific input, the focus would doubtless have been on the fact that in the scientific literature to date no actual risk from transgenesis has been identified. For a meta-analysis, see: in the EU, A decade of EU-funded GMO research (2001–2010), https://op.europa.eu/en/publication-detail/-/publication/d1be9ff9-f3fa-4f3c-86a5-beb0882e0e65/language-en; in the US, National Academies of Science, Engineering, medicine, Genetically Engineered Crops: Experiences and Prospects, https://www.nationalacademies.org/our-work/genetically-engineered-crops-past-experience-and-future-prospects.

the GMO Directive and those that pre-date it, and thus between mutagenesis techniques which, according to the case-law of the Court, are and are not exempted from the application of this Directive. This turns out to be rather complicated. In proceedings against the failure to reach a decision, the French court threw the case back over the fence to the CJEU. In response the CJEU (in Paysanne II) takes the view that the issue is (also) not whether a result is achieved that could have been achieved with traditional mutagenesis techniques. Thus, it will really have to be determined on a case-by-case basis whether or not the technique used dates from after 2001. However, the mere fact that an in vitro technique was used is not sufficient in this regard. And with that, the case goes back over the fence of the French court. Someone will have to write a history of mutagenesis.

15.10 A task for legislators

To deal with the tension created by the CJEU's interpretation of the GMO definition, intervention by the legislator is deemed necessary. On 5 July 2023, the European Commission submitted to the European Parliament a proposal for a new regulation on plants obtained using certain genomic techniques and food and feed derived therefrom.[129]

The Commission proposal is without prejudice to the GMO definition and its interpretation by the CJEU. This means that crops resulting from new breeding techniques will continue to be classified as GMOs. However, as the Commission considers that the authorisation procedure for GMOs is less appropriate for these crops, it proposes to create a separate regime for them by way of lex specialis. However, the Commission has not designed a coherent regime but a series of derogations from Directive 2001/18 and Regulations 1829/2003 and 1830/2009. From a legislative technique point of view, this approach certainly won't win any beauty contests. Interested parties will have to piece together the applicable standards each time from the GMO regulations and derogations. It is anything but user-friendly.

The new regime separates modified crops into two categories. Category 1 is considered equivalent to conventional plants. It is exempt from the authorisation procedure – although under circumstances the Novel Foods Regulation may apply.[130]

---

129 Proposal for a Regulation of the European Parliament and of the Council on plants obtained by certain new genomic techniques and their food and feed, and amending Regulation (EU) 2017/625 COM(2023) 411 final, https://www.europarl.europa.eu/RegData/docs_autres_institutions/commission_europeenne/com/2023/0411/COM_COM(2023)0411_EN.pdf. At the time of writing amendments proposed by the European Parliament on 17 September 2025 are the latest document in the legislative process: https://eur-lex.europa.eu/legal-content/EN/TXT/PDF/?uri=OJ:C_202503751.

130 In particular, category iv, food consisting of, isolated from or produced from plants or parts of plants that were obtained using non-traditional breeding techniques, causing significant changes to the composition or structure of the food that impact their nutritional value, the way in which they are metabolised or their levels of undesirable substances.

Category 2 includes all crops not covered by category 1. For the criteria, the proposed Regulation refers to an annex. According to the proposed annex, plants fall under category 1 when all mutations come from the same gene pool and the number of mutations is limited – less than 20, or there is only switching off of genes.[131] In addition, there is a substantive criterion. A mutation that makes the plant resistant to herbicides such as glyphosate fall under category 2. Category-2 products are essentially subject to GMO regulation, both in terms of non-food use, food use and labelling requirements. The main difference the proposed regulation would bring is the introduction of a category of crops that the legislator does classify as GMOs but exempts from the burden of proof of safety and the duty to inform consumers.

In my estimation, the proposal succeeds in making everyone unhappy. New breeding techniques are not freed from the GMO stigma. Opponents of GMOs are going to be faced with a category of GMOs that have been removed from both safety assessment and transparency.

My preference would be to clarify the GMO definition focusing on transgenesis (the introduction of genes from outside the gene pool which would not happen in nature) and conduct the safety assessment of cisgenesis (combining genes within the gene pool) under the Novel Foods Regulation (category iv)[132]. Labelling obligations can be imposed on a case-by-case basis under the Novel Foods Regulation.

Bernd van der Meulen
Visiting Professor of Global Food Law at the University of Parma

131 The use of categories conjured up associations with Chinese law on GMOs. There four categories are distinguished. The number of mutations plays no part in the classification. Instead they look at the safety profile of the plant to which the modifications are made and the nature of the modifications. Some plants are already high risk before modification, because for example they are naturally toxic. This is an indicator for inclusion in a higher risk category. Some modifications can exacerbate existing risks, some can reduce them. This too influences classification. See B.M.J. van der Meulen, Genetically modified food, Chapter 7 in Buijs/Van der Meulen/Sun Chinese food law. A legal-systematic analysis, Wageningen Academic Publishers/Brill.

132 'The following definitions also apply: a) 'novel food' means any food that was not used for human consumption to a significant degree within the Union before 15 May 1997, irrespective of the dates of accession of Member States to the Union, and that falls under at least one of the following categories: [...](iv) food consisting of, isolated from or produced from plants or their parts, except when the food has a history of safe food use within the Union and is consisting of, isolated from or produced from a plant or a variety of the same species obtained by: traditional propagating practices which have been used for food production within the Union before 15 May 1997; or *non-traditional propagating practices* which have not been used for food production within the Union before 15 May 1997, where those practices do not give rise to significant changes in the composition or structure of the food affecting its nutritional value, metabolism or level of undesirable substances.' (Article 3(2)(a)(iv) Regulation (EU) 2015/2283)

CHAPTER 16

Movizin complex – On hold claims

Annotated by Karin Lenssen

Court of Justice of the European Union 10 September 2020, C-363/19, ECLI:EU: C:2020:693 (*Konsumentombudsmannen v Mezina AB*)

Main considerations:

> 45 Thus, by providing, in both Article 5(1) and Article 6(1) of Regulation No. 1924/2006, that health claims must be substantiated by 'generally accepted scientific evidence', the EU legislature has determined the standard of proof required in that regard.
> 46 The use of the expression 'generally accepted scientific evidence' means that such evidence should not be limited to beliefs, hearsay derived from popular wisdom, or the observations or experiences of persons outside the scientific community.
> 47 On the contrary, the use of such an expression means that health claims should be based on objective and scientific evidence and that, in particular, there should be sufficient scientific agreement as to the benefits of the substances to which the health claims relate, as stated in recital 14 of Regulation No 1924/2006. In addition, and as required in recital 17 of that regulation, health claims must be 'scientifically substantiated by taking into account the totality of the available scientific data and by weighing the evidence'.
> 50 In that regard, it must be noted that, although Article 5(1) and Article 6(2) of Regulation No 1924/2006 do not impose such a requirement, they nonetheless require the food business operator concerned to be able to justify the health claim it uses.
> 51 The evidence put forward may be that contained in the file prepared in support of the application for entry in the list provided for in Article 13(3) of Regulation No 1924/2006 or come from other sources, provided that that evidence is of sufficient scientific value.

16.1 Background

In a dispute between the Swedish Consumer Ombudsman and Mezina, a manufacturer of natural remedies and food supplements, the Swedish court has asked for a preliminary ruling on Articles 5 and 6 of Regulation (EC) No 1924/2006. In the dispute between the consumer ombudsman and Mezina, the ombudsman is seeking a ban on the use of both the specific and general health claims on the food supplements 'Movizin complex', 'Macoform' and 'Vistvital'. These products contain, among others, ginger, dandelion and blueberry; ingredients that are currently cov-

 | HTTPS://DOI.ORG/10.1163/9789004751699-017

ered by the transitional measures of Article 28(5) of Regulation (EC) No 1924/2006. Health claims under these transitional measures, claims that have not yet been assessed by the European Food Safety Authority (EFSA) and for which it has therefore not yet been determined whether they are permitted, may be used only if they comply with the conditions set out in the earlier provisions of the regulation and with national provisions. The submitted health claims on the so-called 'botanicals' also fall within these transitional measures. Botanicals are defined by the EFSA as foods, including food supplements, containing (parts of) plants, algae, fungi or lichens. The evaluation of these health claims by the EFSA was halted in 2010 after a debate arose on the substantiation of these claims. The European Commission compiled a list of the submitted health claims that have not yet been evaluated, the so-called 'on-hold' claims.[133] The claims on this list may be used until their evaluation is completed, provided they meet the other conditions of the regulation.

16.2 Question

The first four questions of the judgment concern the interpretation of Articles 5 and 6 of Regulation (EC) No 1924/2006.[134] The first question asks whether the conditions set out in Articles 5 and 6 of the regulation apply to health claims that have not yet been authorised, including those falling under the transitional measures of Article 28(5). The second question is whether the food business operator or the authority applying to a court for an injunction should bear the burden of proof to substantiate health claims.

Questions three and four deal with the standard of proof, asking first whether Articles 5 and 6 of Regulation (EC) No 1924/2006 also determine the standard of proof or whether this is set by national law. It then asks in the fourth question what the requirements for the proof are.

Finally, question five looks at the application of Regulation (EC) No 1924/2006 and Directive 2005/29/EC on unfair business-to-consumer commercial practices in the internal market and which of these prevails in the event of a conflict between them.

133 List of on-hold claims: https://www.efsa.europa.eu/sites/default/files/2021-06/questions-on-hold-botanical-claims.xlsx

134 Regulation (EC) No 1924/2006 on nutrition and health claims made on foods. Art. 5(1)(a): '[…] has been shown to have a beneficial nutritional or physiological effect, as established by generally accepted scientific data'; Art. 6(1): 'Nutrition and health claims shall be based on and substantiated by generally accepted scientific data'; Art. 6(2) 'A food business operator making a nutrition or health claim shall justify the use of the claim.'

16.3 Discussion

Need for scientific evidence for claims falling under the transitional measures

According to the CJEU, Regulation (EC) No 1924/2006 accommodates the first four questions raised. Botanical claims that are on hold, and therefore fall under the transitional measures, must be substantiated by 'generally accepted scientific evidence', as described in Articles 5 and 6 of the Regulation. Looking at the standard of proof as requested, the CJEU goes two steps further. Firstly, the Court specifically mentions that one must look at 'the totality of the available scientific data and by weighing the evidence'. This would mean that it is insufficient to grant a licence based solely on studies with a positive outcome. One would have to conduct a thorough systematic literature review of all available scientific studies published on the submitted health claim. However, it is not specified in this judgment who is responsible for conducting the analysis or checking whether the documents submitted in the file actually contain all available scientific evidence. Although the Court concludes that a food business operator is responsible for substantiating a health claim, so far it has not been described anywhere, either in the technical guidance of the EFSA,[135] or in the scientific literature on Regulation (EC) No 1924/2006, that a dossier must contain a full systematic analysis of all literature. Substantiation requirements are usually considered to be the human clinical studies required to substantiate a health claim.[136]

Furthermore, from a scientific perspective, it is important to note that 'available' is a complicating factor, to say the least. Publication bias and 'outcome report' bias can, to a greater or lesser extent, result in an overly positive picture of the available scientific studies because the negative results are simply not published.[137] Increasing requirements in terms of transparency and 'open science' do aim to reduce this.

Secondly, the court mentions that generally accepted scientific evidence 'should not be limited to beliefs, hearsay derived from popular wisdom, or the observations or experiences of persons outside the scientific community'. An interesting statement given the ongoing debate on the substantiation of health claims on botanicals with evidence of traditional use. Evidence of traditional use as a substantiation of

135 EFSA NDA Panel in 2016 (EFSA Panel on Dietetic Products, Nutrition and Allergies), Turck D, Bresson J-L, Burlingame B, Dean T, Fairweather-Tait S, Heinonen M, Hirsch-Ernst KI, Mangelsdorf I, McArdle HJ, Naska A, Neuhäuser-Berthold M, Nowicka G, Pentieva K, Sanz Y, Sjödin A, Stern M, Tomé D, Van Loveren H, Vinceti M, Willatts P, Martin A, Strain JJ, Heng L, Valtueña Martínez S, and Siani A, 2021. Scientific and technical guidance for the preparation and presentation of a health claim application (Revision 3). EFSA Journal 2021;19(3):6554, 35 pp. doi:10.2903/j.efsa.2021.6554.

136 A. de Boer, E. Vos & A. Bast, 'Implementation of the nutrition and health claim regulation – the case of antioxidants', Reg. Tox & Pharm., 2014, 475–487.

137 MJE Urlings. 'Selective citations and its consequences', thesis Maastricht 2019, https://cris.maastrichtuniversity.nl/en/publications/selective-citation-and-its-consequences.

health claims often cites sources different from studies derived from the nutritional sciences, such as old books and recipes, documents from archives and other historical information.[138] And even though history is indeed a scientific discipline, it concerns something different from nutritional sciences. Two important questions remain. First: does historical research into a health claim fall under 'beliefs, hearsay derived from popular wisdom or the observations or experiences of persons outside the scientific community'? If the answer is no, it remains to be determined whether historical research into evidence on traditional use would be sufficient to substantiate a claim.

The 'burden of proof' and the 'methods of obtaining that evidence' are matters that are dealt with in national law.[139] For health claims covered by the transitional measures, the lack of an overall assessment of the underlying scientific dossier leads to a precarious sitatuation. Different competent authorities in Member States may translate 'generally accepted scientific evidence' needed to substantiate a health claim under the transitional measures differently. This may lead to health claims being allowed in one Member State but not in another, or at least only under different conditions.[140] To prevent this, there should be clearer guidance for competent authorities on the evidence requirement, or the centralised assessment of the botanical health claims by the EFSA would need to continue. Only in this centralised approach can the potential difference between Member States be prevented or resolved.

Conflict between Regulation (EC) No 1924/2006 and Directive 2005/29/EC

The last question concerns a possible conflict between Regulation (EC) No 1924/2006 and Directive 2005/29/EC. According to the court, Directive 2005/29/EC should be interpreted in a broadly manner. It applies to all commercial practices before, during and after a commercial transaction, where a commercial practice includes anything involving the promotion, sale or supply of a product to consumers.[141] In Article 3(3) of the directive, it is stated that in case of conflict between the

138 R. Anton, M. Serafine, L. Delmulle, 'Traditional knowledge for the assessment of health effect for botanicals – a framework for data collection', Eur Food Feed law Rev, 2012, 74–80 and P. Coppens, L. Delmulle, O. Gulati, D. Richardson, M. Ruthsatz, H. Sievers and S. Sidani, 'Use of botanicals in food supplements – regulatory scope, scientific risk assessment and claim substantiation', Annals of Nutr. Met, 2006, 538–554.

139 Rec. 53.

140 CJEU 23 November 2017 cases C-596/15 P and C-597/15 P, paras 89 & 90 and Commission staff working document evaluation of the Regulation (EC) No 1924/2006 on nutrition and health claims made on foods with regard to nutrient profiles and health claims made on plants and their preparations and of the general regulatory framework for their use in foods, 20.5.2020, SWD (2020) 96 final.

141 CJEU 16 July 2015 cases C-544/13 and C-545/13 ECLI:EU:C:2015:481.

directive and other provision of Union law and even national rules, the Union law and national law will prevail. Thus, the court sees this directive as providing for situations for which there is no more specific legislation within the European Union.

On the issue of the preliminary questions for the CJEU on the substantiation of health claims, Regulation (EC) No 1924/2006 suffices. This regulation sets requirements for the information given to consumers, when this information can promote sales. For example, in Article 3(a) it is stated that nutrition and health claims may not be misleading. Article 5(1)(a) states that health claims may only be used if it has been established by 'generally accepted scientific evidence' that 'the presence, absence or reduced content [...] of a nutrient or other substance [...] has a beneficial nutritional or physiological effect'. This should therefore adequately protect a consumer from misleading commercial practices.

Dr. Karin Lenssen
Food Claims Research Centre

CHAPTER 17

Spermidine – New production process

Annotated by Karin Verzijden

Court of Justice EU 19 January 2023, ECLI:EU:C:2023:38 (*TLL The Longevity Labs GmbH v Optimize Health Solutions mi GmbH*)
Opinion of the Advocate General M. Campos Sánchez-Bordona of 19 January 2023 ECLI:EU:C:2023:38 (case C 141/22)

Novel foods, new production process, history of safe use.

Main considerations:

22. As the Advocate General observed in point 34 of his Opinion, the concept of a 'history of safe food use in a third country', as defined in Article 3(2)(b) of that regulation, may be transposed to the concept of a 'history of safe food use within the Union', within the meaning of Article 3(2)(a)(iv) of that regulation. There is nothing to support a finding that the concept of 'history of safe food use' should have a different meaning depending on whether it is used with reference to a third country in the context of Article 3(2)(b) of Regulation 2015/2283 or to a country of the European Union in the context of Article 3(2)(a)(iv) of that regulation.
26. As the Advocate General observed, in essence, in point 36 of his Opinion, propagating practices for the production of new plants by reproduction are to be distinguished from practices covering the entire production process of a food.
27. It is apparent from the information in the file sent to the Court that the use of an aqueous solution of spermidine for the cultivation of buckwheat seedlings is not a plant propagation technique within the meaning of the preceding paragraph and Article 3(2)(a)(iv) of Regulation 2015/2283, but a production process for enriching the seedlings in order to achieve a high spermidine content. In such a scenario, which it is for the referring court to ascertain, such a production process would be irrelevant for the purpose of examining the second of the cumulative conditions.

Main considerations of the conclusion:

52. Having established the meaning of 'production process' and determined that it applies to the product at issue, it is necessary to ascertain whether the technique used is a new one and causes 'significant changes in the composition or structure of the food affecting its nutritional value, metabolism or level of undesirable substances'.
53. The referring court does not seem to be in any doubt that the technique used in this case is new. Nor does it refer to evidence of the use, prior to 15 May 1997, of an

 | HTTPS://DOI.ORG/10.1163/9789004751699-018

aqueous solution containing a high level of spermidine to cultivate buckwheat seeds hydroponically, in the manner described above.
54. That cultivation technique gives rise to significant changes in the composition or structure of the food affecting its nutritional value. The level of spermidine in the product at issue, according to the analyses provided by TLL, is 3.51 mg/g, which is 106 times higher than the level of that substance under normal cultivation conditions in buckwheat seed, which is 0.03305 mg/g.
55. Spermidine bio-enrichment alters the composition and nutritional value of sprouted buckwheat flour to such an extent that it becomes a novel food. According to the scientific studies referred to, although increased spermidine intake may have health benefits, too much might be harmful for cells. That is why prior authorisation of the product at issue under Regulation 2015/2283 is essential to ensure food safety and prevent risks to the consumer.

17.1 Introduction

On 25 May 2023, the Court of Justice of the European Union (CJEU) rules in a dispute arising in Austria between two manufacturers of food supplements. This is the first decision on the interpretation of the current Novel Food Regulation (in force since 1 January 2018) and thus a first in the industry. The dispute concerns the production method of the functional ingredient spermidine and the industry had hoped to receive some guidance on what exactly constitutes a 'new manufacturing process' under the Novel Food Regulation. The CJEU does not rule on this topic, but shares two other considerations. And the Advocate General (AG) does provide his view on what a 'new manufacturing process' actually is.

17.2 Why do people take spermidine?

Spermidine supplementation takes place with a view to supporting cellular autophagy, or cell renewal. This could promote the prevention of cardiovascular disease, prevent food allergies and combat the symptoms of diabetes. It has even been suggested that spermidine could extend human life span by as much as 5 to 7 years. So says AG Campos Sánchez-Bordona in his Opinion on the case dated 19 January 2023, citing various scientific sources (see footnotes 12 and 13 of this Opinion).

17.3 Background to this dispute

The dispute in Austria was brought by The Longevity Labs (TLL)[142], which markets the spermidine supplement spermidineLIFE. TLL extracts spermidine from unger-

142 https://spermidinelife.us/pages/fasting.

minated wheat germ by means of a complex and expensive chemical process. TTL applied for and obtained Novel Food authorisation for this (see 'spermidine-rich wheat germ extract' in the Union list).[143] Then a competitor, Optimize Health Solutions, enters the market with its own spermidine supplement. This is produced using a much simpler and therefore cheaper production process based on hydroponic cultivation of buckwheat germ with synthetic spermidine. TLL is of the opinion that Optimize Health Solutions also needs Novel Food authorisation for marketing purposes. TLL therefore initiates proceedings seeking an injunction against further marketing of Optimize Health Solutions' product without such an authorisation.

Optimize Health Solutions argues that this is not necessary, as its product is not covered by the Novel Food Regulation. The germination of the wheat germ from which its product is made would be a primary production process, to which the Hygiene Regulation applies.[144] In addition, the Novel Food Regulation does not apply because Optimize Health Solutions' products qualify as 'plants before harvest' and these do not count as food under the General Food Law.[145] And anyway, spermidine has been traded in the EU for over 25 years according to Optimize Health Solutions. The Austrian court decides that clarification of European legislation is needed to resolve this dispute and puts five questions to the CJEU.

17.4 Questions referred by the Austrian referring court

National judges putting questions to the CJEU usually go for as many anchors as possible. If a follow-up question is required based on a particular interpretation by the CJEU, they also ask that question immediately. The flipside of this system is that when the answer to the first question signals the end of the matter, the remaining questions no longer need to be answered. That is what we call procedural economy. In summary, the referring Austrian court asks the following questions:

1. Should a food consisting of flour from buckwheat germ with a high spermidine content be classified as Novel Food in the category 'foods isolated from (parts of) plants'?
2. If not, is it perhaps a Novel Food because a new production process has been used, and does that term include primary production processes?
3. If there is a new production process, does it matter whether that process has not been used at all before or only that it has not been used for spermidine?
4. If primary production processes are not covered by the concept of new production process, is it correct that the process of germinating buckwheat seeds

143 Implementing Regulation 2017/2470 ELI: http://data.europa.eu/eli/reg_impl/2017/2470/2023-06-06.
144 Regulation 852/2004 ELI: http://data.europa.eu/eli/reg/2004/852/2021-03-24.
145 Regulation 178/2002 ELI: http://data.europa.eu/eli/reg/2002/178/2022-07-01.

in a nutrient solution containing spermidine is not covered by the Novel Food Regulation because it does not apply to plants before harvest?

5. Does it make a difference whether the nutrient solution contains natural or synthetic spermidine?

17.5 Decision of the CJEU

The CJEU answered question 1 – with some reservation – in the affirmative. Optimize Health Solutions' product is a Novel Food because there is no evidence that this product was used for human food to any significant extent within the EU before 1997. This short answer is somewhat disappointing, as it makes answering the remaining questions irrelevant. Still, the CJEU does share two interesting considerations regarding the current Novel Food Regulation. It points out that the concept of 'history of safe use within the Union' is not defined with respect to Novel Foods that must undergo the full authorisation process. However, it *is* defined with respect to Novel Foods that are allowed to take a so-called short cut. These are products that have been used as food outside the EU for a long time, such as chia seeds. These products are subject to the requirement that *the safety of the food has been confirmed by compositional data and by experience of continued use for at least 25 years for the usual diet of a significant number of people in at least one third country.* The CJEU holds that this requirement should be applied by analogy to the spermidine in question and finds that said data were not provided.

17.6 Propagation methods v full production process

Another consideration of the CJEU concerns the Novel Food category 'food isolated from (parts of) plants'. This category is *not* triggered by the application of non-traditional propagation methods that do not result in significant changes to the composition or structure of the food in question. The CJEU rules that a distinction must be made between:

1. propagation processes aimed at producing new plants, and
2. processes involving the entire production process of a food product.

The process applied to Optimize Health Solutions' product to achieve high spermidine content falls into the second category. In other words, a production technique to enrich a product is not the same as a propagation technique. The CJEU instructs the Austrian court to take this into account when deciding the case at national level. The likelihood that the exception to the Novel Food category 'food isolated from (parts of) plants' will apply is thus reduced and the reservation applied by the CJEU to its decision is put into perspective. There's a good chance, therefore, that the

Austrian national court will indeed rule that Optimize Health Solutions' product is a Novel Food.

17.7 New production process according to the AG

With the above answer, the Austrian court's question has been answered and the CJEU will not get to the remaining four questions. It is of course unfortunate that we will not know the CJEU's decision with respect to these. It is therefore interesting to see how the AG rules on these questions. And here we have it: according to the AG, enriching wheat germ with spermidine should be classified as a new production process. He argues that the bio-enrichment with spermidine changes the composition and nutritional value of buckwheat germ flour. Indeed, its spermidine content becomes 106 times higher than unenriched wheat germ. The AG cites studies according to which a higher spermidine content may be beneficial to health, but it also follows that too high a content could be harmful to cells.

The AG therefore concludes that prior authorisation of Optimize Health Solutions' product is indispensable to ensure food safety and prevent risks to consumers. To illustrate this, he refers to selenium-enriched mushrooms and mushrooms that are treated with ultraviolet light after harvest to increase their vitamin D2 content. These products have also received authorisation under the Novel Foods Regulation. Furthermore, the AG argues that the effects and hence the safety of a new production process should be assessed in each individual case and not in general. One and the same production process may affect one foodstuff differently from another.

17.8 Relevance to legal practice

This spermidine case clarifies the criterion on the basis of which it has to be determined whether a food has a history of safe use and should therefore be classified as a Novel Food or not. By the way, this is not entirely new – there has already been a European Commission guidance document 'Human Consumption to a Significant Degree'[146] since 1997, which argues essentially the same thing. So the gain is somewhat meagre. However, when this judgment is considered in conjunction with the AG's opinion, it does provide relevant information for legal practice in establishing how to determine if a new production process has been applied to a food.

This is the case if it is established that an applied process significantly alters the composition and nutritional value of a food compared to a food in which such a process has not been applied. Thus, in such a case, a food should obtain authori-

146 https://food.ec.europa.eu/system/files/2018-07/novel-food_leg_guide_humn-consumption.pdf

sation under the Novel Food Regulation. The AG acknowledges that the question of whether it is relevant whether the process has been previously applied to any foodstuff cannot simply be answered on the basis of the legal text. So that requires another interpretation of the Novel Food Regulation. For now, we do not yet know whether the CJEU supports the interpretation set out above by the AG. Hopefully, we will find out soon in another case.

As a sequel to this matter at national level, we were informed that the case between The Longevity Labs and Optimize Health Solutions was settled. But in another case initiated by The Longevity Labs against another competitor, the Austrian Supreme Court issued a decision[147] almost identical to the current CJEU decision.

Karin Verzijden
AXON Lawyers, Amsterdam

147 *Oberste Gerichtshof*, 12 September 2023, ECLI:AT:OGH0002:2023:0040OB00068.23P.0912.000.

CHAPTER 18

Nasal spray – Product demarcation; the relationship between different legal regimes

Annotated by Irene Verheijen

Court of Justice EU 19 January 2023, cases C-495/21 and C-496-21 (*L. GmbH & H. Ltd v Bundesrepublik Deutschland*) ('Nasal spray'), ECLI:EU:C:2023:34

Medical device, medicinal product by function, medicinal product by presentation, relationship legal regimes, priority of medicines legislation over other product legislation. If a product clearly falls within the definition of other product categories, the Medicinal Products Directive does not apply.

Main considerations:

> 25 By its fourth question, which it is appropriate to examine first, the referring court asks, in essence, whether Article 2(2) of Directive 2001/83 must be interpreted as meaning that it applies not only to 'medicinal products by function', as referred to in Article 1(2)(b) of that directive, but also to 'medicinal products by presentation', as referred to in Article 1(2)(a) of that directive.
> 26 As a preliminary point, it should be noted that Article 2(1) of Directive 2001/83 provides, in essence, that that directive is to apply to medicinal products for human use which are intended to be placed on the market in Member States and which are manufactured industrially.
> 27 The scope of Directive 2001/83 is thus limited to products which are industrially manufactured medicinal products, to the exclusion of products which do not correspond to one or other of the definitions of 'medicinal product' listed, respectively, in Article 1(2)(a) of that directive, that is, 'medicinal products by presentation', and in Article 1(2)(b) of that directive, that is, 'medicinal products by function' (see, to that effect, judgment of 13 March 2014, Octapharma France, C512/12, EU:C:2014:149, paragraph 30)
> 28 However, first, the wording of Article 2(2) of Directive 2001/83, which expressly refers to 'medicinal product', does not permit a distinction to be drawn between the two definitions of the concept of 'medicinal product' adopted in points (a) and (b) of Article 1(2) of that directive, unless the very wording of that provision is to be disregarded.
> 29 Second, the exclusion of 'medicinal products by presentation' from the primacy of application of the legal regime applicable to medicinal products is incompatible with the stated intention of the legislature, which sought, by means of the obligation laid down in Article 2(2) of Directive 2001/83, to reconcile the requirements of legal cer-

 | HTTPS://DOI.ORG/10.1163/9789004751699-019

tainty for economic operators with those of quality, safety and efficacy of medicinal products for human use.

30 Directive 2004/27, which introduced that provision into Directive 2001/83, states, in recital 7, that 'the definitions and scope of Directive 2001/83/EC should be clarified in order to achieve high standards for the quality, safety and efficacy of medicinal products for human use' and that, 'in order to take account both of the emergence of new therapies and of the growing number of so-called "borderline" products between the medicinal product sector and other sectors, the definition of "medicinal product" should be modified so as to avoid any doubt as to the applicable legislation when a product, whilst fully falling within the definition of a medicinal product, may also fall within the definition of other regulated products'.

31 In that regard, it must nevertheless be pointed out that recital 7 also states that 'where a product comes clearly under the definition of other product categories, in particular food, food supplements, medical devices, biocides or cosmetics, [Directive 2004/27] should not apply'.

32 However, that exception is subject, in accordance with the wording of recital 7, to the conditions laid down by another definition being clearly satisfied, as was submitted at the hearing by the European Commission, according to which the legislature did not in any way intend to call into question the rule on Directive 2001/83 taking precedence.

33 In the present case, it does not seem to be clear that the conditions of the concept of 'medical device' are satisfied, which is, however, a matter for the referring court to verify.

34 Thus a product that falls within the definition of a 'medicinal product', as referred to in Article 1(2)(a) or (b) of Directive 2001/83, is covered by the legal regime established by that directive and accordingly may not be classified as a 'medical device', within the meaning of Directive 93/42 (see, to that effect, judgment of 3 October 2013, Laboratoires Lyocentre, C109/12, EU:C:2013:626, paragraph 41).

35 In the light of all of the foregoing considerations, the answer to the fourth question is that Article 2(2) of Directive 2001/83 must be interpreted as meaning that it applies not only to 'medicinal products by function', as referred to in Article 1(2)(b) of that directive, but also to 'medicinal products by presentation', as referred to in Article 1(2)(a) of that directive.

[...]

37 In the first place, it should be noted that it follows from Article 1(2)(a) of Directive 93/42 that a substance may be classified as a 'medical device' only where the principal intended action in or on the human body is not achieved by pharmacological, immunological or metabolic means (see, to that effect, judgment of 3 October 2013, Laboratoires Lyocentre, C109/12, EU:C:2013:626, paragraph 44).

38 In that regard, it must be pointed out that Articles 3 and 4 of that directive require a manufacturer wishing to market a product as a 'medical device' to demonstrate that that condition has been satisfied.

39 That approach is further reinforced by the general scheme of Directive 93/42,

which does not provide for the same level of consumer protection as that provided for by Directive 2001/83. The reason for that difference is the negative requirement to which medical devices are subject, in the sense that the principal intended action is not to be achieved by pharmacological, immunological or metabolic means, in accordance with Article 1(2)(a) of Directive 93/42. The presumption that those products are less dangerous justifies the release of goods on a declaratory basis, unlike the legal rules applicable to medicinal products, whether by function or by presentation, in respect of which Article 6 of Directive 2001/83 requires the prior grant of marketing authorisation.

40 In the second place, Article 1(5)(c) of Directive 93/42 specifically requires the competent authorities to take 'particular' account of the principal mode of action of the product concerned. Such wording cannot, however, be interpreted as allowing the national authorities to take account of other criteria, since it follows from Article 1(2)(a) of that directive that the mode of action of any 'medical device' must necessarily be other than pharmacological, immunological or metabolic.

41 *Therefore, in the absence of scientific knowledge that would make it possible to establish that the principal intended action in or on the body is not achieved by pharmacological, immunological or metabolic means, Article 1(2)(a) of Directive 93/42 does not enable a product to be classified as a 'medical device'.*

42 In the third place, as regards classification as a 'medicinal product by function', as referred to in Article 1(2)(b) of Directive 2001/83, it must be noted that, according to settled case-law, unlike the concept of 'medicinal product by presentation', as referred to in Article 1(2)(a) of that directive, a broad interpretation of which is intended to protect consumers from products that do not have the efficacy which they would be entitled to expect, the concept of 'medicinal product by function' is intended to encompass products whose pharmacological properties have been scientifically established (see, to that effect, judgment of 6 September 2012, Chemische Fabrik Kreussler, C308/11, EU:C:2012:548, paragraph 30 and the case-law cited).

43 In addition, as regards determining that a product is a 'medicinal product by function', as referred to in Directive 2001/83, it must be recalled that the national authorities, acting under the supervision of the courts, must decide on a case-by-case basis, taking account of all the characteristics of the product concerned, in particular its composition, its pharmacological, immunological or metabolic properties, to the extent to which they can be established in the present state of scientific knowledge, the manner in which it is used, the extent of its distribution, its familiarity to consumers and the risks which its use may entail (judgment of 15 January 2009, Hecht-Pharma, C140/07, EU:C:2009:5, paragraph 39).

44 *However, in the absence of available scientific knowledge, a product cannot meet the definition of the concept of 'medicinal product by function', which requires, in accordance with Article 1(2)(b) of Directive 2001/83, a pharmacological, immunological or metabolic action.*

45 In the fourth place, as regards classification as a 'medicinal product by presenta-

tion', as referred to in Article 1(2)(a) of Directive 2001/83, it must be recalled that a product is 'presented [as having properties] for treating or preventing disease' within the meaning of Directive 2001/83 when it is expressly 'indicated' or 'recommended' as such, possibly by means of labels, leaflets or oral representation.

46 A product is also 'presented [as having properties] for treating or preventing disease' whenever any averagely well-informed consumer gains the impression, which, provided it is definite, may even result from implication, that the product in question should, having regard to its presentation, have the properties concerned (judgment of 15 November 2007, Commission v Germany, C319/05, EU:C:2007:678, paragraph 46 and the case-law cited).

47 In that regard, account should be taken of the attitude of an averagely well-informed consumer, in whom the form given to a product may inspire particular confidence similar to that normally inspired in him or her by proprietary medicinal products, having regard to the safeguards normally associated with their manufacture and marketing. Although the external form given to the product may serve as strong evidence of its classification as a medicinal product by presentation, the 'form' must be taken to mean not only the form of the product itself but also that of its packaging, which may, for reasons of marketing policy, tend to make it resemble a medicinal product (see, to that effect, judgment of 15 November 2007, Commission v Germany, C319/05, EU:C:2007:678, paragraphs 44, 46 and 47 and the case-law cited).

48 Thus, the factors referred to by the referring court, such as the presentation of the product concerned as having properties for treating or alleviating disease, references to medicinal interactions and adverse effects, and pharmacy-only distribution, are factors which, taken as a whole, are capable of making the products concerned appear to an averagely well-informed consumer to have the properties of a medicinal product, which will, however, be for the referring court to verify.

49 Having regard to all of the foregoing considerations, the answer to the second and third questions is that Article 1(2)(a) of Directive 93/42 and Article 1(2) of Directive 2001/83 must be interpreted as meaning that, where the principal mode of action of a product is not scientifically established, that product cannot meet the definition of the concept of 'medical device', within the meaning of Directive 93/42, or that of 'medicinal product by function', as referred to in Directive 2001/83. It is for the national courts to assess, on a case-by-case basis, whether the conditions relating to the definition of the concept of 'medicinal product by presentation', as referred to in the latter directive, are satisfied.

18.1 The case

Two companies, L. GmbH and H. Ltd, manufacture and sell nasal drops and nasal spray, respectively, as 'medical devices', on the German market and in several other European Union Member States. The regulator in Germany, the *Bundesinstitut für*

Arzneimittel und Medizinprodukte (Federal Institute for Drugs and Medical Devices, Germany) (BfArM), is of the opinion that both products are medicinal products (both by function and by presentation) and require a prior marketing authorisation.

The referring court (the *Bundesverwaltungsgericht*, the highest federal administrative court in Germany) asks the Court to interpret the definitions medical device[148] and medicinal product[149] and how they relate to each other, and the provisions of Article 2(2) of the Medicinal Products Directive, the so-called 'hierarchy provision'. This means, in short, that in case of doubt, when there is a product that falls under the definition of medicinal product, but also under the definition of another product (food product, medical device, cosmetic product), the medicinal product legislation takes precedence.

18.2 Importance of the ruling

Interestingly, referring to recital 7 of Directive 2004/27/EC, the Court stressed that if a product clearly falls within the definition of other product categories, notably food, food supplements, medical devices, biocidal products or cosmetic products, the Medicinal Products Directive does not apply.[150]

18.3 Review

The Court's emphasis on the evidentiary nature of a product category (no doubt) is perfectly in line with the Medicinal Products Directive, after all, this directive states as a condition for applying the hierarchy of legislation that there must be doubt about the nature of the product. When there is evidently a medical device, a foodstuff or a cosmetic product (even if medical information is provided about it, which is not allowed – under the legislation of the relevant product categories), the product does not actually become a medicinal product just like that. However, this is regarded differently by the Dutch Food and Consumer Product Safety Authority,

148 Article 1(2)a of Council Directive 93/42/EEC of 14 June 1993 on medical devices (PB 1993, L 169, p. 1), as amended by Directive 2007/47/EU of the European Parliament and of the Council of 5 September 2007 (PB 2007, L 247, p. 21).

149 Article 1(2)a of Directive 2001/83/EC of the European Parliament and of the Council of 6 November 2001 on establishing a Community code relating to medicinal products for human use (PB 2001, L 311, p.67), as amended by 2004/27/EC of the European Parliament and of the Council of 31 March 2004 (PB 2004, L136, p. 34) (the Medicinal Products Directive).

150 See also: E.H. Hoogenraad, M. Teeuw, L. Snellenberg, B. Duivenvoorde, Kroniek over 2022, part X. Reclamerecht, IER 2023/8; S.M. Wertwijn, Kroniek over 2022, part IX. Geneesmiddelenrecht, IER 2023/8; Barbara Mutsaers and Roeland Carolus, Kroniek jurisprudentie levensmiddelenrecht 2022 up to and including first quarter 2023, Waar & Wet, no. 4, June 2023.

and Dutch judges also often take the view that if (in these cases) a food product contains medical claims, this makes it (by definition) a medicine.[151]

The question is: where does the test start, to assess what kind of product it is? Do you start with the definition of medical device (or of foodstuff, or of cosmetic product) and, if the product meets that definition, do you no longer test the definition of medicinal product? Or do you start with the definition of medicinal product and, if it falls under that, do you check whether the product also meets other definition(s)? The Court considers in marginal 31 of the judgment that for an exception to the priority rule of the Medicinal Products Directive, it is required that the conditions set out in another definition are clearly met. This seems to indicate that the Court is leaning towards saying that when a product meets the definition of medical device, foodstuff, or cosmetic product, the Medicinal Products Directive does not apply and, therefore, there is no further need for a Medicinal Products Directive review.

Thus, the doubt in fact only arises when a product does not meet the requirements of the product category under which it is marketed. The Court considers that this is the case in this instance (marginal 33). If subsequently (marginal 34) a product meets the definition of medicinal product, it cannot be classified as a medical device.[152]

The Court then first tests the products against the definition of 'medical device' and concludes that the products do not meet this definition. The Court considers (marginal 41) that if no scientific knowledge is available on the basis of which it can be established that the principal intended function in or on the body is not achieved by pharmacological, immunological or metabolic means, a product cannot be classified as a medical device.

As the product does not meet the requirements to qualify as a medical device, the Court tests whether it is a medicinal product by function. The Court rules (marginal 44) that if no scientific knowledge is available, a product does not meet the definition of medicinal product by function, which requires a pharmacological, immunological or metabolic effect.

Since there is no question of a medicinal product by function, the Court examines whether there is a medicinal product by presentation. The Court assesses this on the basis of the criteria developed in the case-law (marginals 46 and 47): Is the product described or recommended as a medicinal product? Does it, by the manner of presentation, implicitly but no less emphatically give the impression to the av-

---

151 See Chapters 5 and 25 of this book.

152 The Court considers in this respect that the hierarchy provision in the Medicinal Products Directive concerns not only medicinal products by function but also medicinal products by presentation. According to the judgment of the Court, medicines by presentation cannot be excluded from the priority of legal requirements applying to medicinal products, given that this is incompatible with the meaning expressed by the legislator, who attempted with the obligation laid down in Article 2(2) of Directive 2001/83 to align the demands of legal certainty for market participants with the requirements in terms of quality, safety and efficacy of medicinal products for human use.

eragely well-informed consumer that it has the properties of a medicinal product? The Court holds that taken as a whole several factors – presentation as products capable of alleviating illness, references to interactions between medicinal products and to undesirable effects, and the fact that the product is sold exclusively in pharmacies – may give the averagely well-informed consumer the impression that the products in question have the properties of medicinal products. This is for further assessment by the Court.

18.4 Conclusion

Indeed, from the mode of review, the Court seems to assume that doubt only arises in cases where a product does not clearly meet the requirements of another product category.

A similar consideration was made by the East Brabant District Court, in an earlier ruling on food supplements.[153] This court also referred to recital 7 of the preamble of Directive 2004/27/EC for the verdict 'no doubt, not a medicinal product'.

I discussed the question as to whether this approach (the mere reference to recital 7) is sufficiently nuanced to arrive at this opinion, in the note to the judgment of the District Court of East Brabant. The conclusion I draw is this: whatever approach is taken, if (in substance) there is no doubt about the nature and identity of a product, and it is not a medicinal product, it cannot be qualified as a medicinal product either. This is also in line with the view frequently expressed in the literature by other food lawyers.[154]

Irene Verheijen
Lawyer – Partner Legaltree, Amsterdam

153 Court of East Brabant 25 March 2022, ECLI:NL:RBOBR:2022:1109; see Chapter 39 of this book.
154 See Chapters 5, 25 and 39 of this book.

CHAPTER 19

Kwizda Pharma: demarcation issue food supplement – Medicine

Annotated by Silvia Gawronski

Court of Justice of the European Union 2 March 2023 Case C-760/21, (*Kwizda Pharma GmbH v Landeshauptmann von Wien*) ECLI:EU:C:2023:143

Food safety, foodstuffs, demarcation issues, Regulation (EU) 609/2013 Article 2(2)(g), foodstuff/food supplement vs medicinal product, concept of 'food for special medical purposes', dietetic foods.

Main consideration:

> Recital 34: '[…] in the event of doubt as to the correct classification of the products in question, Article 2(2) of Directive 2001/83 gives priority to the application of EU law relating to medicinal products, which, by reason of the higher requirements deriving from the law relating to medicinal products for the placing of products on the market, is also consistent with the objective of a high level of protection of human health pursued by Article 168 TFEU.'

19.1 Introduction

This contribution concerns a judgment of the Court of Justice of the European Union (CJEU) of 2 March 2023 in a reference for a preliminary ruling filed by the *Verwaltungsgericht Wien* in Austria (Referring Judge) in the case between Kwizda Pharma GmbH (Kwizda Pharma) and the *Landeshauptmann von Wien* (Competent Authority). A company wanted to market a number of products – whose main ingredients were D-mannose and cranberry – as 'food for special medical purposes' within the meaning of Article 2(2)(g) of Regulation (EU) 609/2013. However, this was refused by the competent authority, which considered it a medicinal product.

In food law, demarcation issues are a familiar problem. Is the cheerful children's packaging for a sweet a food contact material or a toy? Is snus tobacco or a foodstuff? Is something an animal by-product or a medicine? Such questions arise with innovative products or production processes in particular, where the European legislator logically did not consider the relevant innovations when drafting the definitions in the various regulations and directives. In this judgment, the CJEU provided an answer in relation to four concrete demarcation questions. But the judgment is

 | HTTPS://DOI.ORG/10.1163/9789004751699-020

also of greater importance, because the way the CJEU answers these preliminary questions may also constitute a precedent for other demarcation issues.

19.2 Background

Kwizda Pharma markets four products whose ingredients it claims prevent bacteria from adhering to the mucous membrane of the urinary tract, and as such the use of these products is recommended in cases of urinary tract infections ('Products Concerned').

The company notified the Competent Authority that it would market these four products as 'food for special medical purposes' within the meaning of Article 2(2)(g) of Regulation (EU) 609/2013.

In four separation decisions, the Competent Authority refused to classify these four products as 'food for special medical purposes'. These decisions were based on the assessment of the government agency responsible for examining the samples of the Products Concerned, according to which these products are not foodstuffs, because the ingredients producing the claimed effect, namely D-mannose and cranberry, do not achieve their effect through absorption during digestion, but through their action on the excretory functions of the kidneys.

Kwizda Pharma is challenging these four decisions before the Referring Judge. The Referring Judge wonders what exactly the concept of 'food for special medical purposes' means and what distinction should be made between this concept and the concepts of 'medicinal product' and 'food supplement'. He also wants to know more about the concept of 'nutrient' and whether it still matters whether a product is available from a pharmacy or other outlet. In that context, in summary, the Referring Judge makes the following observations and asks the following questions.

First, the Referring Judge expresses a desire to learn what characteristics a food must have in order to satisfy the particular nutritional requirements of the persons for whom it is intended ('dietetic foods') within the meaning of Regulation (EU) 609/2013. Indeed, the Referring Judge considers that a product must be classified as 'food for special medical purposes' if it has the declared medical effect exclusively to meet the nutritional needs that the person consuming it wishes to satisfy for health reasons. According to the Referring Judge, Kwizda Pharma advocates a different approach. In its view, the concept of 'dietetic food' encompasses all cases in which the consumption of a particular nutrient is indicated because of a disease or condition. Thus, in the present case, those nutritional requirements are met by the consumption of cranberry and D-mannose, which do not derive their action from a necessary change in diet and which are not absorbed or metabolised through digestion, but whose consumption is indicated to promote renal excretion and, consequently, the healing process of urinary tract infections.

Second, the Referring Judge wonders what distinction should be made between the concept of 'food for special medical purposes' on the one hand, and the concepts of 'medicinal product' and 'food supplement' on the other. According to Kwizda Pharma, the composition of food supplements, the Referring Judge notes, can lead to them also being classifiable as 'food for special medical purposes' because certain substances, such as cranberry or D-mannose, are deemed to meet nutritional needs. Moreover, Kwizda Pharma argues that a product can be classified as 'food for special medical purposes' when it contains a substance – in this case cranberry and D-mannose – that may be conducive to the progression of disease or healing. According to the Referring Judge, however, such an argument blurs the distinction between food for medical purposes and medicinal products.

Thirdly, the Referring Judge wishes to know the scope of the requirement of Article 2(2)(g) of Regulation (EU) 609/2013, which states that, for classification as a food for special medical purposes, determining ingredients must produce their effects in the context of nutritional needs that cannot be met by modification of the normal diet alone. It notes that, according to Kwizda Pharma, cranberry or D-mannose can only be ingested as part of the normal diet with great difficulty, so this requirement is met for the Products Concerned.

Finally, for the purposes of assessing the argument put forward by Kwizda Pharma concerning the divergent definitions of the concept 'nutrient' in EU law – according to which, for the purposes of Regulation (EU) 609/2013, any food and any substance that may also be contained in a food or food supplement must be regarded as a nutrient within the meaning of Article 2(2)(g) of that Regulation – the Referring Judge wishes to know from the Court what is to be understood by the concept 'nutrient' within the meaning of that regulation.

19.3 Prejudicial questions

With this in mind, the Referring Judge asked a large number of preliminary questions. The CJEU deals with the questions clustered into a number of topics, thus providing answers to various issues of demarcation. I would like to single out one of those answers in this annotation: the difference between a foodstuff (food for medical purposes) and a medicinal product.

Food and medicinal product

The CJEU first addresses the third – and most important – question of this judgment. This question was: 'What criteria should be used as a basis for distinguishing a medicinal product from a food for special medical purposes [...]?' An answer to this question involves the concepts of 'medicinal product' within the meaning of Article 1(2) of Directive 2001/83 and 'food for special medical purposes' within the meaning of Article 2(2)(g) of Regulation (EU) 609/2013.

In answering this question, the CJEU first refers to the judgment of 27 October 2022 (Orthomol, C 418/21, EU:C:2022:831, paragraph 37) in which it has already been emphasised that food for special medical purposes is distinct from medicinal products, and that the two categories of products, in view of their specific characteristics, are subject to different definitions and regulations that are mutually exclusive.

The CJEU continues that the characteristics and functions of food for medical purposes are different from those of medicinal products, meaning that they are products that purport to have therapeutic or prophylactic properties in relation to human diseases, or any single or compound substance that may be used in or administered to human beings either to restore, improve or modify physiological functions by exerting a pharmacological, immunological or metabolic effect or to make a medical diagnosis.

However, food for special medical purposes is different, the CJEU stresses. Indeed, it concerns food intended to meet the nutritional needs of patients and not to prevent or cure human diseases, or to restore, improve or modify physiological functions by producing a pharmacological, immunological or metabolic effect, or to make a medical diagnosis. Food for special medical use cannot, as such, combat diseases, disorders or ailments, but is characterised by its nutritional function.

Consequently, if a patient generally benefits from the ingestion of a product to the extent that the substances it contains help to prevent, alleviate or cure a disease, the purpose of that product is not to nourish that patient, but to care for him, prevent a disease or restore, improve or modify physiological functions by producing a pharmacological, immunological or metabolic effect, which argues in favour of classifying that product as other than 'food for special medical purposes'.

So far, nothing new. But then the CJEU applies it to the present case, ruling that it is clear from the information provided by the Referring Judge that Kwizda Pharma markets the products in question and indicates that the consumption of these products in cases of urinary tract infection is conducive to the elimination of the pathogens in question.

Even though it is for the competent national public authorities to determine, on the basis of a case-by-case assessment and taking into account all the characteristics of those products, whether those products can be marketed as food for special medical purposes, the fact remains, according to the CJEU, that products presented as having therapeutic properties in relation to a disease, but which are *not intended* to meet particular nutritional needs of patients, cannot be marketed as food for special medical purposes but as medicinal products. Kwizda Pharma's argument therefore fails.

Indeed, the following consideration also comes into play here, according to the CJEU. Article 2(2) of Directive 2001/83 states, namely, that in case of doubt as to the correct classification of the products concerned, priority is given to the application of EU law on medicinal products, which, because of the stricter requirements resulting from pharmaceutical law for the marketing of products, is also consistent with the objective pursued by Article 168 TFEU of a high level of human health protection.

The CJEU therefore answers the third question, stating that the distinction between concepts of 'medicinal product' and 'food for special medical purposes' must be assessed – in the light of the nature and characteristics of the product in question – whether it is a foodstuff intended to satisfy particular nutritional needs, or a product intended to prevent or cure diseases in human beings, to restore, improve or modify physiological functions by exerting a pharmacological, immunological or metabolic effect, or to make a medical diagnosis, or possibly presented as such. The destination is thus of decisive importance.

19.4 Conclusion: demarcation – a common problem

In food law, demarcation issues are a familiar problem. Is the cheerful children's packaging for a sweet a food contact material or a toy? Is snus tobacco or a foodstuff? Is something an animal by-product or a medicine? Such questions arise with innovative products or production processes in particular, where the European legislator logically did not consider the relevant innovations when drafting the definitions in the various regulations and directives. In this judgment, the CJEU answered two concrete demarcation questions. However, the judgment is also of greater importance because the way in which the CJEU answers these questions may also constitute a precedent for other demarcation issues. In it, the CJEU gives the following basic principles:

- As a first step, it should be carefully examined which definitions – and thus within the scope of which Regulation or Directive – a particular product falls under. This does not rule out the possibility of overlapping definitions.
- It is then necessary to consider whether the two European regulations have a provision determining the rules of precedence.
- If not, the concepts will have to be juxtaposed and an assessment made of where the concepts and the resulting qualifications are necessarily mutually exclusive.
- In this context, it can then be determined on a case-by-case basis which concept a product falls under. Whereby, on the one hand, reference can therefore be made to the CJEU judgment of 19 January 2023 (C-495/21 and C-496/21 – ECLI:EU:C:2023:143) indicating that if a product clearly meets the conditions set out in another definition, then the Medicinal Products Directive does not apply, but that in case of doubt as to the correct classification of the products in question, the Medicinal Products Directive should take precedence, as this legal framework imposes stricter requirements which, in view of the objective of a high level of protection of human health pursued by Article 168 TFEU, is deemed appropriate.

It will still often be a grey area, and in special cases – as here – a national court will not be able to avoid asking preliminary questions. Nevertheless, in my view, the roadmap suggested by the CJEU in this judgment is often sufficient to answer most questions on demarcation.

Silvia Gawronski
partner Food & Commodities at Van Traa law firm, Rotterdam

CHAPTER 20

Food v environment – Crumbling precautionary principle

Annotated by Jacqueline Mailly, Irene Verheijen, Bernd van der Meulen

Court of Justice of the European Union 19 January 2023 (case C-162/21) *PAN Europe v Belgium*, ECLI:EU:C:2023:30
Court of Justice of the European Union 7 February 2023 *Confédération paysanne II* ECLI:EU:C:2023:75
Court of Justice of the European Union 22 June 2023 *Taminco and Arysta LifeScience Great Britain v Commission* ECLI:EU:C:2023:513

Under the precautionary principle it must be ensured that active substances or products placed on the market do not adversely affect human or animal health or the environment. Under the precautionary principle, it cannot have been the legislator's intention to exclude products of mutagenesis techniques that did not exist at the time of the legislation from prior safety assessment as GMOs. Under the principle of proportionality, the less intrusive of the available alternatives must be chosen. The Communication on the precautionary principle does not impose any obligations on the Commission; in particular, it does not impose any obligation to prepare a written cost-benefit analysis.

Main considerations:

Court of Justice of the European Union 19 January 2023 PAN Europe v Belgium

> The precautionary principle should be applied and this Regulation should ensure that industry demonstrates that substances or products produced or placed on the market do not have any harmful effect on human or animal health or any unacceptable effects on the environment.
> The provisions of this Regulation are underpinned by the precautionary principle in order to ensure that active substances or products placed on the market do not adversely affect human or animal health or the environment.
> In that regard, it should be borne in mind that those provisions are based on the precautionary principle, which is one of the bases of the policy of a high level of protection pursued by the European Union in the field of the environment, in accordance with the first subparagraph of Article 191(2) TFEU, in order to prevent active substances or products placed on the market from harming human or animal health or the environment.

 | HTTPS://DOI.ORG/10.1163/9789004751699-021

Court of Justice of the European Union 7 February 2023 Confédération paysanne II

44 In addition, the interpretation of Article 3(1) of Directive 2001/18, read in conjunction with point 1 of Annex I B thereto, must be made in the light of the objective of that directive, as set out in Article 1 thereof, namely, in accordance with the precautionary principle, to protect human health and the environment when, first, GMOs are deliberately released into the environment for any purpose other than placing on the market within the European Union, and, secondly, when GMOs are placed on the market within the European Union as or in products (see, to that effect, judgment of 25 July 2018, Confédération paysanne and Others, C528/16, EU:C:2018:583, paragraph 52).

45 An interpretation of Article 3(1) of Directive 2001/18, read in conjunction with point 1 of Annex I B thereto, according to which organisms obtained by means of techniques/methods of mutagenesis would be excluded from the scope of that directive, without any distinctions, would compromise the objective of the protection of human health and the environment pursued by that directive and would fail to respect the precautionary principle which it seeks to implement (see, to that effect, judgment of 25 July 2018, Confédération paysanne and Others, C528/16, EU:C:2018:583, paragraph 53).

46 In the light, in particular, of the foregoing considerations, the Court has held that Article 3(1) of Directive 2001/18, read in conjunction with point 1 of Annex I B to that directive and in the light of recital 17 thereof, must be interpreted as meaning that only organisms obtained by means of techniques/methods of mutagenesis which have conventionally been used in a number of applications and have a long safety record are excluded from the scope of that directive (see, to that effect, judgment of 25 July 2018, Confédération paysanne and Others, C528/16, EU:C:2018:583, paragraph 54).

47 It is important to point out, in that regard, that the limitation of the scope of the exemption provided for in Article 3(1) of Directive 2001/18, read in conjunction with point 1 of Annex I B thereto, by reference to the dual criterion of conventional use in a number of applications and with a long safety record, is closely linked to the very objective of that directive, set out in paragraph 44 of the present judgment.

48 The application of that dual criterion thus makes it possible to ensure that, because of age and the variety of uses of a technique/method of mutagenesis and the information available as to its safety, organisms obtained by that technique/method may be released into the environment or placed on the market within the European Union, without it being necessary, in order to avoid adverse effects on human health and the environment, to subject those organisms to the risk assessment procedures laid down in Part B and Part C respectively of Directive 2001/18.

Court of Justice of the European Union 22 June 2023 Taminco and Arysta Life-Science Great Britain v Commission

127 It is necessary, as a preliminary point, to examine the Commission's complaints concerning paragraphs 186 to 194 of the judgment under appeal, in so far as those paragraphs impose on it a separate obligation to carry out a cost/benefit analysis of the regulation by means of which it decides on an application for renewal of the approval of an active substance.
128 In that regard, it should be noted that Article 20(1) of Regulation No 1107/2009, read in conjunction with Article 14(1) thereof, provides that, when the Commission receives such a request, it may adopt a regulation authorising the renewal requested, but it may also refuse or authorise such renewal subject to conditions and restrictions.
129 Thus, by providing for the non-renewal or renewal of the approval of an active substance subject to conditions and restrictions, those provisions expressly incorporate the principle of proportionality, which is one of the general principles of EU law. That principle requires that acts of the EU institutions should not exceed the limits of what is appropriate and necessary for the achievement of the legitimate objectives pursued by the legislation in question, it being understood that, where there is a choice between several appropriate measures, the least restrictive should be used and that the inconvenience caused must not be disproportionate to the aims pursued (see, by analogy, judgment of 6 May 2021, Bayer CropScience and Bayer v Commission, C-499/18 P, EU:C:2021:367, paragraph 166).
130 In addition, Article 20(1) of Regulation No 1107/2009 does not impose any particular form or detailed rules to ensure compliance with the principle of proportionality (see, by analogy, judgment of 6 May 2021, Bayer CropScience and Bayer v Commission, C-499/18 P, EU:C:2021:367, paragraph 169). Article 14 of that regulation, read in conjunction with Article 4 and points 3.6.2 to 3.6.4 of Annex II, entitled 'Procedure and criteria for the approval of active substances, safeners and sygergists pursuant to Chapter II', of that regulation, requires the Commission to carry out a risk assessment, such as tests of genotoxicity, carcinogenicity or reproductive toxicity, only when it is called on to decide on an application for renewal of the approval of an active substance (see, by analogy, judgment of 6 May 2021, Bayer CropScience and Bayer v Commission, C-499/18 P, EU:C:2021:367, paragraph 172).
131 Those provisions do not therefore require the Commission to draw up a document specifically intended to present its examination of the benefits and costs resulting from its action or inaction. Nor can such an obligation arise from the Communication on the precautionary principle, since, as stated in Section 2 thereof, 'The guidelines outlined in this Communication ... [must] in no way ... modify or affect the provisions of the Treaty or secondary Community legislation'.
132 It therefore follows from paragraphs 128 to 131 of the present judgment that, when it receives an application for renewal of the approval of an active substance under Article 14 of Regulation No 1107/2009, the Commission is not bound by the guidelines contained in the Communication on the precautionary principle. While it is admittedly for the Commission to demonstrate that it has complied with the principle of proportionality when ruling on such an application for renewal, the Commission

is in no way required to base its assessment on a calculation of the respective costs associated with the renewal or non-renewal of the approval of the active substance concerned.

133 It must therefore be held that the General Court erred in law by imposing on the Commission the obligation to examine separately the benefits and costs of the non-renewal of thiram for use as a seed treatment, in accordance with point 6.3.4 of the Communication on the precautionary principle.

134 That being said, the judgment under appeal cannot be set aside, since the General Court rightly found, in that judgment, that, in the present case, the Commission had not failed to observe the precautionary principle, on which the provisions of Regulation No 1107/2009 are based.

135 The General Court has justified to the requisite legal standard that the Commission had fulfilled its obligation to examine the benefits and costs of non-renewal of the approval of an active substance and, therefore, that it had complied with the principle of proportionality, by referring, in paragraph 197 of the judgment under appeal, to the information that the Commission had provided to the Standing Committee at the meetings of 24 and 25 May 2018, to which reference is made in paragraph 23 of this judgment. Moreover, the appellant has not shown that the extract from the summary of that meeting, which is cited in the same paragraph of the judgment under appeal, was distorted by the General Court.

136 In addition, the appellant's arguments relating to paragraphs 196 and 200 of the judgment under appeal, according to which the General Court did not require the Commission to comply with point 6.3.4 of the Communication on the precautionary principle, must be declared ineffective, since it follows from paragraphs 128 to 132 of the present judgment that the Commission is not bound by the guidelines contained in that communication when it acts under Regulation No 1107/2009.

20.1 Introduction

For readers who base their impression of the content of the precautionary principle on the text of Article 7 of Regulation 178/2002, the above judgments will no doubt come as a surprise. For those readers, this book contains several more surprises elsewhere. Sometimes the Court draws conclusions from the precautionary principle that are difficult to grasp in the light of Article 7 GFL. Those considerations reproduced above are examples of this. One possible explanation is a mutual influence between food law and environmental law.

In this annotation, we want to do two different things. We begin with some critical reflections following the surprises in the relatively short sections 20.2 to 20.6. We then try to bring some order to the chaos by listing in the much longer paragraph 20.7 what did become clear over time about the application of the precautionary principle under Article 7 GFL.

20.2 Sources of the precautionary principle

Perhaps some of the surprises can be explained by the different sources of the precautionary principle. The highest source is Article 191(2), first sentence TFEU: 'Union policy on the environment shall aim at a high level of protection taking into account the diversity of situations in the various regions of the Union. It shall be based on the precautionary principle and on the principles that preventive action should be taken, that environmental damage should as a priority be prevented at source and that the polluter should pay.' Then there is Article 7 of Regulation 178/2002, and finally there is the Commission's 2000 Communication on the Precautionary Principle.[155] References in other legislation cast their light and shade on the principle.

In Court of Justice of the European Union (CJEU) 22 June 2023 *Taminco and Arysta LifeScience Great Britain v Commission* ECLI:EU:C:2023:513, the CJEU gives the General Court a slap on the wrist for allegedly wrongly holding the Commission to the Communication on the Precautionary Principle. The Court's direction is open to question. Of course, the court is right that policy rules as such are not binding.[156] The same, of course, applies to this communication. On the other hand, however, they can provide legal certainty.[157] In our view, the Court should not have skipped the step of assessing to what extent the communication's interpretation of the precautionary principle is correct. If that were the case, and that interpretation could be applied to Article 191 TFEU, a very different story emerges. On the one hand, the Court positions the proportionality principle in that case at treaty level, but the precautionary principle at policy rule level. That is not entirely satisfactory.

20.3 In which situations does the precautionary principle apply?

Regulation 178/2002 places the precautionary principle in the context of the principle of risk analysis. Risk analysis is a method by which risk management measures are based on scientific risk assessment. According to Article 7, the risk manager can fall back on the precautionary principle when the risk assessment is inconclusive.

A preliminary question in this context is when the principle of risk analysis applies, i.e. in which cases we want to base decisions to be made on science. In part, this question is answered in secondary legislation. In particular, consider the Additives Regulation, the Novel Foods Regulation, the Claims Regulation, GMO leg-

155 COM 2000(1) def.

156 In this sense, for example: CJEU 22 September 2016 (C-113/15 ECLI:EU:C:2016:718) Breitsamer and Ulrich (recital 78).

157 General Court 12 June 2015, (case T-296/12 ECLI:EU:T:2015:375) Health Food Manufacturers' Association and Others v European Commission, recital 151).

islation, et cetera. These regulations stipulate that decisions on the authorisation of certain foods must be based on science. However, secondary legislation does not always provide the answer.

In case-law prior to Regulation 178/2002, the Court derives the requirement of scientific assessment from the principle of proportionality. Member States may disrupt interstate trade only to the extent necessary. What is necessary for the life and/or health of consumers is a scientific question. Therefore, that question must indeed be put to science. Scientific uncertainty can be invoked as a basis only after an assessment of all available data has been carried out.

CJEU: 'Where it proves to be impossible to determine with certainty the existence or extent of the alleged risk because of the insufficiency, inconclusiveness or imprecision of the results of studies conducted, but the likelihood of real harm to public health persists should the risk materialise, the precautionary principle justifies the adoption of restrictive measures, provided they are non-discriminatory and objective.'[158]

In Court of Justice of the European Union, 28 January 2010, ECLI:EU:2010:44 (*Commission v France*),[159] the Court requires a scientific risk assessment before a Member State may introduce an authorisation procedure (for processing aids). A reference to inherently deficient knowledge does not suffice. Reliance on the precautionary principle is rejected.[160]

A successful invocation of the precautionary principle requires that scientific uncertainty arises from the risk assessment. These are therefore situations where risk assessment is required, has been carried out and is inconclusive. In the case of a conclusive risk assessment, it cannot be set aside by invoking scientific uncertainty.

See, for example, the Fidenato-case:[161]

> in order to prevent Article 7 of Regulation No 178/2002 failing to reduce the level of uncertainty required by the rule laid down in Article 34 of Regulation No 1829/2003 for the adoption of emergency measures, such an autonomous application of the precautionary principle, as laid down in Article 7 of Regulation No 178/2002, without the substantive conditions laid down by Article 34 of Regulation No 1829/2003 being

158 CJEU, 2 December 2004, C-41/02, ECLI:EU:C:2004:762, rec. 54 (*Commission v Netherlands*), CJEU 23 September 2003, C-192/01, ECLI:EU:C:2003:492, rec. 52 and 53 (*Commission v Denmark*), CJEU 29 April 2010, C-446/08, ECLI:EU:C:2010:233, rec. 70 (*Solgar France*) and CJEU 2 December 2004, C-41/02, ECLI:EU:C:2004:762, rec. 93 (*Commission v Netherlands*). See also the first NVLR 5th anniversary edition: Kenneth Defares and Bernd van der Meulen, *Het voorzorgsbeginsel. Preadvies voor de Nederlandse Vereniging voor Levensmiddelenrecht*, Sdu 2009.

159 See chapter 7 of this book.

160 For comparison, CJEU 27 April 2017, C-672/15, ECLI:EU:C:2017:310 (*Noria*) and CJEU 29 April 2010, C-446/08, ECLI:EU:C:2010:233 (*Solgar*).

161 CJEU 13 September 2017, C-111/16 ECLI:EU:C:2017:676. See Chapter 14 of this book.

satisfied in light of the adoption of the emergency measures set out in that article, cannot be accepted.

that Article 34 of Regulation No 1829/2003, read in conjunction with the precautionary principle as set out in Article 7 of Regulation No 178/2002, must be interpreted as meaning that it does not give Member States the option of adopting, in accordance with Article 54 of Regulation No 178/2002, interim emergency measures solely on the basis of that principle, without the conditions set out in Article 34 of Directive No 1829/2003 being satisfied.[162]

20.4 Burden of proof

In situations where risk assessment is required, the question of burden of proof arises. Who should ensure that the necessary research is carried out, collected and made available? First and foremost, it should be the government which is seeking market intervention. In several pieces of legislation regulating application procedures, the burden of proof is reversed and placed on companies wishing to market a product. According to the Communication on the Precautionary Principle, this too is an application of the precautionary principle. Regulation 1829/2003 also sees it this way.[163]

In any case, if so, this is not the precautionary principle as subsequently codified in Article 7 of Regulation 178/2002.

It is perhaps the precautionary principle from the TFEU. The pesticide legislation infers from this precautionary principle that it must be ensured that the industry demonstrates that the substances or products produced or placed on the market do not have any harmful effect on human or animal health or any unacceptable effect on the environment. The Court quotes this in the PAN Europe judgment of 19 January 2023 and then considers accordingly.

20.5 Precautionary obligations

More generally, the above leads to the question of whether legal obligations can arise from the precautionary principle. Article 7 of Regulation 178/2002 appears to grant only discretionary powers to the risk manager, not obligations.

In case-law, we do encounter obligations.

See, for example, Court of First Instance (CFI) Malagutti:[164] (Chapter 1 of this book):

162 Recital 53 and 54.
163 Regulation (EC) 1829/2003, recitals 2, 3, 5.
164 CFI 10 March 2004, T-177/02 ECLI:EU:T:2004:72. See chapter 1 of this book

> If any doubt remains in that regard, it should be pointed out that under the precautionary principle prevailing in the matter of the protection of public health the competent authority may be obliged to take appropriate measures to prevent certain potential risks for public health without having to wait until the existence and seriousness of those risks has been fully demonstrated.

In the above ruling on PAN Europe v Belgium, both the pesticides legislation cited by the Court and the Court itself infer an obligation on the legislator to ensure that the industry only puts safe products on the market. This would have made the reversal of the burden of proof referred to above a legal obligation. In Confederation II, also, the Court derives from the precautionary principle a duty on the legislator. In this case, it concerns the scope of the group of products to be placed under prior safety assessment. Through interpretation, the Court ensures that the legislation complies with this duty.

As far as we can see, there has not yet been a structured discussion on the existence and scope of precautionary duties. In our opinion, part of that discussion should concern whether the precautionary principle is the right place to invest such obligations. The difference between the precautionary principle in the sense of the TFEU and the precautionary principle in the sense of the GFL would then become very large.

20.6 Distribution of burden

In Malagutti, the CFI assessed whether the Commission had acted unlawfully. Since the Commission's action in a situation of uncertainty – factual rather than scientific – could be justified by the precautionary principle, it had not, and the damage caused by the state action remained the victim's responsibility. Invoking the precautionary principle, the Court considered that '[t]he applicant, which is a victim of that alert system introduced in order to protect human health, must accept its adverse economic consequences, since the protection of public health must take precedence over economic considerations.'

This would have been an ideal case that would have lent itself to administrative compensation if the EU had such a doctrine. The idea that it would have something to do with precaution, whereby adverse consequences must be borne by the affected company, is hard to follow.

20.7 Established CJEU case-law on precautionary and proportionality principles

Over the years, CJEU case-law has refined and clarified the conditions under which EU institutions or Member States can invoke the precautionary principle in order to take protective measures.

I. *Application requirement: the identification of an uncertain public health risk and the obligation to carry out a risk assessment*

With regard to the two 'activating criteria' mentioned in Article 7 GFL and in the 2000 communication, several CJEU rulings have indicated the level of uncertainty required to take measures based on the precautionary principle.

Case-law recognises that uncertainty is inseparable from the concept of precaution: if the risk is certain, measures must be taken, not as a precaution, but to reduce the risk. Uncertainty, however, does not mean that a risk has not been identified at all. Protective measures cannot legally be based on a purely hypothetical approach. In particular, in its judgment of 8 September 2011, the CJEU held that: 'Protective measures adopted under Article 34 of Regulation No 1829/2003 cannot validly be based on a purely hypothetical approach to the risk, founded on mere assumptions which have not yet been scientifically verified.'[165]

However, once it has been established that the risk is likely to occur, and is not merely hypothetical, the EU institutions are entitled to take measures based on the precautionary principle, even if the studies carried out are still inconclusive as to the existence or extent of the risk. In several judgments, the CJEU recalls, 'Where it proves to be impossible to determine with certainty the existence or extent of the alleged risk because of the insufficiency, inconclusiveness or imprecision of the results of studies conducted, but the likelihood of real harm to public health persists should the risk materialise, the precautionary principle justifies the adoption of restrictive measures.'[166]

Although the risk may remain uncertain, it follows from the case-law that the Union institutions or the Member States must endeavour to ascertain whether a risk exists and assess the significance of that risk through a proper risk assessment. This assessment must be carried out in the specific case. As considered in the Monsanto judgment of 8 September 2011, '[...] such protective measures, notwithstanding their temporary character and even if they are preventive in nature, may be

165 CJEU 8 September 2011, Monsanto SAS and Others v Ministre de l'Agriculture et de la Pêche, case C-58/10, ECLI:EU:C:2011:553, rec. 77.

166 CJEU 17 December 2015, *Neptune Distribution SNC v Ministre de l'Economie et des Finances*, case C-157/14, ECLI:EU:C:2015:823, rec. 82, and CJEU 9 June 2016, *Pesce e.a. c/ Presidenza del Consiglio dei Ministri*, case C-78/16 and C-79/16 (added cases), ECLI:EU:C:2016:428, rec. 47.

adopted only if they are based on a risk assessment which is as complete as possible in the particular circumstances of an individual case, which indicate that those measures are necessary.'[167]

In its judgment of 19 January 2017 regarding food supplements containing amino acids, the CJEU ruled that, as required by Article 6 of the GFL: '[...] the risk assessment, upon which food legislation must be based, is based on available scientific evidence and is undertaken in an independent, objective and transparent manner.'[168]

In the same judgment of 19 January 2017 and in a 2010 judgment on processing aids in foodstuffs, the Court also ruled that '[...] a correct application of the precautionary principle presupposes, first, identification of the potentially negative consequences for health of the substances or foods concerned, and, second, a comprehensive assessment of the risk to health based on the most reliable scientific data available and the most recent results of international research.'[169]

Therefore, CJEU case-law confirms a two-step approach of 'first, identification of the potentially negative consequences for health' and 'second, a comprehensive assessment of the risk to health' as required for the application of the precautionary principle.

II. *Temporary measures and scientific developments*

The case-law of the CJEU has also established criteria relating to the period of time granted to precautionary measures. According to Article 7(2) GFL, preventive measures taken by the institutions of the Union under the precautionary principle must be temporary.

More recent CJEU rulings have confirmed this requirement: 'In that regard, it should be noted that Article 7 of Regulation No 178/2002 defines the precautionary principle in the area of food law. Article 7(1) of that regulation states that, in specific circumstances where, following an assessment of available information, the possibility of harmful effects on health is identified but scientific uncertainty persists, provisional risk management measures necessary to ensure the high level of health protection chosen in the Union may be adopted, pending further scientific information for a more comprehensive risk assessment.'[170]

In the Afton judgment of 8 July 2010,[171] which concerned the public health is-

167 CJEU 8 September 2011, *Monsanto SAS e.a. v Ministre de l'Agriculture et de la Pêche*, case C-58/10, ECLI:EU:C:2011:553, rec. 77.

168 CJEU 19 January 2017, *Queisser Pharma GmbH &; Co. KG v Bundesrepublik Deutschland*, case C-282/15, ECLI:EU:C:2017:26, rec. 52

169 Ibid. 56

170 CJEU 13 September 2017, Criminal proceedings against Giorgio Fidenato e.a., case C-111/16, ECLI:EU:C:2017:676, rec. 44. See in this book, Chapter 14.

171 CJEU 8 July 2010, *Afton Chemical Limited v Secretary of State for Transport*, case C343/09, ECLI:EU:C:2010:419.

sue arising from the use of metallic additives in fuels, the measure adopted was applied temporarily during the development of test methods for a more thorough assessment of the product. The temporal criterion also relates to the issue of risk assessment, since in this case the Court 'acknowledged that the European Union legislature may, under the precautionary principle, take protective measures without having to wait for the reality and the seriousness of those risks to be fully demonstrated ...'[172]

Measures taken under the precautionary principle may be temporary, either because (i) they require a review of the requirements set out therein and/or (ii) they are time-limited by regulation.

Another ruling relating to spongiform encephalopathies,[173] dated 12 January 2006, ruled that 'when new elements change the perception of a risk or show that that risk can be contained by less restrictive measures than the existing measures, it is for the institutions and in particular the Commission, which has the power of legislative initiative, to bring about an amendment to the rules in the light of the new information.'

It therefore follows from the CJEU case-law that the legislator or regulator must take into account new information or new scientific evidence and must reconsider temporary measures taken under the precautionary principle.

III. *Simultaneous application of the precautionary and proportionality principles*

In the area of food and feed, the CJEU has stated on several occasions: 'The principle of proportionality, which is one of the general principles of Community law, requires that measures implemented through Community provisions must be appropriate to achieve the objective pursued and must not go beyond what is necessary to achieve that objective.'[174]

With regard to the application of the principle of proportionality and assessment of the relevance of the measures adopted by the authorities, the Court recognises that its judicial review is limited because 'the legislature of the European Union, in the exercise of the powers conferred on it, must be allowed a broad discretion in

172 Ibid. 62

173 CJEU 12 January 2006, *Agrarproduktion Staebelow Gmbh v Landrat des Landkreises Bad Doberan*, case C-504/04, ECLI:EU:C:2006:30, rec. 40.

174 CJEU 8 June 2017, *Dextro Energy GmBH & Co. KG v European Commission*, case C-296/16, ECLI:EU:C:2017:437, Rec. 48 (translated from the French); see also in agriculture cases CJEU 17 October 2013, *Schaible v Land Baden-Württemberg*, case C-101/12, ECLI:EU:C:2013:661, rec. 48: 'In the area of agriculture, the European Union legislature enjoys, inter alia, such a broad discretion, corresponding to the political responsibilities given to it by Articles 40 TFEU to 43 TFEU. Consequently, review by the Court is limited to verifying whether that legislature has manifestly exceeded the limits of its discretion.'

areas which involve, on its part, political, economic and social choices and in which it is called upon to undertake complex assessments.'[175]

In this regard, the Court recognises that judges may not be able to fully assess the proportionality of the measures challenged before the Court and may have to apply limited judicial review: 'the legality of a measure adopted in those fields can be affected only if the measure is manifestly inappropriate having regard to the objective which the competent institution is seeking to pursue'[176]

The assessment of the proportionality of the measures taken usually includes a cost-benefit analysis, an assessment that is also necessary in the context of applying the precautionary principle. The Court stated in a 2013 ruling on agricultural issues that 'when there is a choice between several appropriate measures recourse must be had to *the least onerous*, and *the disadvantages caused must not be disproportionate to the aims pursued*.'[177] (own emphasis)

However, public health considerations, which are crucial in food products, tend to outweigh economic arguments, as mentioned in a 2015 General Court ruling on the detection of toxins in edible sea scallops: 'Protection of public health takes precedence over economic considerations and may therefore justify adverse economic consequences, even those which are substantial, for certain traders.'[178]

This principle applies equally to measures taken by Member States, as the CJEU has held that 'in exercising their discretion relating to the protection of public health, the Member States must comply with the principle of proportionality. The means which they choose must therefore be confined to what is actually necessary to ensure the safeguarding of public health; they must be proportional to the objective thus pursued, which could not have been attained by measures which are less restrictive on intra-Community trade.'[179]

According to Article 7(2) GFL, measures taken by the legislator or regulator based on the precautionary principle must also comply with the principle of proportionality.[180] This provision thus establishes the crucial link between the precautionary and proportionality principles in the field of food safety.

---

175 CJEU 17 October 2013, *Schaible v Land Baden-Württemberg*, case C-101/12, ECLI:EU:C:2013:661, rec. 47.

176 From a recent example of this case law: CJEU 13 March 2025, Shuvalov v. Council, case C-271/24 ECLI:EU:C:2025:180

177 CJEU 17 October 2013, *Schaible v Land Baden-Württemberg*, case C-101/12, ECLI:EU:C:2013:661, rec. 29.

178 CFI 11 February 2015, *Kingdom of Spain v European Commission*, case T-204/11, ECLI:EU:T:2015:91, rec. 141. See in this book also the Malagutti case discussed in Chapter 1.

179 CJEU 23 September 2003, *Commission of the European Communities v Kingdom of Denmark*, case C-192/01, ECLI:EU:C:2003:492, rec. 45.

180 'Measures adopted on the basis of paragraph 1 shall be proportionate and no more restrictive of trade than is required to achieve the high level of health protection chosen in the Community, regard being had to technical and economic feasibility and other factors regarded as legitimate in the matter under consideration.'

Articles 53 and 54 GFL on emergency measures and the safeguard mechanism also incorporate the principle of proportionality. Article 53(1) explicitly calls for emergency measures by the Commission in cases where 'it is evident that food or feed originating in the Community or imported from a third country is likely to constitute a serious risk to human health, animal health or the environment, and that such risk cannot be contained satisfactorily by means of measures taken by the Member State(s) concerned'.

In this context, the CJEU carried out a case-by-case assessment, taking into account the importance of the risks and issues at stake. In particular:

- In a 2017 judgment, the CJEU ruled that EU Member States' legislation banning the manufacture and sale of food supplements containing amino acids goes beyond what is necessary to achieve the objective of the GFL.[181] In that case, the national legislation challenged before the Court provided for a general ban with only temporary derogations, even in cases where the safety of a substance had been established. Such a rule was deemed disproportionate by the Court.
- In a 2002 judgment, the CJEU upheld a zero tolerance policy on the presence of pathogens in food products (smoked fish) as compatible with the objective of the Treaty, and ruled with regard to the permissible level of pathogens that 'national legislation seeking to ensure that the pathogen cannot be detected in 25 g of fish product, simply because, for certain products, even low concentrations of Listeria monocytogenes may constitute a risk to the health of particularly susceptible consumers, must be considered compatible with the requirements of the Treaty.'[182]
- In the Afton judgment[183] (relating to public health issues arising from the use of additives in fuels), the CJEU recognised that, given the nature of the harm that could be caused by the hazardous substance, there could not be more precise information making it possible to establish a maximum level of such a limit for hazardous substances that could ensure that the harm was adequately prevented. The CJEU therefore held that setting limit values did not appear manifestly disproportionate in relation to the economic interests of producers of the substance in question to ensure a high level of health and environmental protection.

---

181 CJEU 19 January 2017, *Queisser Pharma GmbH & Co. KG v Bundesrepublik Deutschland*, Case C-282/15, ECLI:EU:C:2017:26.

182 CJEU 24 October 2002, Criminal Proceedings against Walter Hahn, case C-121/00, ECLI:EU:C:2002:611, rec. 46.

183 CJEU 8 July 2010, *Afton Chemical Limited v Secretary of State for Transport*, case C343/09, ECLI:EU:C:2010:419.

20.8 Conclusion

The battle over the precautionary principle is not yet over. Many questions deserve further elaboration. As an environmental court, the Court seems to be using a different concept of the precautionary principle from the one it uses as a food court. However, there is no real difference between the two. The Court has no environmental chamber or food chamber. Therefore, it will often be the same people who consider different issues. In doing so, they bring their own representations of what words such as 'precautionary principle' mean. The jurisdictions themselves also overlap. This is to a large extent the case, for example, in the world of genetic modification of crops that can be used for food production. Some degree of mutual influence is therefore inevitable.

However, none of this alters the fact that clear lines of case-law also emerge under Article 7 GFL. No precaution without scientific uncertainty, so no precaution without science. Measures must be proportionate and temporary.

Jacqueline Mailly
Senior EU Regulatory Affairs Advisor at Hogan Lovells in Brussels

Irene Verheijen
lawyer/partner at Legaltree, president of the European Food Law Association (EFLA)

Bernd van der Meulen
president of the Dutch Food Law Association (NVLR)

PART II

Dutch case-law

A short introduction to Food Law in the Netherlands

Oswald Jansen

As is the case in most EU Member States, EU regulations are directly applicable in the national legal order. In addition, national food law is in place, both implementing EU directives and complementing EU law. Most procedural matters are governed by national administrative law. Moreover, the self-regulatory codes on advertising play an important role in food law in the Netherlands as well. The Dutch system of self-regulatory codes is explained in more detail in the introduction of Part III, on page 293.

EU food law in the Netherlands

The General Food Law (Regulation 178/2002) provides for the applicable rules on food safety and food quality. This regulation was the first regulation resulting from the 'radical new approach'[184] of the European Commission following from the 'Mad Cow'-crisis or BSE-crisis. Within ten years the complete reform of the EU legal framework on Food Law and Food Safety was completed.[185]

Next to the General Food Law this framework covers among others the following Regulations and directives:

- Directive 2002/46 on the approximation of the laws of the Member States relating to food supplements;
- Regulation 1829/2003 on genetically modified food and feed;
- Regulation 1830/2003 concerning the traceability and labelling of genetically modified organisms and the traceability of food and feed products produced from genetically modified organisms;
- Regulation 852/2004 on the hygiene of foodstuffs;
- Regulation 853/2004 laying down specific hygiene rules for food of animal origin;
- Regulation 1935/2004 on materials and articles intended to come into contact with food;
- Regulation 183/2005 laying down requirements for feed hygiene;
- Regulation 1924/2006 on nutrition and health claims made on foods;
- Regulation 1925/2006 on the addition of vitamins and minerals and of certain other substances to foods;
- Regulation 1331/2008 establishing a common authorisation procedure for food additives, food enzymes and food flavourings;

184 According to the European Commission in the executive summary of the White Paper on Food Safety, COM (1999), 719 final, 12 January 2000.

185 Bernd van der Meulen and Emma Stok, Levensmiddelenrecht, Nijmegen: Ars Aequi Libri 2025, p. 13.

HTTPS://DOI.ORG/10.1163/9789004751699-022

- Regulation 1332/2008 on food enzymes;
- Regulation 1333/2008 on food additives;
- Regulation 1334/2008 on flavourings and certain food ingredients with flavouring properties for use in and on foods;
- Regulation 1169/2011 on the provision of food information to consumers;
- Regulation 2015/2283 on novel foods;
- Regulation 2017/625 on official controls and other official activities performed to ensure the application of food and feed law, rules on animal health and welfare, plant health and plant protection products.

National food law in the Netherlands

The main act on foods and other consumer products in the Netherlands is the *Warenwet* (Commodities Act).[186] It is a framework act that provides some key principles but mostly forms a legal basis for delegated legislation with more detailed rules. Delegated legislation is provided at two different levels; there are royal decrees (government level) (*Algemene maatregel van Bestuur: AMvB*) and ministerial regulations (*Ministeriële Regeling*). Delegated legislation under the Commodities Act is known as the *Warenwetbesluiten* (*Commodities Act Decrees*) and *Warenwetregelingen* (*Commodities Act Regulations*). The main authority in the field of food law is the Netherlands Food and Consumer Product Safety Authority (*Nederlandse Voedsel- en warenautoriteit; NVWA*).[187]

An overview of Food Law regulation in the Netherlands

In this book the following Decrees and Regulations are mentioned:
- Warenwetbesluit Bereiding en behandeling van levensmiddelen (Decree on preparation and treatment of foodstuffs);
- Warenwetbesluit Toevoeging microvoedingsstoffen aan levensmiddelen (Decree on food fortification);
- Warenwetregeling Vrijstelling vitaminepreparaten (Regulation on food fortification);
- Warenwetbesluit Hygiëne van levensmiddelen (Decree on food hygiene);
- Warenwetbesluit uitvoering verordening officiële controles en andere officiële activiteiten (Decree implementing the EU Official Controls Regulation);
- Warenwetbesluit Etikettering van levensmiddelen (Decree on food labelling);
- Warenwetbesluit Voedingswaarde-informatie (Decree on nutrition information);

186 The act to implement Regulation 178/2002 by changing national acts such as the Commodities Act was the act of 13 april 2004 (Staatsblad 2004,164; see for the parliamentary history *kamerstukken 29317*). This act was enacted on 1 january 2005.

187 See www.nvwa.nl.

- Warenwetbesluit Gereserveerde aanduidingen (Decree on reservered designations);
- Warenwetbesluit Zuivel (Decree on dairy).

Other legislation mentioned in this book includes the Wet dieren (animals act) and the Geneesmiddelenwet (medicines act).

Access to Dutch law

Legislation and case law in the Netherlands are available online in Dutch. Case law can be found at www.rechtspraak.nl. Only the Council of State has its own website: https://www.raadvanstate.nl/. All rulings disclosed on these websites have been provided an ECLI. Official documents of the executive are published at: www.overheid.nl. Legislation is available at https://wetten.overheid.nl/. Pieces of legislation are identified only by their Dutch name. There is no number coding like in EU law nor is the ELI system applied. Unlike Eurlex, the database takes the user directly to the most recent consolidated version. Earlier versions can be found as well.

The Netherlands Food and Consumer Product Safety Authority (NVWA) and administrative decision making on issues of food law

The NVWA monitors animal and plant health, animal welfare, the safety of food and consumer products and enforces nature legislation. It is a field service (*buitendienst*)[188] and departmental agency (using accrual accounting) (*baten-lastendienst*)[189] of the Ministry of Agriculture, Fisheries, Food Security and Nature (*ministerie van Landbouw, Visserij, Voedselzekerheid en Natuur*).[190] Together with the Health and Youth Care Inspectorate (*Inspectie Gezondheidszorg en Jeugd*)[191] the NVWA forms the State Supervision on Public Health (S*taatstoezicht op de volksgezondheid*).[192]

The NVWA is either an administrative authority itself or it acts on behalf of an administrative authority, for example the Minister of Agriculture, Fisheries, Food Security and Nature. The acts of the NVWA are regulated by administrative law, it

188 Besluit mandaat, volmacht en machtiging LNV 2019 (bijlage)

189 Instellingsbesluit baten-lastenagentschap Nederlandse Voedsel- en Warenautoriteit, Stcrt. 2013, 24446. The legal term has changed from baten-lasten*dienst* (*see* article 10, first paragraph, of the Comptabiliteitswet 2001 (Government Accounts Act 2001) to baten-lasten*agentschap* (see article 2.2 of the Comptabiliteitswet 2016 (english translation: https://english.rekenkamer.nl/binaries/rekenkamer-english/documenten/publications/2018/01/01/government-accounts-act-2016/CW+2016+vertaling+WR.pdf).

190 See https://www.government.nl/ministries/ministry-of-agriculture-fisheries-food-security-and-nature.

191 See https://english.igj.nl.

192 Article 36 Gezondheidswet (*Public Health Act*) and article 1(b) Besluit Staatstoezicht op de volksgezondheid (Decree on supervision on public health).

is part of the law field of economic administrative law, and by the Dutch General Administrative Law Act (Algemene wet bestuursrecht: Awb).[193] This codification of general administrative law applies to all administrative acts of all administrative authorities of all levels of government.[194]

Legal protection and the court system in the Netherlands

Several of the upcoming chapters show examples of administrative law procedures on food law issues.[195] Food law disputes will regularly be reviewed in civil law procedures or criminal law procedures as well. Chapters 23 and 38 discuss criminal law cases on a food law item, whereas chapter 26, 30, 31, 36 analyse civil law cases. Most of the rest is administrative law.

In the Dutch law on administrative procedure, it is the general pattern that someone who wants to take up arms against a decision must first lodge an objection (*bezwaar*) with the administrative authority that has taken the contested decision. After the decision on the objection (*beslissing op bezwaar*) has been taken by the administrative authority, appeal (*beroep*) can be lodged with the court. The regional courts of first instance are divided into a number of divisions, for example the criminal division and the administrative division. The administrative division is responsible for administering administrative law. The judges sitting on this bench are all independent of authorities and legislature and are members of the judiciary. Appeals against the judgment of one of these regional courts in the Netherlands can be lodged with one of the highest administrative judges: the Administrative Jurisdiction Division of the Council of State, the Central Appeals Tribunal or the Trade and Industry Appeals Tribunal.[196] The choice of which is the highest administrative court to address with an appeal depends on the matter of the appeal.

---

193 The last English translation I could find was published on 28 February 2013 (https://www.acm.nl/sites/default/files/old_publication/publicaties/15446_dutch-general-administrative-law-act.pdf). Although outdated this translation gives a good impression. The latest version in Dutch can be found via wetten.overheid.nl.

194 A few sources in English discussing this General Administrative Law Act (GALA) are the following: Oswald Jansen, A few characteristics of the Law of Administrative Procedure in the Netherlands, in: J.B. Auby (éd.), Droit comparé de la procédure administrative / Comparative Law of Administrative Procedure, Brussels: Bruylant 2016, p. 101-124.

195 See chapters 21, 22, 24, 25, 27, 28, 29, 32, 33, 34, 35, 37, 39 and 40.

196 It is sufficient here to mention the three most important administrative courts which fit in with the standard model described. Appeal in tax matters is entrusted to the tax chamber of the Courts of Appeal. The highest administrative court in tax matters is the tax division of the Netherlands Supreme Court. In the law on traffic fines, there is also a deviation from standard procedure: there is administrative appeal to the public prosecutor, appeal to the sub district court and appeal to the Court of Appeal in Leeuwarden. The system of legal protection against administrative fines imposed by municipalities following infringements relating to nuisance in public areas [*overlast in de openbare ruimte*] follows the system of traffic fines.

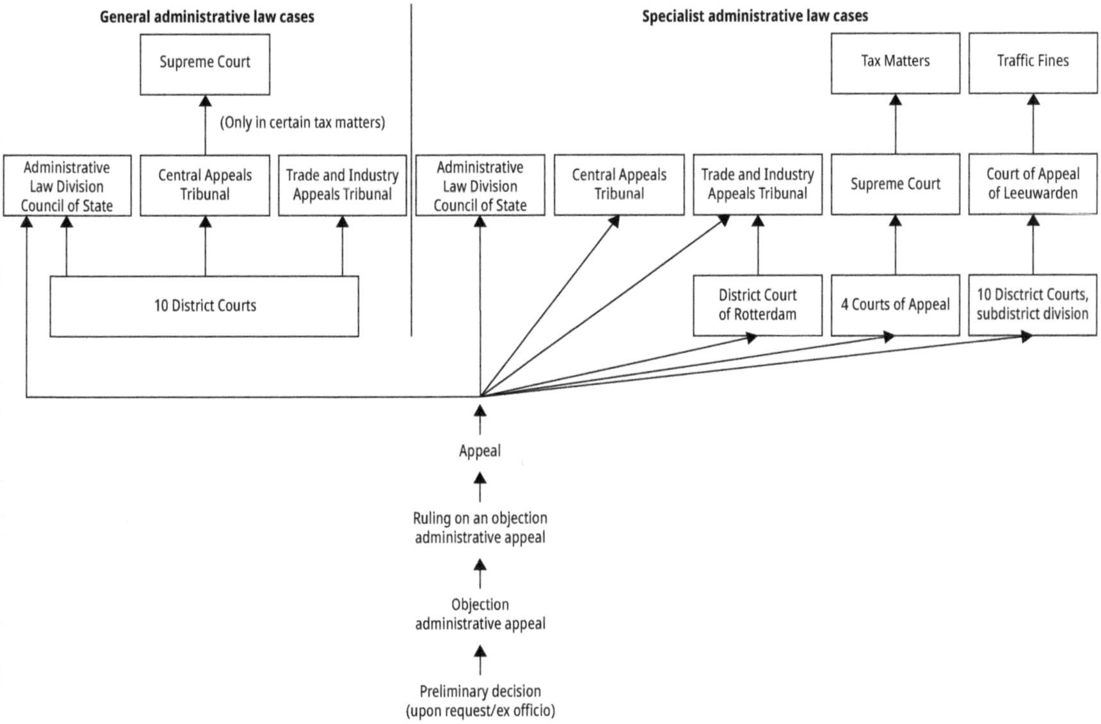

Since the enactment of the Act on the modification of administrative procedural law (*Wet aanpassing bestuursprocesrecht*) on 1 January 2013, an appendix lists the acts that are within the competence of the Central Appeals Tribunal and the Trade and Industry Tribunal. The remaining acts automatically fall within the competence of the Council of State.[197] The Central Appeals Tribunal (*Centrale Raad van Beroep*) decides on social questions (benefits, pensions and civil service matters), the Trade and Industry Appeals Tribunal (*College van Beroep voor het bedrijfsleven*) on economic administrative matters and the Administrative Jurisdiction Division (*Afdeling bestuursrechtspraak*) often on environmental matters and environmental planning. In food law matters the Trade and Industry Appeals Tribunal is the competent court.

Chapters 6, 7 and 8 of the General Administrative Law Act contain the provisions for administrative procedure. Chapter 6 contains general provisions relating to objections, administrative appeals and appeals to the administrative courts. Chapter 7 contains provisions concerning objections and administrative appeals and Chapter 8 outlines appeal to the administrative courts.[198] Almost all of these provisions also apply to appeal to a higher court (8:108 GALA).

197 Article 9,10 and 11 of Appendix 2. This Appendix is based upon the Articles 8:5, 8:6, 8:7, 8:105 and 8:106 GALA.

198 This applies for the special procedures mentioned in the previous footnote.

The absence of a constitutional court and review of the constitutionality of the legislation enacted by national parliaments is interesting from a comparative administrative law point of view. In short, the European Convention on Human Rights (ECHR) plays an important role, as national courts can review the conformity of legislation with the ECHR and, in many cases, the European Court of Human Rights (ECtHR) fulfils the role of a constitutional court. More and more the ECJ applying the EU Charter of Fundamental Rights (CFR) is playing a similar role. Also, article 8:2 GALA limits appeal against generally binding provisions (*algemeen verbindende voorschriften*) and policy rules (beleidsregels), but this important provision does not limit the possibility of review of these rules indirectly via the appeal against the acts based upon these general rules (*exceptieve toetsing*).

Self-regulation

An important part of food law in the Netherlands, is based in self-regulation. The business sector has set up a series of advertisement codes which set behavioural requirements with regard to advertisement – including food labelling. In connection with this an Advertisement Code Commission has been established. The Advertisement Code Commission provides a complaints mechanism.[199] The procedure is simple and accessible. Everybody can lodge a complaint (action popularis) against an advertisement. This opportunity is mostly used by individual consumers and consumer organisations. On the basis of a complaint the advertisement is assessed for compliance with the applicable advertisement code and with the law. As regards food products, many complaints relate to compliance with the Nutrition and Health Claims Regulation.

Complaints may be declared inadmissible or unfounded. In other situations they result in a recommendation by the Advertisement Code Commission (or its president). Interested parties can appeal such recommendation. Recommendations under this self-regulatory system are not legally binding. This notwithstanding, levels of compliance are high.

The third part of this book is dedicated to recommendations under the Advertisement Codes.[200] The Dutch system is explained in more detail there.

Prof. dr. Oswald Jansen
Partner and attorney at Libertas Lawyers in Rotterdam and professor of European and Comparative Administrative Law at Balekesir University, Türkiye. Former professor of European Administrative Law and Public Administration at Maastricht University and adjunct professor American University Washington College of Law (AUWCL).

199 See https://www.reclamecode.nl/engels/stichting-reclame-code/about-stichting-reclame-code
200 The English translation of the Dutch Advertising Code can be found at https://www.reclamecode.nl/engels/dutch-advertising-code/

CHAPTER 21

Knorr Meals with iodised salt – Precautionary principle

Annotated by Bernd van der Meulen

Rotterdam District Court, 6 August 2004, ECLI:NL:RBROT:2004:AQ6954 (*Unilever Bestfoods*)

Risk analysis, precautionary principle
Proper application of the precautionary principle requires, on the one hand, identification of the adverse health effects that the addition of iodine to instant meals may have and, on the other hand, a comprehensive health risk assessment based on the most reliable scientific data and on the most recent results of international research.

Main considerations:

> It must be stated at the outset that Community law, partly in the absence of harmonisation in this area, does not in principle preclude the Netherlands from deciding in its laws and regulations the extent to which it wishes to protect the health and life of individuals.
> However, when using this discretion, the Netherlands must, according to the established case law of the European Court of Justice, comply with the (European law) principle of proportionality. This implies in the present case that, since the defendant invokes the exception laid down in Article 30 of the EC Treaty, it is for them to demonstrate in this particular case, against the background of the dietary habits of the Dutch population and of the results of the internationally accepted scientific research known at the time of the adoption of the contested decision, that the Dutch legislation and regulations and the exemption policy pursued on the basis thereof, under which the contested exemption cannot be granted, are necessary for the effective protection of the health and life of humans and, in particular, that the marketing of the instant meals poses a real risk to public health. To that end, the defendant must demonstrate, on the basis of a thorough assessment, the possible effects or risks of adding iodine to the instant meals (lawfully manufactured and marketed in another Member State). Such an assessment may reveal that there is as yet no scientific certainty as to the existence or extent of real risks to human health. In such circumstances, it is acceptable for a Member State such as the Netherlands to take protective measures on the basis of the precautionary principle without waiting until it is fully established that those risks do indeed exist and are significant. The correct application of this precautionary principle requires, first, that the adverse health effects which may result from the addition of iodine to instant meals be identified and, second, that a comprehensive health risk assessment be carried out on the basis of the most reliable scientific data

 | HTTPS://DOI.ORG/10.1163/9789004751699-023

> and the most recent results of international research. In this regard, if it proves impossible to determine with certainty the existence or extent of the alleged risk because the results of the studies are insufficient, inconclusive or imprecise, but real harm to public health remains likely in the event that the risk materialises, the precautionary principle justifies the adoption of restrictive measures, provided that they are objective and non-discriminatory.

21.1 Importance of the ruling

This ruling provides a nice and neat example of how the precautionary principle, as currently codified in Article 7 of Regulation 178/2002, should be used.

21.2 The facts and the proceedings

Unilever Bestfoods BV has requested the Minister of Health, Welfare and Sports (Volksgezondheid, Welzijn en Sport: VWS) to grant an exemption under 16(2) of the Dutch Commodities Act (Warenwet) to the prohibition (from section 4(1) opening words and under a) of the Commodities Act, in conjunction with sections 2(1) and 9a of the Commodities Act Decree on the Addition of Micro-nutrients to Foodstuffs) of the presence of iodine compounds in instant meals. The exemption was requested for the marketing of instant meals produced by its German sister company Unilever Bestfoods Deutschland GmbH under the names Knorr Spaghetteria and Knorr Orienty. Iodised salt has been added to these instant meals.

On behalf of the Minister of Health, Welfare and Sport, the exemption was refused on the grounds that granting it would be contrary to policy. The refusal was upheld in administrative review after which Unilever Bestfoods BV appealed to the Rotterdam District Court as special administrative judge in Commodities Act cases.

According to the ruling, it is the policy of the Netherlands that only a very limited number of daily food products (bread salt, kitchen salt, bread substitutes, kitchen salt substitutes and meat products) may have iodine added within a certain bandwidth (and at a relatively high level), so that no iodine may be added to other food products such as instant meals. Underlying this policy is the idea that, according notably to the report of the Scientific Committee on Food on the Tolerable Upper Intake Level of Iodine of 7 October 2002, commissioned by the European Commission, the recommended daily intake of iodine for an adult is 100 mcg per day, that according to simulation studies (carried out by TNO in 1993–1995 on the basis of the second Food Consumption Survey of 1992) this is met by iodine enrichment of the aforementioned products, and that intake of more than 600 mcg iodine per person per day, barring an occasional excess, is detrimental to health (in particular thyroid function).

Dutch policy can thus be characterised as a policy of 'a lot in a few' (a relatively large amount of iodine in relatively few products is allowed), while in other Member States (such as Germany) a policy of 'a little in many' is advocated.[201]

21.3 Risk analysis and precautionary principle

The precautionary principle provides part of the answer to the question of what food safety authorities can do in situations of scientific uncertainty. One might question the relevance of the presence or absence of scientific certainty. In a democratic society, is it not much more about what democratically legitimised majorities think or want?

In some ways, (choice) freedom can be compared to money. Making choices limits the options. In situations where the government has committed not to interfere with the freedom of citizens or businesses beyond what is necessary, it imposes on itself the obligation to demonstrate objectively that its action is necessary. Where objective demonstration is required, science soon comes into play. This can also be explicitly established as has been done internationally in the World Trade Organization Agreement on the Application of Sanitary and Phytosanitary Measures (the SPS Agreement). In the European Union, it is enshrined as a principle of food law in Article 6 of Regulation 178/2002.

More implicitly, it is also anchored in the prohibition of trade barriers within the European single market with the corresponding principle of mutual recognition. At the time of the case, this was still enshrined in Article 28 TEC. Today, it is Article 34 TFEU: 'Quantitative restrictions on imports and all measures having equivalent effect shall be prohibited between Member States.' Article 36 TFEU opens up the possibility of exceptions: 'The provisions of Articles 34 and 35 shall not preclude prohibitions or restrictions on imports, exports or goods in transit justified on grounds of public morality, public policy or public security; the protection of health and life of humans, animals or plants; the protection of national treasures possessing artistic, historic or archaeological value; or the protection of industrial and commercial property. Such prohibitions or restrictions shall not, however, constitute a means of arbitrary discrimination or a disguised restriction on trade between Member States.'

It is established case-law of the Court of Justice of the EU that measures of equivalent effect to quantitative restrictions on imports may only be taken if prior

201 For readers interested in the causes and consequences of iodine deficiency in the Netherlands, see: R.H. Verheesen and C.M. Schweitzer, 'Het jodiumtekort is terug', Medisch Contact 21 October 2008, https://www.medischcontact.nl/nieuws/laatste-nieuws/artikel/het-jodiumtekort-is-terug.htm. ' [Dutch language only].

risk assessment shows a danger (in our context) to public health.[202] The Court of Justice derives this from the principle of proportionality.

In situations where the government has the burden of proof to demonstrate necessity scientifically, the precautionary principle provides relief from that burden of proof. Where scientific risk assessment, while providing indications that a risk might exist, is not completely conclusive, the government may assume a worst-case scenario and act as if the risk assessment had scientifically demonstrated a risk. It could be said that 'a beginning of scientific evidence' is sufficient to be allowed to take risk mitigation measures[203] (see Article 7 of Regulation 178/2002). However, the measures to be taken must be proportionate and non-discriminatory. They are also, in principle, limited in time. That time should be used to deepen scientific understanding.

Not every doubt is scientific uncertainty within the meaning of the precautionary principle. Uncertainty is scientific only when an assessment has been conducted and when that assessment has not produced unambiguous results.

21.4 Before and after science

The question with regard to what kinds of situations we would like to have scientific certainty about, and thus require scientific assessment, is different from the question of what to do in case of scientific uncertainty. Nevertheless, the expression 'precautionary principle' is also used in this context, also by the CJEU. In this book, CJEU 25 July 2018, case C-528/16 ECLI:EU:C:2018:583 (*Confédération paysanne and Others v Ministre de l'Agriculture, de l'Agroalimentaire et de la Forêt*) is included. The question there is whether crops obtained by site-directed mutagenesis should be classified as genetically modified organisms. The Court answers this question in the affirmative, referring, inter alia, to the argument that the risks of site-directed mutagenesis would be comparable to those of transgenesis and to the precautionary principle that also underlies the GMO Directive. The issue here is whether the crops in question should be subject to risk analysis, not what to do in the event of scientific uncertainty. If that is an application of the precautionary principle, it is a precautionary principle in a different sense from that of Article 7 of Regulation 178/2002 (but, for example, in that of Article 191 TFEU). Against a requirement for prior proof of safety, it is sometimes argued that it is not possible to prove something negative such as the absence of risk. The value of this argument is only rela-

202 For an analysis of this ruling, see K.J. Defares and B.M.J van der Meulen, *Het voorzorgsbeginsel. Preadvies voor de Nederlandse Vereniging voor Levensmiddelenrecht*, Sdu 2009.

203 Hence the title: K.J. Defares and B.M.J. van der Meulen, *Een begin van wetenschappelijk bewijs. Het voorzorgsbeginsel in het levensmiddelenrecht*, Sociaal-economische wetgeving: tijdschrift voor Europees en economisch recht (SEW) 2009, pp. 462–482.

tive. It only shows that it is important to establish the scientific standards by which judgments should be made. Toxicology, for example, is well placed to judge whether a certain substance at a certain dose is toxic or non-toxic to humans, or whether there is scientific uncertainty about it.[204]

An example in this book of an application by the Court of the precautionary principle in cases of scientific uncertainty (i.e. within the meaning of Article 7 of Regulation 178/2002) is CJEU 13 September 2017, case C-111/16 ECLI:EU:C:2017:676 discussed by Nathan Meijer (Chapter 14).

Both issues come together in CJEU 28 January 2010 case C-333/08 ECLI:EU:C:2010:44 (*Cie. v France*) discussed by Victor van Ahee in Chapter 7. There, the Court ruled that Member States are not allowed to subject categories of food to an authorisation requirement whereby scientific safety must be demonstrated if the Member State has not first carried out a risk analysis showing the existence of a risk.

Bernd van der Meulen
Visiting Professor of Global Food Law at the University of Parma

---

204 A very useful introduction to toxicology for lawyers is: Aalt Bast and Jaap C. Hanekamp, *Toxicology: What Everyone Should Know. A Book for Researchers, Consumers, Journalists and Politicians*, Elsevier Academic Press 2017. See also his contribution in Bart Wernaart and Bernd van der Meulen (red.) *Applied Food Science*, Wageningen Academic Publishers/Brill 2022 https://brill.com/edcollbook-oa/title/68695; (open access).

CHAPTER 22

Ghanaian children – Right to food in the United Nations Convention on the Rights of the Child

Annotated by Bart Wernaart

Central Appeals Council, 8 August 2005, ECLI:NL:CRVB:2005:AU0687

Articles 2, 3 and 27 United Nations Convention on the Rights of the Child (UNCRC); Article 16 Dutch Work and Social Assistance Act (WWB). Linking principle, very urgent reasons, assistance for minors of non-beneficiary parents.

Main considerations:

> In the opinion of the preliminary relief judge, there is a reasonable likelihood that the Council will have to disapply the latter provision [Article 16 of the Work and Social Assistance Act] in the cases on the merits at issue here in respect of the appellants. In doing so, it relies on the text of Articles 2(1) and (2), 3(1) and (2) and 27(3) of the UNCRC and on the comments of the Committee on the Rights of the Child published on these articles. The words 'without discrimination of any kind, irrespective of the child's or his or her parents or legal guardian's [...] status' in Article 2(1) of the UNCRC, viewed in conjunction with the other provisions just mentioned, indicate that the principle of linkage cannot provide sufficient justification for completely excluding the possibility of providing assistance solely for the benefit of minors in a situation where their non-beneficiary parents, requesting such assistance, are themselves unable to meet the costs of food, clothing and other essential expenses necessary for the minors.

22.1 Introduction

In these preliminary relief proceedings, the municipality of Zaandam is being sued by two children with Ghanaian nationality, born in the Netherlands in 1998 and 2003 (the ruling dates from 2005). The children's parents – also with Ghanaian nationality – have applied for a residence permit and are allowed to await the course of these proceedings in the Netherlands. However, they have insufficient means to care for their children and are not entitled to social assistance because of their residence status. The central issue is the possible application of the WWB, Article 16, in force at the time, on the basis of which the Municipal Executive may, in cases of extreme urgency, grant assistance to persons not entitled to assistance. However, paragraph two of this article – in implementation of the so-called linking principle – stipulated that this assistance could not be granted to, among others, illegally residing aliens, which is the category to which the parents belonged at the time.

22.2 The case

In the present case, the judge in preliminary relief proceedings assesses whether the application of the coupling principle violates mainly Article 27 UNCRC. Among other things, the court examines the legislator's view of the right to an adequate standard of living, as set out in Article 27 UNCRC. It concludes that in the parliamentary debate for the bill to ratify this convention, the legislator established that the maintenance of the child is the primary responsibility of the parents, but when the parents cannot fulfil their duty of care, the State has a responsibility, as also expressly recognised in Article 27(3) UNCRC. In particular, this responsibility can be met by providing assistance under the General Assistance Act (Abw). The nature of this assistance depends on the specific situation and needs of the child in question, and may be provided to parents in the form of family assistance.[205]

The preliminary relief judge further examined whether the legislator had examined whether the *Koppelingswet* (Dutch Linking Act)[206] was compatible with Articles 2, 3 and 27 UNCRC when it could lead to withholding assistance to families with children residing illegally in the Netherlands. The answer is that the legislator did not do so explicitly during the parliamentary debate on this act, but only established that withholding assistance to this group of children could not violate the UNCRC.[207]

The preliminary relief judge then concluded that the Linking Principle in general does not raise concerns, but that, in the specific application,

> [it] cannot constitute a sufficient justification for completely excluding the possibility of granting assistance solely for the benefit of minors in a situation where their non-beneficiary parents requesting such assistance are themselves unable to pay the cost of food, clothing and other essential expenses necessary for the minors.

In doing so, the preliminary relief judge expressly used Articles 2, 3 and 27 UNCRC as a compass, as well as the accompanying General Comments of the UN Committee on the Rights of the Child (CRC). The judge decided to grant the requested remedy by awarding the difference between the assistance for a single parent and the assistance for a single person to the parents for the maintenance of their children.

205 *Kamerstukken II*, 1992-1993, 22 855 (R 1451), no. 3 and 6.

206 Stb. 1998, 203.

207 *Kamerstukken I*, 1996-1997, 24 233, no. 76c, pp. 17 and no. 149, pp. 11.

A similar reasoning was also later used by the Central Appeals Tribunal in the Appeals Court against the contested refusal of assistance, extending the assistance granted in the preliminary injunction until six weeks after the municipality of Zaandam announced the new decision on the original objection to refuse assistance.[208]

Although some of the above legislation is now no longer in force (the Work and Social Assistance Act has been replaced by the Participation Act in 2015), this ruling is important for legal scholarship for several reasons.

First, we see a rare mindset of one of the Netherlands' highest courts whereby an international treaty article containing a socio-economic human right ('ECOSOC law') is used as an interpretative standard in assessing the validity of national legislation.

In the second place, we see in this judgment a wonderful example of a balancing judge who, on the one hand, does not want to sit in the legislator's chair, and, on the other hand, wants to administer justice in a distressing case. There is an important role for parliamentary history in the trial judge's finding of law. This corresponds to a trend we have observed more broadly in the issue of the direct effect of internationally enshrined ECOSOC laws.[209] Only when the Dutch legislator (in the explanatory memorandum to the ratification act) has not expressly excluded direct effect from an international treaty provision, and the compatibility of new legislation with these treaty provisions has not been expressly discussed, does the court feel free within the separation of powers to proceed to its own further review.

Third, we see in this judgment a rare application of the General Comments of a UN committee. Dutch courts (and courts worldwide) are not easily inclined to use such authoritative commentaries as part of determining the law.

Finally, this line of thinking has had far-reaching consequences. Whereas before this ruling there was mainly a lack of clarity about the legal status of Article 27 UNCRC, since this ruling we see that Article 27 UNCRC – usually in combination with Articles 2 and 3 UNCRC – is used with some regularity as an interpretative norm by Dutch courts.[210]

In addition, this view also led to a policy change. The then Minister of Security and Justice amended a ministerial regulation, implementing a broader granting of assistance to families with children without residence permits.[211]

Dr. Bart Wernaart
Professor of Moral Design Strategy; Chair of the Centre of Expertise Sustainable and Circular Transitions at Fontys University of Applied Sciences

208 CRVB 24-01-2006 ECLI:NL:CRVB:2006:AV0197.

209 B. Wernaart and B. van der Meulen, 'The Right to Food in International Law, with case studies from the Netherlands and Belgium'. In: G. Steier and K. Platel (red), International Food Law and Policy. New York: Springer 2017.

210 See for a more detailed summary: B. Wernaart, *The enforceability of the human right to adequate food, a comparative study*. Wageningen Academic Publishers 2013, p.112–118.

211 *Wijziging regeling verstrekking bepaalde categorie.n vreemdelingen*, ruling of the Minister of Justice of 22 December 2006, no. 5458886/06/DVB.

CHAPTER 23

Fat & Burn – Criminal law: Food company must keep up with scientific developments

Annotated by Bernd van der Meulen

Leeuwarden Court of Appeal, 30 November 2006, ECLI:NL:GHLEE:2006:AZ3591

The open general food safety clause can be criminally enforced when the substantive standard is sufficiently apparent from substantive regulations and policy principles.

Main considerations:

With regard to the accused, it was proven that on 10 September and 22 November 2002, she stocked pre-packaged dietary supplements designated as Fat & Burn No. 3, which, due to their unsoundness, could endanger human health or safety, [Court: after all, upon examination, the level of benz(a)pyrene found in those dietary supplements was found to be (approximately) 22.5 and 28.4 micrograms per kilogram of product, respectively].
The indictment is based on Section 18 opening words and under a) of the Commodities Act (Warenwet), which at the time of the periods charged – as far as relevant here – read as follows:

> Without prejudice to the provisions under or pursuant to the preceding articles, it is prohibited:
> a. to deal in food or drink products which, because of their unsoundness, may endanger human health or safety.

It is clear that this section of the Commodities Act does not assume a concrete or hard standard but merely sets the requirement of soundness. It should be noted that setting a more or less vague standard is not unusual in law. This method of regulation, i.e. to describe an offence in general terms, may be necessary or unavoidable to prevent conduct worthy of punishment from falling outside the scope of the offence definition (Supreme Court (*Hoge Raad:HR*) 31 October 2000, NJ 2001, 14). It may also be unavoidable because it cannot always be foreseen in what way the interests to be protected will be violated in the future (HR 28 May 2002, NJ 2002, 483).
Whether vagueness is permissible in a concrete case must then be judged on the basis of whether the norm addressed to the addressee of the norm is sufficiently knowable in substantive regulations and policy principles to be able to adjust behaviour accordingly.

 | HTTPS://DOI.ORG/10.1163/9789004751699-025

[...]

From what counsel has argued in his pleading, the court of appeal concludes that at the time when the products described in the indictment were stocked by the defendant, she was aware of the contents of the letter of 15 December 1999, in which the Minister of Health, Welfare and Sport (*Volksgezondheid, Welzijn en Sport: VWS*) wrote to the House of Representatives that the acceptable safe standard (of intake by humans) is between 20 and 40 nanograms of b(a)p per kilogram of body weight per day. By letter of 8 May 2000, also addressed to the House of Representatives, this standard was subsequently tightened by the minister to 5 nanograms of benzo(a)pyrene per kilogram of body weight per day.

[...] According to the Explanatory Memorandum (House of Representatives, session 1981–1982, 17495, no. 3) to the Commodities Act, the legislator also assumes this. Page 48 of the Explanatory Memorandum states that every trader may 'be expected to keep himself informed to the best of his ability about new developments, also in the field of health and safety, with regard to products in his branch.' In doing so, the trader should heed warnings from institutions charged with enforcing the Commodities Act. In this context, the aforementioned letter of 8 May 2000 refers to the National Institute for Public Health and the Environment (*Rijksinstituut voor Volksgezondheid en Milieu: RIVM*) , so that it would have been logical to seek their opinion on the permissibility of b(a)p and the extent to which b(a)p may occur in foodstuffs (and therefore also in food supplements).

In the court of appeal's opinion, this applies to the accused to an even greater extent, as the one who, in addition to putting these products on the market, also gave the order for their production and set requirements regarding their composition, as the accused's representative stated at the court session.

In this respect, the court of appeal considers that the certificate that the manufacturer of the contested products – according to the statement of the defendant's representative at the court session – provided with the products and from which the defendant deduced that the products were sound and suitable for human consumption, contrary to the defendant's representative's opinion, did not relieve the defendant of the duty to independently form an opinion about the soundness of the product distributed by her.

The court also takes into account that the court's hearing showed that the defendant had little or no knowledge of the concrete (chemical) composition of her products, how they work, or the possible risks they pose to public health.

23.1 Importance of the ruling

The ruling makes it clear that an open food safety standard as laid down in Article 18 of the Dutch Commodities Act at the time and now in Article 14 of Regulation

(EC) 178/2002 entails a responsibility for food companies to actively keep abreast of scientific developments and that this can be enforced by criminal law.

23.2 The facts and the proceedings

In 2002, batches of a food supplement were found at the premises of the accused, a legal entity that marketed food supplements, which were contaminated with aromatic hydrocarbons. At that time, there was no legal limit in this respect. The following is an extract from the ruling of the court at first instance (Leeuwarden District Court 28 May 2004 ECLI:NL:RBLEE:2004:AP1271): 'Based on a RIVM report of November 2001, the Netherlands Food and Consumer Product Safety Authority (NVWA) has set its standards such that products containing more benzo(a)pyrene than the lower measurement limit of 1 µg/kg are deemed unfit for use. At levels up to 1.8 µg/kg, no action is taken, in order to exclude measurement uncertainties. At even higher levels up to 2 µg/kg, a written warning follows. Between 2 and 5 µg/kg, a statement of requirements follows with urgent advice for recall, and above 5 µg/kg a European rapid alert follows. The consulted NVWA experts examined stated that this lower limit of 1 µg/kg has been set in such a way that it is achievable in practice but does not give any false security.' The defendant is convicted of levels of between 20 and 30 microgram/kg of product. Fines of €4,000 each are imposed for six different offences, half of which are suspended. During the course of the proceedings, the defendant went bankrupt and therefore ceased to exist. The court noted that trading continued under a different company name.

The Court describes in some detail the development of thinking on the harmfulness and occurrence of benzo(a)pyrene in edibles. From the overview, the following:

- the Minister of VWS (Dutch Ministry of Public Health, Welfare and Sport) wrote in a letter dated 15 December 1999 to the Dutch House of Representatives that the acceptable safe standard (of human intake) is between 20 and 40 nanograms of benzo(a)pyrene per kilogram of body weight per day;
- the Minister of VWS, in a letter of 8 May 2000 to the Dutch House of Representatives, tightened this acceptable safe standard to 5 nanograms benzo(a)pyrene per kilogram of body weight per day (Dutch House of Representatives, session year 1999–2000, 26991, no. 6).

Ultimately, the Court arrives at a standard of 0.5 nanograms of benzo(a)pyrene per kilogram of body weight per day. From expert testimony, the Court deduces that this standard is exceeded when consuming the food supplements in question according to instructions for use.

The Court does not address the question of how a standard per kilogram of body weight leads to a standard per kilogram of product weight in food supplements.

Only that, according to the experts, it was exceeded. The Court's reference to the admission on behalf of the defendant of familiarity with the letter of 15 December 1999, but not with the letter of 8 May 2000, legitimises the presumption that the standard that would follow from the insights of late 1999 would have been met, but not that of spring 2000. That reading accuses the defendant of relying on outdated science. According to the Court, this is at the food company's risk.

23.3 Food safety in food law

The food legislature – especially the European one – sets maximum limits for an increasing number of microorganisms, residues of veterinary medicines and plant protection products and for chemical contaminants. For situations not specifically provided for by the legislator, the general food safety standard ('food may not be placed on the market if it is unsafe') applies as a safety net. At the time of the offences charged, the Commodities Act had not yet been adapted to Regulation 178/2002. The general food safety standard was then to be found in section 18 of the Commodities Act. It can now be found in Article 14 of Regulation 178/2002. There is an editorial difference. The Commodities Act required unsoundness as a reason for health risk. The regulation juxtaposes health risk and unfitness for consumption. An important similarity is that both provisions focus on health risk as a natural science concept. There is no reason to suppose that the Court would have ruled differently under Article 14 of Regulation 178/2002 than it did under Article 18 of the Dutch Commodities Act.

Article 14 of Regulation 178/2002 sets out in paragraphs 7 and 9 the relationship to legal food safety standards such as maximum limits. Where such standards are set at European (paragraph 7) or national (paragraph 9) level, compliance with these standards gives rise to a legal presumption of safety (subject to progressive understanding, see paragraph 8).

According to the Standing Committee on the Food Chain and Animal Health (SCoFCAH – now Standing Committee on Plants, Animals, Food and Feed – SCoPAFF), a contrario, the opposite also applies: in case of non-compliance, a rebuttable legal presumption of unsafety applies.[212] The Dutch Trade and Industry Appeals Tribunal (CBb) follows the Standing Committee in this line of thinking.[213] The question is whether such a legal fiction has added value. Violations of legal

212 Standing Committee on the Food Chain and Animal Health (SCoFCAH) now Standing Committee on Plants, Animals, Food and Feed (SCoPAFF) Guidelines for the implementation of Articles 11, 12, 14, 17, 18, 19 and 20 of Regulation (EC) No 178/2002 on General Food Law, 26 January 2010 https://food.ec.europa.eu/system/files/2016-10/gfl_req_guidance_rev_8_en.pdf/.

213 CBb 3 July 2018 ECLI:NL:CBB:2018:326.

limits can be enforced directly without the need to take the intermediate step via Article 14.[214]

The importance of Article 14 as a general food safety clause is that it includes a provision for all those situations not specifically foreseen by the legislature. In those cases, food law turns out to be so 'science-based' that scientific insights become a legal standard. A legal standard that, according to the above judgment, can even be enforced under criminal law.

With legal limits, new scientific findings, e.g. from EFSA or JECFA (Joint FAO/WHO Expert Committee on Food Additives), do not have legal effect until the law is changed. The current state of the law is knowable from the law. The law is accessed in a fixed place and according to a fixed structure. Transitional arrangements are often made. In the case of the open food safety standard, new scientific findings have an immediate effect. EFSA (or the RIVM) opinion thus acquires almost the force of law.

By what means can the current state of science be known? We see in the judgment diffuse sources such as letters to Parliament and reports of the National Institute for Public Health and the Environment (*Rijksinstituut voor Volksgezondheid en Milieu: RIVM*). The court at first instance refers to the enforcement policy of the NVWA. This reference is not found in the court's judgment. Still, there is something to be said for putting the naming of the prevailing state of science somewhere central in a government body – in the form of law-interpreting policy. For that matter, because of the major impact, I would prefer to place such policy with the politically responsible minister rather than the regulator.

The Court places the responsibility with individual companies. While it does impose the requirement that the substantive standard is sufficiently knowable in substantive regulations and policies, it does not set high standards for this. On the contrary, companies have to make efforts to stay scientifically informed. This is also a criminal responsibility. The food company cannot hide behind supplier certificates. The latter is in line with CJEU 13 November 2014, case C-443/13 (*Ute Reindl v Bezirkshauptmannschaft Innsbruck* ECLI:EU:C:2014:2370)[215].

Bernd van der Meulen
Professor of EU Food Law at University of Copenhagen

214 See the official controls regulation; at the time Regulation 882/2004; now Regulation 2017/625. See also B.M.J. van der Meulen, 'Enforcement of EU agri-food law', ERA Forum, https://link.springer.com/article/10.1007/s12027-018-0532-5.

215 This ruling is discussed in Chapter 4 of this book.

CHAPTER 24

LL Rice 601 – Application for enforcement, interested party and discretion

Annotated by Silvia Gawronski and Kirsten Lenssen

Rotterdam District Court 23 June 2008, ECLI:NL:RBROT:2008:BD5791 (*Greenpeace*);
Trade and Industry Appeals Tribunal (CBb) 15 April 2010, ECLI:NL:CBB:2010: BM3228

Position of complainant, interested party. When can a legal entity qualify as an interested party within the meaning of Article 1:2(3) of the Dutch General Administrative Law Act (GALA)[216]*? The evolution of the concept. The Netherlands Food and Consumer Product Safety Authority (NVWA) has discretion to act or not to act against unauthorised GMOs.*

Main considerations:

> District Court
> Regarding interested parties:
> 2.4 [...] Under Article 1:3(3) of the Awb, application means a request by an interested party to make a decision.
> [...] Furthermore, in the court's opinion, the requirement in Article 1:3(3) of the Awb that the request comes from an interested party has been met. In the court's opinion, the claimant qualifies as an interested party within the meaning of Article 1:2(3) of the Awb. To this end, it considered the following.
> In view of its articles of association, the claimant's objective is to promote nature conservation in a broad sense, while the claimant's actual activities also demonstrate this advocacy, as was also considered in the decision of the Administrative Jurisdiction Division of the Council of State (*Afdeling bestuursrechtspraak Raad van State* (ABRvS)) of 21 February 2007 (LJN: AZ9048; LJN: JM 2007/58). It is true that the legislation in question is mainly concerned with food safety, but in view of Article 2(a) and (f) of Directive 2003/4/EC, the claimant can be followed in her argument at the hearing that no close distinction can be made between nature conservation and food safety. In addition, in its decision of 14 June 2007 (LJN: BA8075), the preliminary relief judge of the Trade and Industry Appeals Tribunal apparently considered the claimant to be an interested party in a case similar to the present one.

216 In Dutch: Algemene wet bestuursrecht: Awb.

 | HTTPS://DOI.ORG/10.1163/9789004751699-026

> Regarding discretionary power:
> [...] Having regard to Article 21(1) of the Commodities Act, if, in its opinion, goods pose a danger to human safety or health, the defendant may, with a view to protecting those interests, order the person who trades or has traded the goods to inform the holders or presumed holders of those goods of the danger without delay and in an effective manner. The defendant therefore has discretionary power in that regard. The Court did not find that the defendant, in view of the NVWA's opinion of 6 September 2006, could not reasonably conclude that the measures already taken were insufficient.
> In addition, the court considers it important that the risk of contaminated rice does not adversely affect public health.

Trade and Industry Appeals Tribunal (CBb):

> Regarding application
> 5.5 The next issue before the Tribunal is whether the District Court was correct in ruling that the Minister's response to the appellant's letter of 15 December 2006, supplemented by a letter of 22 December 2006, constituted a decision within the meaning of Article 1:3 of the Awb, against which an objection may be lodged under the Awb. Contrary to the appellant's apparent view, that issue is of public policy, so that the Tribunal must assess *ex officio* whether the court reached a correct verdict.
> 5.6 The Tribunal first of all points out that an interested party who applies to an administrative body for an enforcement decision must sufficiently specify its request. Anyone requesting the administrative body to take a decision to enforce statutory provisions cannot be content with expressing that wish in general terms, but must outline more specifically the case to which the request relates.
> 5.15 In the opinion of the Tribunal, the appellant's request of 15 December 2006, supplemented by a letter of 22 December 2006, cannot therefore be regarded as a request for a decision as referred to in Article 1:3 of the Awb. Consequently, the letter of 21 December 2006, in which the Minister responded to the appellant's request, does not contain a (refusal of a) decision, so that the court's opinion that the Minister wrongly declared the appellant's objection inadmissible cannot legally stand.

24.1 Importance of the ruling

This case touches on all the key issues of whether a person can claim that the Netherlands Food and Consumer Product Safety Authority (the "NVWA") acts against food law violations.

If people or organisations approach the NVWA with a complaint about breaches of food law, should they resign themselves to waiting to see if the NVWA does anything about it, or do they have a right to take action or at least a decision that they can challenge if necessary?

Under the Awb (the "Dutch General Administrative Law Act" or "GALA"), it is mainly the person who submits an application as an interested party who has standing to apply for a legal remedy. The court addresses the question of when an application for enforcement is an application, who is an interested party and to what extent the NVWA is obliged to do something when a violation has been established. The Trade and Industry Appeals Tribunal ("Tribunal" or "CBb") corrects the court on the issue of application.

24.2 The case

On 18 August 2006, the authorities of the United States of America informed the European Commission that contamination with genetically modified rice, which has not been authorised for placing on the market in the European Community, was found in samples taken in the US market of long grain rice from the 2005 crop. The rice in question is 'LL RICE 601' from producer Bayer CropScience. This has prompted the European Commission, in application of Article 53 of Regulation (EC) No 178/2002, to take emergency measures in relation to food imported from a third country in order to protect human health, animal health or the environment.

Greenpeace (the claimant) requested the competent authority (hereafter referred to as the respondent) to take enforcement action against the sale of this US-origin rice contaminated with 'LL Rice 601'. Greenpeace was informed by letter that the Netherlands Food and Consumer Product Safety Authority's (NVWA) approach to the control of 'LL Rice 601' provided sufficient guarantees to protect Dutch consumers. Greenpeace objected to the respondent's letter. In Greenpeace's view, the letter could qualify as a written refusal to take enforcement action. However, the respondent was of the opinion that Greenpeace's request was so poorly formulated that it didnot qualify as a request for enforcement. This also means that the respondent's letter is considered to be merely informative and not a decision within the meaning of Article 1:3(1) GALA (Awb) which can be objected to. Greenpeace's objection is therefore inadmissible. Greenpeace appealed against this. The court then addressed the requirements arising from Article 1:3(3) GALA (Awb). Pursuant to Article 1:3(3) GALA, an application means 'a request by an interested party to take a decision'. The question is whether Greenpeace's request can be qualified as a request for the respondent to take a decision (of enforcement).

24.3 The interested party[217]

The court dissected the term 'interested party'. An interested party is 'the person whose interest is directly affected by a decision'. Since Greenpeace is a legal entity,

217 See on this subject also Chris Backes, 'Access to court', in: Chris Backes and Mariolina Eliantonio (eds.), *Cases, Materials and Text on Judicial Review of Administrative Action*, Hart Publishing 2019, p. 191ff

it must also meet the requirements of Article 1:2(3) GALA (Awb). This article reads as follows:

> In respect of legal entities, their interests shall include the general and collective interests which they specifically promote by virtue of their objectives and as shown by their actual activities.

The court ruled that Greenpeace met the requirements of Article 1:2(3) GALA (Awb). It looked at the statutes and actual activities of the foundation. In addition, it follows from previous case-law that Greenpeace was also found to be an interested party in a similar case. Furthermore, the court did not say anything about whether a legal entity is an interested party or not. The reason for this is probably that the court came to the conclusion that Greenpeace's appeal cannot succeed in demonstrating that the competent authority is obliged to take enforcement action:

> Doing what the defendant ought to have done, the court has itself (...) decided the case and declared the objection as yet unfounded.

Over the years, case-law has developed further interpretations of Article 1:2(3) GALA (Awb). Below, we will give a brief account of the main developments. This representation only relates to cases in which legal entities stand up for general and collective interests and not those in which they stand up for their own direct interests. The latter is the case, for example, if a new construction project is started in the vicinity of a legal entity's head office.

24.4 Developments in case-law

To qualify as an interested party, there must be an *objectively* determinable, *personal, individual*, and sufficiently *current* interest, *directly* affected by the decision:[218]

- Objectively: Is the importance actually objectively determinable or is it just a personal perception?
- Personal: Is the interest distinctive from random other interests? Two characteristics that may come into play here are *proximity* and *competition*. With regard to consumers, it is generally assumed that they do not have a distinctive interest outside the issue of personal injury.

218 CBb 22 December 2017, ECLI:NL:CBB:2017:508, rec. 2.1. Referred to in Dutch as the OPERA criteria. In English this would probably be FOPERA (F (Formal):The decision must be a formal act of the government, such as a permit or a ban. O (Objective): The decision must have an objective basis, meaning it's not based on subjective preferences. P (Personal): The person claiming to be a "belanghebbende" (interested party) must be affected in their own personal interests, not just in a general interest. E (Effect): The decision must have a real effect on the person, such as impacting their property or business. R (Relevant): The effect of the decision must be relevant to the person's specific interests. A (Assessment): The government must consider and assess the impact of the decision on the person's interests.

- Individual: This is the most important criterion for a legal entity. As discussed above, there must be a general and/or collective interest. In the case of a general interest the question is whether there is a sufficiently specific idealistic interest. For a collective interest the question is whether the interest affects a larger group of natural persons or legal entities.
- Current: Is the interest currently relevant or does it relate to possible developments in the future?
- Directly: Are the interests invoked affected by the decision or is it a derivative interest? If there is a derivative interest, there cannot be an interested party.
  The OPERA criteria apply to any potential interested party; not just legal entities.

However, in the case of a legal entity, the 'individual interest' is coloured by Article 1:2(3) GALA. For a legal entity to qualify as an interested party, the following cumulative criteria must be met:
- a legal entity representing general and collective interests
- these interests must arise from the statutory objective; *and*
- from actual operations
- interests need to be specifically promoted

Legal entity

The first element implies that it must actually be a legal person as described in Book 2 of the Dutch Civil Code. However, based on case-law, an informal association can also be an interested party if it meets three cumulative requirements:
- membership
- an organisational association established for a specific purpose, so there must be regular membership meetings, a board and cooperation aimed at some continuity
- the organisation should participate in legal transactions as a unit[219]

General and collective interests

General and collective interests of a legal entity derive from the statutory objective and actual activities of a legal entity. General interests refer to specific idealistic interests pursued by an organisation, e.g. environmental interests. Often, a legal entity has a statutory objective that is too broadly formulated, so that it no longer qualifies as a specific idealistic interest.[220] Collective interests refer to the pooling of individual interests. Case-law assumes the existence of a collective interest un-

219 ABRvS 12 March 2008, ECLI:NL:RVS:2008:BC6406 and ABRvS 12 May 2016, ECLI:NL:RVS:2016:1370.

220 ABRvS 15 October 2008, ECLI:NL:RVS:2008:BF8953.

less the contrary is shown.[221] A legal entity may also act on behalf of a limited number of members, but to do so, the legal entity must be authorised by the relevant members.[222]

Statutory objective and actual activities

On 28 May 2008, the Administrative Jurisdiction Division of the Council of State (the "Council" or the ABRvS) considered that the statutory objective of the *Stichting Monumentenbehoud Nederland* (Dutch Society for the Preservation of Monuments) was formulated too broadly in both functional and territorial terms.[223] The foundation's objective read as follows: 'the protection of the beauty and historical-spatial character of the Netherlands in general and of cultural monuments in particular'. The case concerned the demolition of a few houses in Amsterdam and thus concerned only a very small part of the Netherlands. In addition, in this ruling, the Council (ABRvS) further elaborates on the foundation's actual activities. The Council ruled that 'merely opposing decisions in court cannot, as a rule, be regarded as actual activities'. It is therefore important to look at both the statutory objective and the actual activities.

The decision of the Council (ABRvS) on 1 October 2008 further elaborates on factual activities.[224] The submission of requests for enforcement action and the submission of views on draft decisions are not regarded as actual activities, as they both serve to prepare for legal challenges to decisions. It is important to look not only at the substance of the actual activities, but also at continuity. A legal entity may lose interested party status if the actual activities they normally carry out are not carried out for an extended period. There must therefore be a certain continuity in the performance of a legal entity's actual activities. This was ruled by the Council (ABRvS) in a 19 December 2014 ruling.[225] Remarkably, the Rotterdam District Court ruling of 23 June 2018 makes a reference to a 2007 ruling[226] of the Council (ABRvS) in which Greenpeace was found to be an interested party based on the actual activities. Based on this ruling, the Court assumes that over 11 years later the advocacy is still evident from the actual activities. Furthermore, no consideration is given to the continuity of Greenpeace's actual activities, even though this seems to be a logical step based on the 2014 ruling.

221 ABRvS 14 March 2007, ECLI:NL:RVS:2007:BA0642. See also ABRvS 1 April 2015 ECLI:NL:RVS:2015:992.

222 ABRvS 1 May 2006, ECLI:NL:RVS:2006:AX2120.

223 ABRvS 28 May 2008, ECLI:NL:RVS:2008:BD2647.

224 ABRvS 1 October 2008, ECLI:NL:RVS:2008:BF3911.

225 ABRvS 19 December 2014, ECLI:NL:RVS:2014:4736.

226 ABRvS 21 February 2007, ECLI:NL:RVS:2007:AZ9048.

Specifically promote

From the Explanatory Memorandum of the Dutch General Administrative Law Act (Awb),[227] it appears that 'specifically promote' is included so that political parties do not qualify as interested parties within the meaning of Article 1:2(3) GALA(Awb): 'The third paragraph speaks of general and collective interests that are 'specifically' promoted by a legal entity. These words have been included in order to make it clear that political parties as such are not interested parties within the meaning of the Awb: the mere fact that a political party's objective is to promote the general interest, as it sees it, does not mean that its interest is directly affected by a decision that it believes is contrary to the general interest.' In a ruling by the Limburg District Court, a political party, despite the presence of legal personality, is also explicitly not considered an interested party, because the words 'specifically promote' oppose this.[228]

24.5 Application for a decision

Regulators prefer to see communications about violations as 'tips' that they can deal with as they see fit without being accountable to the person raising the issue. We therefore regularly see the view that what has been requested is not (sufficiently clearly) a decision or that the person asking is not an interested party.

On appeal, the Trade and Industry Appeals Tribunal (CBb) considered that there was no question of an application and therefore no decision, as Greenpeace's request was not sufficiently concrete. At least, it did not concern a request for enforcement in a concrete case, but rather the adjustment of the enforcement policy in general. The Tribunal subsequently set aside the court ruling and declared Greenpeace's appeal unfounded.

All in all, even given the ruling on appeal, the Court's ruling looks like a Pyrrhic victory for Greenpeace. Especially where the Tribunal (CBb) initially indicates that Greenpeace does have a litigation interest in a (more principled) ruling on the substance, but then omits the requested opinion on this – given the conclusion of the inadmissibility of objection.

From an administrative procedural point of view, however, it is a nice – indirect – victory for Greenpeace in both rulings: it can be an interested party in enforcement of food safety legislation. A conclusion that may not make much sense at first glance, but is an understandable one, given (a) the broad scope of the term 'environment' and (b) the broad objective of European food safety regulations. This means that once Greenpeace submits a sufficiently concrete and substantiated enforcement

227 *Kamerstukken II* 1988/89, 21.221, no. 3 (MvT).
228 Court of Limburg 16 April 2013, ECLI:NL:RBLIM:2013:BZ7281.

request, the competent authority – given the duty of principle to enforce[229] – will have to do something with it. We do wonder, however, whether Greenpeace would not in this day and age run aground on the relativity requirement of Article 8:69a GALA (Awb), or could perhaps still be 'saved' by the Widdershoven correction.[230] Something for further investigation!

24.6 Discretion and policy-making powers

The NVWA also won on an important point in these proceedings. In this case, it was sufficiently established that an unauthorised GM food was on the market. Is the Netherlands Food and Consumer Product Safety Authority (NVWA) obliged to act in such a situation? The Court answered this question in the negative. The Court referred to the text of the then Article 21(1) of the Dutch Commodities Act (Warenwet), according to which, 'in its opinion, goods pose a danger to human safety or health, the defendant may, with a view to protecting those interests, order the person who trades or has traded the goods to inform the holders or presumed holders of those goods of the danger without delay and in an effective manner.' According to the court, the minister/NVWA has discretion in this regard. The Court performs a spot-check, i.e. the Court considers whether the minister/NVWA could reasonably have come to the opinion that the measures already taken were sufficient. In doing so, the court considers it important that the risk of contaminated rice does not adversely affect public health. By doing so, the Court seems to be saying that risk analysis is a prerequisite for the use of what it calls 'discretion'. Now that an unlawful situation has been established, the Court seems in fact to grant policy-making powers regarding the matter of whether to enforce or tolerate.

GMOs that are not authorised, and in respect of which the European Food Safety Authority (EFSA) has also not issued a favourable safety opinion, are in principle subject to zero tolerance. In Chapter 3 of this book, a judgment of the Court of Justice of the European Union[231] is discussed in which zero tolerances with regard to antibiotic residues are a matter for debate. In that case, the Court rejects the existence of any discretion or policy-making powers. According to the Court, even in

229 See, for example, ABRvS 24 January 2018, ECLI:NL:RVS:2018:209.

230 ABRvS 28 December 2016, ECLI:NL:RVS:2016:3451, AB 2017/110, annotated by B.J.P. Stramrood under AB 2017/112, Gst. 2017/64, annotated by W.P. Adriaanse (Gall Gemak). The "Widdershoven correction" is referring to the opinion of Advocate-General Rob Widdershoven according to which the relativity requirement of Article 8:69a GALA should be corrected. The opinion of 2 December 2015 of AG Rob Widdershoven, ECLI:NL:RVS:2015:3680 was followed by the Council of State in a decision of 16 March 2016, ECLI:NL:RVS:2016:732.

231 CJEU EU 28 September 2006, cases C-129/05 and C-130/05, ECLI:EU:C:2006:613 (*NV Raverco, Coxon & Chatterton Ltd. v Minister van Landbouw, Natuur en Voedselkwaliteit*).

cases of contamination limited to eight parts-per-billion, it is mandatory to proceed to destruction of the foodstuffs in question.

Silvia Gawronski
partner International Trade & Commodities at Van Traa law firm, Rotterdam

Kirsten Lenssen
lawyer at Solv law firm, Amsterdam

CHAPTER 25

Cooking oil – Designation as medicine or prohibited medical claim?

Annotated by Irene Verheijen

Court of The Hague 14 November 2012, ECLI:NL:RBSGR:2012:BY5486, IER 2013/32, annotated by E.H. Hoogenraad and C.H.E. Fontaine, JGR 2013/2, annotated by Lisman, JW 2013/2, WW 2013/1.

Medicine or foodstuff? Presentation criterion and administration criterion. Priority Medicines Act over Commodities Act. Correct application of hierarchy provision? Ban on medical claims foods.

Main considerations:

> The court considers that the defendant was able to use the presentation criterion when answering the question whether the products are medicines or not. The presentation criterion has its origins in Article 1, paragraph 2 (a) of the Medicinal Products Directive. According to established case-law of the European Court of Justice (ECJ), the presentation criterion must be interpreted broadly and, when applying it, what matters is the impression that the manner of presentation makes on an averagely well-informed consumer (cf. ECJ in Van Bennekom, 30 November 1983, case C-227/82, paragraph 18). Given that standard, the court endorses the defendant's view that, because of the text on the information sheets reproduced above, the products were presented as medicinal products.

25.1 The case

An inspector from the Netherlands Food and Consumer Product Safety Authority (*Nederlandse Voedsel- en Warenautoriteit: NVWA*) found that The Natural Health Company (TNHC) was selling pre-packaged coconut oil, omega 3 oil mix and krill oil in its shop. In all cases, these are oils for cooking and baking. Information accompanying the products states or suggests that they have a healing or health-promoting or disease-preventing effect. In the NVWA's opinion, the products are therefore presented as medicines within the meaning of Article 1(1)b of the Dutch Medicines Act. Such medical claims may only be made for products for which a marketing authorisation has been granted as medicine. This is not the case here.

 | HTTPS://DOI.ORG/10.1163/9789004751699-027

The NVWA imposes a fine of EUR 26,382 (being 10% of its annual turnover) on the company for stocking and offering for sale a medicine for which no marketing authorisation has been issued and for advertising that unlicensed product.[232] In doing so, the NVWA takes the view that the products are medicines because of the medical claims and therefore fall outside the scope of the Commodities Act.

The court considers that, given the presentation of the products and the information provided with them, it is not evident that the oil products in question are foodstuffs. Given the doubt that may exist as to how to define the products, in the court's opinion the Netherlands Food and Consumer Product Safety Authority (*NVWA*) correctly applied the hierarchy provision from the Medicines Act and could apply the presentation criterion when answering the question of whether or not the products are medicines.

25.2 Relationship between Medicines Act and Commodities Act

The relationship between (the application of) the Medicines Act and the Commodities Act is regulated in Article 1, paragraph 6 of the Medicines Act. Based on this so-called hierarchy provision, which is based on Article 2 of the Medicinal Products Directive,[233] in case of doubt as to whether a product should be qualified as a foodstuff or a medicine, the Medicines Act takes precedence over the Commodities Act.[234] The General Food Law (GFL[235]) explicitly excludes medicines from the application of food laws and regulations. It logically follows from the above that, from a legal point of view, a foodstuff can never be a medicine and vice versa.[236]

Briefly, a product qualifies as a medicinal product if it:

b. has prophylactic or therapeutic action: the *administration* criterion;[237]
a. is presented as having prophylactic or therapeutic properties: the *presentation* criterion.[238]

---

232 Violation of Article 40(2) and Article 84(1) of the Dutch Medicines Act.

233 Directive 2001/83/EU.

234 The interests that are protected by the Medicinal Products Directive are regarded as being of higher order; see CJEU 21 March 1991, case C-369/88 (*Delattre*), ECLI:EU:C:1991:137, rec. 20-22; CJEU 21 March 1991, case C-60/89 (*Monteil and Samanni*), ECLI:EU:C:1991:138, rec. 15-17. See G.A. van der Wal and F. van Schaik, *De afbakening tussen geneesmiddelen en levensmiddelen: een onderscheid van elastiek?, JGRplus*, 2007/2, p. 45.

235 Regulation (EU) 178/2002.

236 See: B.M.J. van der Meulen, *Medicinal Claims. Prohibition, Enforcement and Delineation: Food in Fact but Medicine in Law?* European Food and Feed Law Review 2017, pp. 392-405.

237 Article 1(2)b of the Medicinal Products Directive.

238 Article 1(2)a of the Medicinal Products Directive.

25.3 Food, medical claims, health claims

Article 7(3) of the Food Information to Consumers (FIC) Regulation[239] and Article 19 of the Commodities Act prohibit attributing to foodstuffs properties regarding the prevention, treatment or cure of a human disease, or alluding to such properties, and marketing them as such. The legislator thereby implies that medical claims can be made on food (other than medicines) but that this is not permitted. The Claims Regulation[240] makes explicit exceptions to that prohibition for permitted health claims referring to disease risk reduction (disease risk reduction claims) and to children's development and health (children's claims).[241]

In the opinion of the Dutch Ministry of Health, however, the ban on medical claims is unnecessary. The position is that any medical claim fulfils the presentation criterion from the Medicines Act and any food product with a medical claim should therefore be classified as a medicine. The government has taken this position for many years, but has not attempted or at least has failed to persuade Brussels to remove the ban on medical claims from food legislation. The position also finds no support in other EU Member States.[242]

25.4 Assessment

The Netherlands Food and Consumer Product Safety Authority (*Nederlandse Voedsel- en Warenautoriteit: NVWA*) , applying the hierarchy provision, did not classify the cooking oils in question as foods with (unauthorised) medical claims, but as (unlicensed) medicines. The NVWA is following its enforcement policy in this matter, whereby 'when assessing medical claims on foodstuffs, the NVWA will first examine whether the product should be classified as a medicine'.[243] The question is whether the path followed by the NVWA and its outcome is correct.

After all, the Medicinal Products Directive states as a condition for applying the legislative hierarchy that there must be doubt about the nature of the product.[244]

239 Regulation (EU) 1169/2011.

240 Regulation (EU) 1924/2006.

241 See E.H. Hoogenraad e.a., *Bijzondere voeding en voedingssupplementen, Praktijkgids*, Waar & Wet 2017/02.

242 B.M.J. van der Meulen and H.J. Bremmers, *Geneesvoedsel; het verbod op medische claims, episode II*, Waar & Wet 2015/6; B.M.J. van der Meulen, Medicinal Claims, European Food and Feed Law Review 5|2017.

243 https://www.nvwa.nl/onderwerpen/voedselveiligheid/voedingsclaims-en-gezondheidsclaims/verbod-op-medische-claims.

244 Although the word 'doubt' is not included in Article 1(6) of the Medicine's Act, I am basing this on the interpretation of this legal article in line with the directive. See nuanced consideration on this subject: Van der Meulen/Bremmers, *Het verbod op medische claims*, Waar & Wet 2015/1, p. 4–5. See also: C.H.E. Fontaine, *Levensmiddel of geneesmiddel?*, Waar & Wet 2014/65.

Doubt does not seem to be an issue here, as it is evident that the product is a foodstuff; and a foodstuff about which medical information is provided does not in fact simply become a medicine, while without this information, it is evident that the product is a foodstuff.[245] The court nevertheless ruled that it was not obvious that the cooking oils were foodstuffs, *given the presentation of the products and the information provided about the products*, basing itself on the presentation criterion in the Medicines Act.

Relevant for the application of the presentation criterion is whether a product is considered a medicinal product based on its presentation and claims. The effect of the product claimed by the manufacturer is relevant in this respect,[246] and there is an important role for the perception of the *averagely well-informed consumer*. As the court considered, the presentation criterion is interpreted broadly in case-law. It is not just about a particular communication, but about the overall image that is evoked in the mind of this consumer, given the totality of elements such as form of the product, packaging, labelling and recommendation.[247]

It is far from likely that averagely well-informed consumers will mistake oil products for medicines, even if certain beneficial effects are attributed to them. This consumer can be expected to distinguish between a medicine and a food product (with admittedly unauthorised medical or health claims). However, the Netherlands Food and Consumer Product Safety Authority (*NVWA*) quickly qualifies such a product as a medicine[248] and the court goes along with it.

The court does not, however, go along with TNHC's reasoning that it is not the presentation criterion that determines doubt about the product's characteristics, but the administration criterion. It can be argued that it logically follows from the connection between the two criteria that the presentation criterion must relate to characteristics that would fulfil the administration criterion.[249] Why should a product that (only) claims to have medicinal properties be classified as a medicinal product under the presentation criterion, while foodstuffs that have medicinal properties are not necessarily classified as medicinal products under the administration criterion?[250]

The court also does not address at all whether the information provided is correct.

---

245 See annotations Hoogenraad and Fontaine (IER 2013/32) and Lisman (JGR 2013/2).

246 CJEU 20 October 1992, case C-219/91 (*Ter Voort*), ECLI:EU:C:1992:414, rec. 26.

247 CJEU 30 November 1983, case C-227/82 (*Van Bennekom*), ECLI:EU:C:1983:354, rec. 17-19; ruling *Monteil and Samanni*, rec. 23-24; and HR 8 March 2002, ECLI:NL:PHR:2002:AD8178, JGR 2002/17. See also: CJEU 16 July 1998, case C-210/96 (*Gut Springenheide*), ECLI:EU:C:1998:369, rec. 31 and 37; CJ EU 13 January 2000, case C-220/98 (*Estée Lauder v Lancaster*), ECLI:EU:C:2000:8, rec. 27.

248 See above under 3 and footnote 11.

249 Van der Meulen/Bremmers, *Het verbod op medische claims*, Waar & Wet 2015/1, p. 9.

250 See CJEU 15 November 2007, case C-319/05 (*Commission v Germany*, Garlic), ECLI:EU:C:2007:678, rec. 65.

This is in line with the policy of the NVWA, which considers it of secondary importance whether the product actually has the claimed medicinal effect, arguing that the legislator thereby wants to protect consumers from inadequately effective medicines, preventing such products from being used instead of adequate remedies.[251]

Finally, TNHC argued that there was arbitrariness and inequality of law.[252] On the basis of factual considerations, the court did not (further) comment on this either. It therefore remains undiscussed in what way the facts and circumstances in the judgments cited by TNHC, in which products were indeed classified as foodstuffs, were different or should have been assessed differently.[253]

25.5 Consequences of the ruling

The policy line adopted by the government has been honoured in several court decisions.[254] As a result, honey and margarine, among others, have been qualified as medicines, based on statements made about these products. Importantly, the presentation criterion relates to the (permanent) classification of a product, while the ban on medical claims relates to communication about the product and is therefore incidental. Assessing the nature of a product on the basis of the presentation criterion for every communication about it leads to an ever-changing (legal) colour of the product, resulting in a continuous demarcation issue between various areas of law and associated legislation.[255]

Qualifying a product as a medicine has major consequences. Not only do these products have strict(er) requirements (with regard to authorisation, efficacy, quality control and advertising), but the fines for non-compliance with the Medicines Act are also significantly higher than those for infringement of the Commodities Act. Remarkably, although the Medicines Act violation in this case resulted in a (high) fine, the products did not have to be taken off the market. I will leave open the question of whether the policy pursued by the NVWA is based on financial motives, as has been suggested.[256]

---

251 See ruling *Van Bennekom*, rec. 17; https://www.nvwa.nl/onderwerpen/voedselveiligheid/voedingsclaims-en-gezondheidsclaims/verbod-op-medische-claims.

252 See also: B. Schwitters, *Medische claims bij levensmiddelen en de schending van het gelijkheidsbeginsel*, Waar & Wet 2014/66.

253 CBb 11 July 2008, ECLI:NL:CBB:2008:BD8238; CBb 20 December 2007, ECLI:NL:CBB:2007:BC2231 and Court of Rotterdam, 1 September 2009, ECLI:NL:RBROT:2009:BJ421.

254 See, among others, Court of Rotterdam 31 October 2013, ECLI:NL:RBROT:2013:8378 (Manuka); Court of Rotterdam 6 March 2014, ECLI:NL:RBROT:2014:1615 (*Rhodiola, Krill-olie, vitamine-D, magnesium*); Court of Rotterdam 23 January 2014, ECLI:NL:RBROT:2014:277 and ECLI:NL:RBROT:2014:278 (*Becel margarine*); see also: RCC 19 April 2018 (*Neprofarm/ v SBZP*), JGR 2018/31.

255 Van der Meulen/Bremmers, *Het verbod op medische claims*, Waar & Wet 2015/1, p. 8-9.

256 See C.H.E. Fontaine, *Levensmiddel of geneesmiddel?*, Waar & Wet 2014/65.

Finally, there is a danger in too broad an interpretation of the presentation criterion: 1) the concept of 'medicinal product' loses its distinctive character; 2) EU regulations on specific categories of foodstuffs (such as novel foods and food supplements) become meaningless; and 3) extending the scope of the Medicinal Products Directive to products that do not belong under it is to the detriment of the free movement of goods.[257] It is questionable whether the Dutch government's broad interpretation of the presentation criterion and its consequences benefit consumers.

25.6 Conclusion

What value does a ban on medical claims in food still have if every time a product is communicated with an allusion to disease, its prevention or cure, food products become medicines? Not surprisingly there is a provision in the FIC Regulation, which prohibits the use of medical claims for foods. This prohibition implies that such claims can be made. By doing this, the legislator (implicitly) indicates that a medical or health claim on a food product does *not* necessarily give the impression to the averagely well-informed consumer that it is a medicine. The legal theory followed by the NVWA is far removed from the actual reality; it does not benefit consumers either. Nevertheless, this has been the consistent policy to date and, despite the sometimes bizarre outcomes, has been followed by case-law.

25.7 Postscript

This annotation was written in 2019. For further development of the distinction between medicinal products and other product categories (food, medical devices), see Chapters 18 and 39 elsewhere in this book.

Irene Verheijen
Lawyer-Partner at Legaltree, Amsterdam

257 See conclusion A-G Geelhoed voor CJEU 9 June 2005, added cases C-211/03, C-299/03, C-316/03 and C-318/03 (*HLH Warenvertrieb and Orthica v Germany*), ECLI:EU:C:2005:78, rec. 36.

CHAPTER 26

Melatonin – enforcement policy on borderline products

Annotated by Karin Verzijden

Judge in preliminary relief proceedings, Gelderland District Court, 24 February 2015, ECLI:NL:RBGEL:2015:1182
Arnhem Court of Appeal, 1 September 2015, ECLI:NL:GHARL:2015:6399, GZR-Updates.nl 2015-0388, JW 2015/47, JGR 2015/27 annotated by Peek
District Court of The Hague, 1 June 2016, ECLI:NL:RBDHA:2016:6019, GZR-Updates.nl 2016-0241 JGR 2016/19 annotated by Lisman

When is a product medicinal by function and how should it be assessed?

Main considerations of the Hague District Court ruling of 1 June 2016:

> A generic, unconditional assessment of similar melatonin products, based on the opinion that at a dosage starting at 0.3 mg melatonin there are physiological processes by means of a pharmacological effect, without having investigated the other individual characteristics of the products named in the Hecht-Pharma judgment, was (and is) therefore not allowed. The IGZ's enforcement policy as announced in its letter of 9 November 2011 is therefore unlawful due to violation of European law. (Rec. 4.6)
> The Court rejects the State's defence that the unlawfulness judgment given above precludes the possibility of conducting (any) enforcement policy in the case of a group of products, in the past tolerated, on the market in the Netherlands containing the same active ingredient (melatonin) and of which the individual products belonging to that group have a resemblance, strong or otherwise, while the IGZ is obliged to conduct an enforcement policy pursuant to the Dutch Medicines Act (Gnw). In this respect, the Court considers the following. (Rec. 4.7)
> The court first notes that the IGZ has policy freedom in the context of its statutory duty to monitor compliance with the Gnw. In this connection, Union law must be observed and therefore – it follows from the foregoing – its policy with regard to a group of products such as the one at issue in this case cannot be accompanied by a generic, unconditional assessment as a medicine of products with a certain dosage of melatonin without a prior assessment of the individual products belonging to the group against the criteria set out in the Hecht-Pharma. In this case, IGZ did not want to opt for an individualised investigation of each individual melatonin-containing product before announcing its enforcement policy. It is free to make that choice and, as far as the court is concerned, it is also justified in itself, if only because the market for

 | HTTPS://DOI.ORG/10.1163/9789004751699-028

this type of product is not static and more or less the same product can easily be put on the market in slightly different capacities, after which it could be asserted against the IGZ that this policy cannot apply in respect of products introduced after the announcement. (Rec. 4.8)

This does not mean, however, that it is not possible for the IGZ to announce and pursue (some) enforcement policy with respect to the aforementioned group of products. In the first place, it is true that the IGZ is entitled to make its own considerations in the context of its authority to determine (the danger of) the pharmacological properties of the active substance melatonin, as long as those considerations find sufficient basis in the current state of science. In doing so, the IGZ is not bound by the considerations of the authorities of other Member States. In this context, the State pointed out that the IGZ's position that a dosage starting at 0.3 mg melatonin involves physiological processes by means of a pharmacological effect is based on an expert statement from 2009, based on submitted professional literature, a conclusion of the Status Determination Working Group (which includes representatives of the IGZ, the Dutch Food and Consumer Product Safety Authority and the Medicines Evaluation Board), also based on such literature, and, finally, written reports of the National Institute for Public Health and the Environment (RIVM). (Rec. 4.9)

26.1 Introduction

This contribution to the collection of leading cases concerns three Dutch rulings, all focusing on the same type of product, namely melatonin supplements consisting of tablets or capsules for oral use. Melatonin is a hormone that plays a role in the human sleep-wake rhythm and has a sleep-inducing effect. The IGZ (Dutch Health Care Inspectorate)[258] had announced its intention to take enforcement action against these supplements, as they were classified as a medicine based on the administration criterion.[259] It based this intention on literature studies, which described pharmacological activity of melatonin after single exogenous oral doses of 0.3 mg. Central to these rulings is the interpretation of the Hecht-Pharma ruling:[260] in what way should the IGZ determine whether a product containing 0.3 mg of melatonin qualifies as a medicinal product by function, so that a marketing authorisation must be obtained prior to marketing[261] of such a product?

258 As of 1 October 2017 the Dutch Health Care Inspectorate (IGZ) was merged with the Youth Care Inspectorate to become the Health and Youth Care Inspectorate (IGJ).

259 See Article 1(1)b 3° Dutch Medicines Act.

260 CJEC 15 January 2009, ECLI:EU:C:2009:5 C-140/07, Hecht-Pharma GmbH v Staatliches Gewerbeaufsichtsamt Lüneburg.

261 See Article 40(2) Dutch Medicines Act.

26.2 Admissibility of the claimants

The Dutch Association of the Pharmaceutical Industry of Self-care Medicines and Health Products (Neprofarm) and *Natuur- en gezondheidsProducten Nederland* (NPN) acted as claimants in the summary proceedings. In the appeal and proceedings on the merits, only the NPN still acted as plaintiff. Neprofarm had waived an appeal and probably therefore did not see the importance of proceedings on the merits.

Based on Article 3:305a of the Dutch Civil Code, a Foundation or Association with full legal capacity may bring a claim that seeks to protect the similar interests of other persons (so-called collective action). Admissibility does require that the claimant has tried to achieve the claimed action on the basis of consultation with the defending party. In addition, the individual members' own remedy before the administrative court must be excluded, according to Court of The Hague on 17 February 2016.[262]

Despite the fact that this series of cases involved a collective action, the admissibility of Neprofarm and NPN was not challenged at any point. This was due to the fact that the IGZ, of its own accord, had chosen Neprofarm and NPN as its point of contact for communicating its enforcement policy regarding melatonin products. Subsequently, the IGZ and Netherlands Food and Consumer Product Safety Authority (*NVWA*) on the one hand, and Neprofarm and NPN on the other, corresponded extensively in the period from 2011 to 2014 on the question of whether melatonin in a dosage of 0.3 mg or more should be considered a medicine. From earlier case-law, the IGZ deduced that it was necessary to set a lower limit of pharmacological activity in order to create clarity for the Dutch market. Neprofarm and the NPN disputed the evidence the IGZ presented for its intended enforcement policy. They saw reason to initiate summary proceedings to prevent the effectuation of this policy.

26.3 Demarcation issue related to food and drugs

Demarcation issues are not new. Back in 2008, the Dutch Minister of Health wrote a letter to the House of Representatives indicating that there was insufficient clarity on the demarcation between medicines and herbal medicines and thus on the applicable legislation. At the time, the Minister did not consider it necessary (nor possible within the EU framework) to take legal measures to solve the identified problem. According to him, the distinction between medicines and foodstuffs would be sufficiently clear based on the administration and presentation criteria. In addition,

262 Court of The Hague 17 February 2016, *Centrale Organisatie voor de Vleessector v. de Staat der Nederlanden*, ECLI:NL:RBDHA:2016:1417.

he expected a positive impact of the list of permitted health claims for food,[263] which was yet to be published at the time. More specifically, it was expected that products marketed with such a (disease risk reduction) claim would not qualify as medicinal products.[264] The series of cases discussed here shows that the impact of the Claims Regulation[265] in this respect was overestimated at the time.

26.4 Background: facts and legal considerations from Hecht-Pharma

The facts of the Hecht-Pharma case concerned a dietary supplement consisting of red fermented rice that Hecht-Pharma had marketed in Germany under the name 'Red Rice 330 mg capsules'. The recommendations for use read: 'as a food supplement, 1 capsule, 1–3 times daily'. The German authorities had qualified this product as a medicinal product by function, but Hecht-Pharma disagreed. It argued that, given the recommended dose, the product in question could not exert a pharmacological effect. In its reference for a preliminary ruling, the *Bundesverwaltungsgerichtshof* sought to clarify whether, following an amendment to the Medicinal Products Directive,[266] criteria previously developed to determine whether a product qualifies as a medicinal product by function were still applicable. Qualification as a medicinal product by function implies that it aims to alter physiological functions through a pharmacological, immunological or metabolic action. The Court of Justice confirmed in its judgment that previously developed criteria were still valid. Consequently, national authorities, under judicial review, must determine on a case-by-case basis whether a product falls within the definition of a medicinal product by function within the meaning of the Medicinal Products Directive. In doing so, account must be taken of all the characteristics of the product, in particular its composition, its pharmacological, immunological and metabolic properties as they can be determined in the current state of scientific knowledge, its methods of use, the extent of its distribution, consumer familiarity with it and the risks which its use may entail. It follows that products containing a substance with physiological action cannot be systematically classified as medicinal products by function.[267]

263 Commission Regulation (EU) 432/2012 of 16 May 2012 establishing a list of permitted health claims made on foods, other than those referring to the reduction of disease risk and to children's development and health.

264 See blogpost 'Dutch Scoop: new Advisory Group on the Status of Borderline Products' FoodHealthLegal: https://www.axonlawyers.com/dutch-scoop-new-advisory-group-on-the-status-of-borderline-products/.

265 Regulation (EC) No 1924/2006 of the European Parliament and of the Council of 20 December 2006 on nutrition and health claims made on foods.

266 Directive 2001/83/EC of the European Parliament and of the Council of 6 November 2001 on the Community code relating to medicinal products for human use.

267 See rec. 39 and 40 CJEC C-140/07 concerning Hecht-Pharma ECLI:EU:C:2009:5.

26.5 Enforcement measures and subsequent summary proceedings regarding melatonin

In early 2015, the IGZ sent a letter to all Dutch manufacturers of melatonin-based products informing them that they needed a market licence to continue marketing products containing 0.3 mg melatonin or more. Each manufacturer was required to inform the authorities by 15 March 2015 for which melatonin-based products it had applied or would apply for such a licence, failing which enforcement action could follow. In summary proceedings, NPN and Neprofarm claimed that the IGZ should refrain from taking enforcement action unless it demonstrated in respect of each melatonin product that it qualified as a medicine according to the applicable legal criteria, as validated in case-law. In these proceedings, NPN argued that the IGZ had not sufficiently demonstrated on the basis of scientific evidence that products containing 0.3 mg melatonin or more could alter physiological functions in the human body, for example through a pharmacological effect. According to NPN, these products were dietary supplements, not medicines.

26.6 Decision in summary proceedings

According to the preliminary relief judge, the assessment of each individual product could be carried out by the Dutch Medicines Agency (CBG) or the European Medicines Agency (EMA), following the submission of an application for a marketing authorisation. It was not necessary for the IGZ to carry out this assessment earlier, as the likelihood of finding individual deviations from the qualification of this group of products was considered very low. However, the preliminary relief judge considered that the communication and application of enforcement measures by the IGZ had not been unequivocal. Even if the manufacturers of the melatonin products had complied with the request to indicate by 15 March 2015 which products they had applied for a marketing authorisation for, it would not be clear when they would know whether the IGZ – pending such an application – would refrain from taking enforcement action. This led to uncertainty in the market and was considered unlawful towards NPN and Neprofarm. The preliminary relief judge therefore ordered the IGZ to set a deadline after 15 March 2015 within which it would decide whether products containing more than 0.3 mg of melatonin, for which a market licence had been applied for, would be tolerated. Furthermore, the IGZ was to refrain from taking enforcement action against products containing more than 0.3 mg but less than 2 mg of melatonin. The upper limit of 2 mg was based on earlier working agreements between the IGZ and Neprofarm.

26.7 Decision reversed on appeal

On appeal, the discussion concerned the correct application of Hecht-Pharma. Contrary to the preliminary relief judge, the Court of Appeal held that a public health authority announcing enforcement measures must at that time justify for itself why a product containing more than 0.3 mg of melatonin qualifies as a medicine by function. A different approach could lead to unjustified market restrictions, for example for products that would not be considered medicinal products when applying for a marketing authorisation. Moreover, the obligation to apply for a marketing authorisation for every product containing more than 0.3 mg of melatonin is not a mere formality. Such an act would require manufacturers of such products to make a significant investment of time and resources. Taking into account that there were no acute health concerns for the continued marketing of melatonin products, at least for products containing up to 5 mg of melatonin, the court ordered that the IGZ should apply all the criteria developed in Hecht-Pharma when considering enforcement action against such borderline products. The upper limit of 5 mg of melatonin comes as a surprise in this ruling. In determining this, the court referred to the remedy granted in summary proceedings, while that concerned products with an upper limit of 2 mg of melatonin.

26.8 Confirmation in proceedings on the merits

The District Court of The Hague ruled in the proceedings on the merits between the NPN and IGZ that the unconditional qualification of a group of products as medicinal products, without an individual assessment, is not in line with EU case-law. The court referred, inter alia, to paragraph 68 of the Advocate General (AG)'s opinion on the Hecht-Pharma judgment. The latter stated there that the insidious extension of the scope of the Medicinal Products Directive by including products that do not belong there would harm the free movement of goods. Thus, until an individual assessment of a borderline product based on the Hecht-Pharma criteria has taken place, public health authorities may not take enforcement action. The Court therefore enjoined the IGZ to carry out the Hecht-Pharma qualification test in any unconditional qualification of melatonin-containing products as medicines on a case-by-case basis and in a transparent and verifiable manner. This injunction also applies to any policy announcement or determination. Until such an individual test is carried out, the IGZ should leave the marketing of such products untouched.

26.9 Importance for legal practice

National health authorities must strike the right balance between the free movement of goods and the optimal protection of public health. While the enforcement policy towards borderline products that pose a threat to public health may be justified, this does not mean that every food product containing a substance with a physiological effect is automatically classified as a medicinal product by function. Lisman rightly pointed out in his note to the judgment of the court on the merits that this reduces the possibility of classifying a product as a medicinal product on the basis of the administration criterion. Food business operators therefore do not have to immediately consider adjustments to their product, such as reducing the active ingredient or even withdrawing it from the market, upon receiving a warning letter from the IGJ (Health and Youth Care Inspectorate, successor of IGZ) announcing enforcement measures. First, IGJ must have unconditionally qualified the product in question as a medicine in a transparent and verifiable manner, and in line with the Hecht-Pharma criteria.

26.10 Advisory Group

The fact that defining the boundaries between food and medicines, and thus determining the applicable legislation, is by no means always straightforward is demonstrated, among other things, by the fact that a special Advisory Group on Status Determination of Medical Products was set up in summer 2018.[268] This Advisory Group consists of representatives of the current IGJ, the Netherlands Food and Consumer Product Safety Authority (*NVWA*), the Medicines Evaluation Board (*College ter Beoordeling van Geneesmiddelen: CBG*) and the CCMO (Central Committee on Research Involving Human Subjects) and focuses on so-called 'borderline products'. Specifically, it aims to advise the enforcing authorities on which legislation applies to a particular product or substance.

26.11 Advancing insight

Legal developments do not stand still. Since the three melatonin judgments discussed above, there has been new case-law blowing in from a whole new direction.

268 Agreement between the Health and Youth Care Inspectorate currently being set up (IGJ), the Netherlands Food and Consumer Product Safety Authority (NVWA), the Dutch Medicines Evaluation Board (CBG) and the Central Committee on Research Involving Human Subjects (CCMO) on the setting up of an advisory group on determining the status of borderline medicinal products, Staatscourant jaargang 2018, No. 37730.

For example, in a judgment of 25 March 2022[269], the Court of East Brabant stipulated that the Medicinal Products Directive does not apply if a product is clearly exclusively a food supplement. This is the case if the content of the product in question must be regarded as a food supplement and not also as a medicine. In such a case, Article 1(6) of the Dutch Medicines Act – which stipulates that in case of doubt as to which legislation applies, the Medicinal Products Directive prevails – does not apply. This case has since been echoed in the decision of the District Court of The Hague of 28 June 2022,[270] but the line is not yet consistent. Moreover, both cases concerned the presentation criterion, not the administration criterion as in the melatonin cases discussed above. Nevertheless, it will be interesting to see whether the application of the 'clearly exclusively a food supplement' criterion will also apply in cases involving the application criterion. I would personally see that as a good thing.

Karin Verzijden
Life sciences lawyer at Axon law firm, Amsterdam

269 ECLI:NL:RBOBR:2022:1109 JGR 2023,3 annotated by Koosje van Lessen Kloeke.
270 ECLI:NL:RBDHA:2022:6294.

CHAPTER 27

Food bank – Who ensures the human right to adequate food?

Annotated by Bernd van der Meulen

Central Netherlands District Court, 30 March 2015, ECLI:NL:RBMNE:2015:2162 (*X. v Ferm Werk*)
Central Appeals Council, 21 June 2016, ECLI:NL:CRVB:2016:2361 (*X. v Ferm Werk*)
Court of Justice of the European Union 12 November 2019, case C 233/18, ECLI:EU:C:2019:956 (*Zubair Haqbin v Federal Agency for the Reception of Asylum Seekers*)

Food bank is not a facility within the meaning of the Dutch Social Support Act (*Wmo*).

Main consideration:

> The court endorses the defendant's position that the claimant has not made it plausible that she belongs to the target group of chronically ill and disabled persons as referred to in the implementing regulations. The claimant argues that she has continuously used a facility under the Wmo for more than a year, but she has not substantiated this position adequately. While she does refer in this context to documents showing that she uses a Food Parcel provision from the food bank, these documents do not show that this is a provision under the Wmo, nor that it is based on a decision by the defendant. Whether a provision falls within the scope of the Wmo may be evident from a granting decision by the defendant on an application for such a provision. Provisions covered by the Wmo and aimed at people with a disability or a chronic mental or psychosocial problem relate to promoting their independent functioning or participation in traffic. A provision of a Food Parcel does not fall into the category of facilities provided under the Wmo to people with a disability or a chronic psychological or psychosocial problem. Thus, it has not become plausible that she makes use of a provision under the Wmo based on a decision of the defendant.

27.1 Importance of the ruling

The ruling raises the question of the place of Food Banks in relation to social security systems. In the context of municipal regulation in Bodegraven-Reeuwijk, that appears to be a place on the sidelines, outside the playing field.

 | HTTPS://DOI.ORG/10.1163/9789004751699-029

27.2 The facts and the proceedings

The municipality of Bodegraven-Reeuwijk, in 'Implementing regulations 2013 Bodegraven-Reeuwijk', allowed under special assistance a contribution (e.g. towards health care premiums) for chronically ill and disabled persons who meet at least one of a list of conditions including: 'Has been continuously using a provision under the Social Support Act (Wmo) for more than 1 year.'

The claimant in this case applied for this contribution. She argued that she enjoyed a Wmo provision in the form of food parcels from the Food Bank. The defendant, the local implementing organisation 'Ferm Werk', rejected the application because a food parcel from the food bank is not a Wmo provision. The court shares this opinion and declares the appeal unfounded. The Central Appeals Council upheld the ruling without any relevant considerations here.

27.3 Foundation of food law

It is not unusual to discuss areas of law on their human rights/constitutional basis.

The right to adequate nutrition could be the human rights foundation of food law. In that approach, the function of food law would be to ensure that consumers have access to sufficient healthy food and are protected from unsafe food and from food exclusion. Food law lacks this foundation mainly because of the almost consistent refusal of courts to assign legal relevance to the right to food.

In the chapter 22 in this book on the Central Council of Appeal decision of 8 August 2005 ECLI:NL:CRVB:2005:AU0687, Bart Wernaart discusses the human right to food. The Netherlands is party to treaties enshrining this right, including the International Covenant on Economic, Social and Cultural Rights (ICESCR) and the United Nations Convention on the Rights of the Child (UNCRC). As Wernaart showed in his doctoral thesis, the courts in the Netherlands – with the exception of cases where the UNCRC can be invoked – fail to enforce the obligations that the Netherlands has accepted as a party to the Convention. In reports to the UN, the government nevertheless paints a rosy picture of the realisation of the right to food in the Netherlands.[271]

It would have been conceivable for hunger and malnutrition to be eliminated in the Netherlands as it is one of the world's richest countries per capita. In that case, the right to food would have been realised for all and food banks would be redundant.

271 Fons Coomans, 'A "rosy picture"? Dutch ICESCR report reviewed by UN Committee', in: Otto Hospes and Bernd van der Meulen (red.) *Fed up with the right to food? The Netherlands' policies and practices regarding the human right to adequate food*, Wageningen Academic Publishers, 2009. https://brill.com/edcollbook/title/68536.

27.4 Poverty and malnutrition in the Netherlands

The Central Bureau of Statistics (CBS)[272] and the Social Cultural Planning Office (SCP)[273] keep statistics on poverty.[274] Over a million people in the Netherlands (out of a population of 18 million) live on or below the poverty line. Over 400,000 are in this situation for the long term.

Poverty affects working people (about 200,000) less than non-working people, women slightly more than men, single-parent families more than families with two parents.

Driven by the Covid pandemic, the number of people living in poverty has increased significantly.

Poverty brings with it the risk of malnutrition. On 9 September 2014, the newspaper NRC headlined with: 'Every day, at least one Dutch person dies of undernourishment'.[275] The newspaper relied on research conducted by the independent economic research institution SEO.[276] Among other things, SEO concludes that '[t]he total cost of disease-related malnutrition in the Netherlands was €1.8 billion in 2011 plus unquantified care costs for people living at home.' In 2005, a Malnutrition Steering Group was launched, focusing on malnutrition in healthcare. 'The prevalence of malnutrition is high in all sectors of the Dutch healthcare system. On average, one in four to five patients in hospitals, care facilities and home care is severely undernourished. Risk groups for malnutrition are frail elderly patients, chronically ill patients, oncology patients, patients undergoing major surgery and patients with severe trauma.'[277]

27.5 Food banks

Private initiatives have taken up the cause of hunger and undernourishment in the Netherlands (and elsewhere). The first food bank was set up in Rotterdam in 2002. Politicians responded by promising that the food bank would be redundant within four years. In 2006, there were more than 40 food banks in the Netherlands.

---

272 See: https://www.cbs.nl/nl-nl/visualisaties/armoede-en-schulden.

273 See: https://www.scp.nl/actueel/nieuws/2024/10/24/voorkom-dat-nog-meer-mensen-in-armoede-vervallen.

274 See: https://digitaal.scp.nl/armoedeinkaart2018/waar-ligt-de-armoedegrens/.

275 See: https://www.nrc.nl/nieuws/2014/09/09/iedere-dag-sterft-minstens-een-nederlander-aan-ondervoeding-a1422530. If these figures are even remotely accurate, then food insecurity in the Netherlands is a far bigger problem than food safety. This however is by no means reflected in policy priorities.

276 See: https://www.seo.nl/publicaties/ondervoeding-onderschat/.

277 See: https://www.kenniscentrumondervoeding.nl/.

By 2025, 179 food banks were operating in 181 municipalities (out of 342) serving 180,579 people.

The rationale behind food banks is simple. In our society, food is being wasted; but on the other side of society, people are struggling with undernourishment. By solving one problem, the other can be addressed. When food is diverted from wastage and provided to those in short supply, both waste and undernourishment are countered. Thus, food banks pursue both social justice and sustainability.

Using the collected food, the food banks put together food parcels with which clients can prepare healthy meals. The packs are distributed weekly at distribution points.

Food banks are neutral and non-discriminatory. In other words, they do not assume any political or religious outlook. Food parcels are provided indiscriminately to any applicant who meets the income criteria. No distinction is made according to gender, age, orientation, origin or philosophy of life.

Food banks are independent of each other. They are usually affiliated with the Association of Dutch Food Banks, where they formulate their joint policies. The Association is affiliated with the European Food Banks Federation (FEBA). According to the FEBA annual report for 2023, food banks are active in 30 European countries.[278]

27.6 Food bank customers

Food banks cannot help everyone living below the poverty line. They provide temporary emergency assistance (maximum three years – though exceptions are possible) mainly to the poorest of the poor to help them through an emergency situation. This is subject to tight financial limits. Eligibility for food parcels is only for households that, after deducting unavoidable fixed expenses, have a disposable income of up to €300 per month for one person, plus €110 for each subsequent household member (2025 figures).[279]

According to a 'facts and figures' document from the Association of Dutch Food Banks, the client base is diverse: self-employed, long-term sick, people with (mortgage) debts and jobseekers. The 50+ share is growing. Over half (51%) of the 180,579 clients are women; about a third (33%) are under 18.[280]

Clients must actively cooperate to solve their problems, for instance through debt restructuring: The policy is 'no package without an assistance programme.' Exceptions occur in situations that are unsolvable, such as elderly people who, because of years spent abroad, have not built up enough state pension rights to make

278 See: https://www.eurofoodbank.org/publications/?_search=annual%20report.

279 See: https://voedselbankennederland.nl/ik-zoek-hulp/kom-ik-in-aanmerking/.

280 See: https://voedselbankennederland.nl/jaarverslagen-feiten-en-cijfers/.

ends meet. Food banks cooperate with agencies that work on problem solving, such as social work. The average period during which people depend on the food bank is about one year.

27.7 Donors

Food banks depend on support from food donors and also financial support. Food is donated mainly by food companies. These are often surpluses and products taken off the shelves towards the end of their shelf life. In addition, food is collected by companies, schools and individuals. Because of the Russia boycott, the EU buys fresh fruit and makes it available for distribution. Donors usually act out of social and sustainability motives. Food companies may have a business interest partly because they reduce waste streams and associated costs. However, this is offset by logistical costs and responsibility for food safety.

Food bank costs are covered by grants, donations from charitable funds, businesses and individuals. Occasionally, a bequest or inheritance also benefits a food bank. The food banks are ANBI-certified, i.e. officially recognised as a charity, which means that donations are tax free.

Food banks in the Netherlands work almost exclusively with volunteers. As a result, staff costs are negligeable. The highest costs relate to transport and logistics: vans, refrigeration, premises for storage and processing and distribution points.

27.8 Its own legal framework

One of the European Parliament's first questions on the scope of the General Food Law (Regulation (EC) 178/2002) concerned the status of charities in relation to the traceability requirement.[281] The European Commission replied that institutions such as food banks are simply 'food businesses' and must comply with traceability. Other requirements also apply, especially hygiene and food safety. For a voluntary organisation, this is quite a challenge.

Both national and European policies have been established to facilitate the collection and distribution of food for charitable purposes.

The Dutch Food Safety Authority (NVWA) has published 'Infoblad 76' dedicated to Charitable institutions and organisations.[282] An important part of this info sheet is the shelf life of products. Products marked with a use-by date should not be offered after this date has passed, while the best before date relates to microbiological

281 Question E-2704/04; answer: E-2704/04DE.

282 See: https://www.nvwa.nl/documenten/eten-drinken-roken/levensmiddelenketen/publicaties/informatieblad-76-verstrekking-levensmiddelen-via-charitatieve-organisaties.

safety in the case of perishable products and mainly with quality in the case of non-perishable products where health risks are less likely to occur. The info sheet contains a non-exhaustive list indicating how long after the best before date products may in principle still be distributed: 1) for very long shelf-life products such as salt, sugar, dry pasta products, canned and frozen products, the guideline is up to one year after best before date; 2) for long-life products such as dry biscuits, salted snacks, hard cheeses, sterilised milk and dairy products and frozen products, up to 2 months after best before date; 3) for perishable products such as bread, fruit and vegetables, it is recommended to stick to the best before date; and 4) for short shelf-life products such as fresh meat, chilled meals and salads, cut vegetables and fruit, it is recommended never to accept or distribute them after the best before date.

The European Commission has published EU guidelines on food donation.[283] It writes: 'In the EU, close to one-quarter of the population – 119,1 million people – were at risk of poverty or social exclusion in 2015 and 42,5 million people were not able to afford a quality meal every second day. At the same time, it is estimated that around 88 million tonnes of food waste are generated annually in the EU with associated costs estimated at 143 billion EUR.

In addition to its important economic and societal impact, food waste places undue pressure on finite natural resources and on the environment. According to the Food and Agriculture Organisation of the United Nations (FAO), approximately one third of all food produced in the world is lost or wasted. Food that is harvested but ultimately lost or wasted consumes about one quarter of all water used by agriculture each year and requires cropland area the size of China. Food waste generates about 8 % of global greenhouse gas emissions annually.'

The EU Guidelines mainly provide explanations for all parties involved about food law and the division of responsibilities and liabilities. There are some nice interpretations in there. For example, it explicitly states that the responsibility that Article 17 GFL places on food business operators should be distinguished from liability. 'The marketing of foods beyond their date of minimum durability (i.e. 'best before') is allowed under EU rules, provided that the foods concerned are still safe and their presentation is not misleading. It is permitted, at each stage in the food supply chain, to place food on the market which has passed the date of minimum durability. It is the food business operator's responsibility (e.g. the retailer) to ensure that the food is still safe for human consumption and that consumers are duly informed that the product concerned is past the 'best before' date (for instance, such products may be marketed separately with signs indicating that the minimum durability date is exceeded). Certain Member States restrict or even prohibit marketing of foods past the 'best before' date leading to avoidable food waste.'

283 Commission notice – EU guidelines on food donation, C/2017/6872, OJ C 361, 25.10.2017, p. 1–29.

27.9 New hope

I began this discussion by expressing disappointment at the low level of recognition of the human right to adequate food. In that respect, there is a startling development to report. Totally unexpectedly, the Court of Justice of the EU has joined the ranks of courts who recognise the human right to adequate food as a legally enforceable right. The case was decided by the Court's Grand Chamber on 12 November 2019.[284] It concerned a minor asylum seeker who had been placed outside reception and all facilities for a certain period of time as a punishment for misconduct. In response to preliminary questions on the compatibility of the punitive measures with Directive 2013/33/EU laying down standards for the reception of applicants for international protection, the Court considered inter alia (recital 46):

> With regard specifically to the requirement to ensure a dignified standard of living, it is apparent from recital 35 of Directive 2013/33 that the directive seeks to ensure full respect for human dignity and to promote the application, inter alia, of Article 1 of the Charter of Fundamental Rights and has to be implemented accordingly. In that regard, respect for human dignity within the meaning of that article requires the person concerned not finding himself or herself in a situation of extreme material poverty that does not allow that person to meet his or her most basic needs such as a place to live, food, clothing and personal hygiene, and that undermines his or her physical or mental health or puts that person in a state of degradation incompatible with human dignity (see, to that effect, judgment of 19 March 2019, Jawo, C163/17, EU:C:2019:218, paragraph 92 and the case-law cited).

The International Covenant on Social, Economic and Cultural Rights explicitly recognises the right to adequate food as part of the right to an adequate standard of living (Article 11). In the Charter of Fundamental Rights of the European Union, the right to nutrition is not mentioned in so many words. Now, the Court explicitly brings it under human dignity in Article 1 of the Charter. This provision reads, 'Human dignity is inviolable. It must be respected and protected.' Potentially, this case law is a powerful push towards recognition of nutrition as a human right.

27.10 Reflection

National and European governments appear to be sympathetic to food banks. One criticism, though, is that they provide governments with an excuse to leave safeguarding the right to food to others. The rulings on Bodegraven-Reeuwijk show that

284 CJEU 12 November 2019, case C233/18, ECLI:EU:C:2019:956 (*Zubair Haqbin v Federaal Agentschap voor de opvang van asielzoeker*).

things can go a step further. In cases where the food bank catches whoever falls through the public law cracks, the public law system helps create new cracks for the same person to fall through yet again. Perhaps Dutch judges will take inspiration from the courageous example of the EU Court of Justice and reconsider their position on socio-economic human rights.

I would like to remind politicians of the promise to make food banks redundant in the Netherlands. That is a moral duty, and now also a legal obligation.

Bernd van der Meulen
Volunteer Food Bank Amsterdam Distribution Centre North Holland – Flevoland (https://voedselbank.amsterdam/)

CHAPTER 28

Arsenic – Food safety requirements

Annotated by Bernd van der Meulen

Rotterdam District Court 5 July 2016 ECLI:NL:RBROT:2016:5619 (*DIS v NVWA*)
Trade and Industry Appeals Tribunal (*CBb*) 3 July 2018 ECLI:NL:CBB:2018:326 (*confirmation*)

With the words 'food safety requirements', Article 19 GFL refers to Article 14 GFL. From the presumption of safety in case of compliance with specific standards, courts derive a contrario a rebuttable presumption of unsafety in case of non-conformity.

Main considerations of the District Court of Rotterdam:

8. Article 19(1) of Regulation (EC) No 178/2002 stipulates, inter alia, that if a food business operator considers or has reason to believe that a food which it has imported, produced, processed, manufactured or distributed does not comply with the food safety requirements, it shall immediately initiate procedures to withdraw the food in question from the market when it has left the direct control of the food business operator. In the court's opinion, the word 'food' in conjunction with the phrase 'does not comply with the food safety requirements' must be understood as meaning that a situation as referred to in the first sentence of Article 19(1) of Regulation (EC) No 178/2002 arises if there is a food which may be presumed to be unsafe within the meaning of Article 14 of Regulation (EC) No 178/2002, which bears the caption 'Food safety requirements'.
9. Article 14(7) of Regulation (EC) No 178/2002 stipulates that food that complies with specific Community provisions governing food safety shall be deemed to be safe in so far as the aspects covered by those specific Community provisions are concerned. A reasonable interpretation of Article 14 of Regulation (EC) No 178/2002 implies that – as noted in the Guidelines – food that does not comply with specific Community safety provisions shall be deemed unsafe to the extent that a risk assessment does not demonstrate otherwise.

Main considerations of the Trade and Industry Appeals Tribunal (*CBb*):

5. The Tribunal finds that it is not in dispute that the additive in question contains a level of arsenic that exceeds the authorised maximum of 1 mg/kg, as stipulated in Article 4(1) of Regulation (EC) No 1333/2008 in conjunction with Annex II to that Regulation and the Annex to Regulation (EU) No 231/2012. Together with the court, the Tribunal is of the opinion that a reasonable interpretation of Article 14 of Regulation

 | HTTPS://DOI.ORG/10.1163/9789004751699-030

(EC) No 178/2002 entails that food that does not comply with specific Community safety provisions is deemed to be unsafe to the extent that a risk assessment does not demonstrate the contrary. In view of this, it may be presumed that the soft and energy drinks produced and distributed by the appellant with the additive in question are unsafe because this additive exceeds the maximum permitted level of arsenic. The appellant, as food business operator responsible for compliance with food safety requirements pursuant to Articles 17 and 19 of Regulation (EC) No 178/2002, would then have the burden of proving that this presumption of unsafety was incorrect. The court therefore correctly considered that the burden of proof to that extent lies with the appellant.

6. Contrary to the respondent's view, the report by [name 3] of the same date, submitted by the appellant by letter dated 9 February 2018, may, in the opinion of the Tribunal, be taken into account in assessing whether the appellant has demonstrated that the aforementioned suspicion of unsafety is incorrect, as this report serves to supplement or elaborate on its position that the finished product is safe in all respects.

13. Given that the purpose of Regulation (EC) No 178/2002 is to provide the basis for ensuring a high level of protection of public health and consumers' interests in relation to food, and having regard to the items listed in Article 14(4) of Regulation (EC) No 178/2002 that must be taken into account when assessing whether a food is injurious to health, in particular long-term effects and suspected cumulative toxic effects, the NVWA [*Netherlands Food and Consumer Product Safety Authority, eds.*] and the RIVM [*the Dutch National Institute for Public Health and the Environment, eds.*]were, in the opinion of the Tribunal, entitled to include the long-term exposure due to the consumption of soft and energy drinks to which arsenic-contaminated trisodium citrate has been added in the risk assessment, and the RIVM also took into account, on good grounds, the exposure to arsenic from other sources.

28.1 The procedure

In 2014, Brenntag, Netherlands BV supplied DIS BV with the additive E331. In mid-2014, Brenntag informed the Netherlands Food and Consumer Product Safety Authority (*NVWA*) that a batch received from China showed excessive levels of arsenic. When soft drinks and energy drinks containing the additive in question were examined, the levels of arsenic remained below the detection limit. The NVWA instructed DIS to block the relevant batches. The *Netherlands Food and Consumer Product Safety Authority* (*NVWA*) subsequently informed DIS that a recall should take place for 'first-tier customers' but not consumers. At the NVWA's request, the RIVM carried out a risk assessment. The RIVM concluded: 'For persons who consume soft and/or energy drinks for long periods, health risks cannot be excluded if these drinks have been produced with contaminated trisodium citrate to a level of 22 µg arsenic/L. No acute health effects are expected.'

A few more reports go back and forth.

An objection is declared unfounded by the Minister of Health (contrary to the advice of the objection advisory committee).

On 15 July 2016, the court concluded that it was not the minister that had jurisdiction, but the NVWA. The lack of jurisdiction was passed over based on Article 6:22 of the Dutch General Adminstrative Law Act (*Awb*) and the appeal was declared unfounded.

On 3 July 2018, the Trade and Industry Appeals Tribunal (*CBb*) confirmed the court's opinion.

Below, I include some brief comments on competence. Central to this discussion is the interpretation given to the term 'food safety requirements'. Finally, I pay brief attention to the guidelines of the Standing Committee on which judges partly base their judgments.

28.2 Enforcement power under the Official Controls Regulation

Enforcement powers under the Commodities Act and the Animals Act such as the power to impose administrative fines, exercise administrative coercion and impose an order for incremental penalty payments fall to the Minister of Health, Welfare and Sport and the Minister of Agriculture, Nature and Food Quality respectively. The Official Controls Regulation (at the time of the case it was Regulation 882/2004, now 625/2017) grants the national competent authority direct powers to take all kinds of measures. Contrary to the national division of powers, in the Netherlands the Food and Consumer Product Safety Authority (*NVWA*) rather than the minister was designated as the competent authority. This was regulated in the Commodities Act Decree on Hygiene of Foodstuffs. In this way, the NVWA (like the tax authority, for example) became partly a so-called devolved office. It is, in other words, an official subordinate administrative body endowed with its own powers. Not surprisingly, this lopsided division of powers leads to confusion. In this case, the decision on objection had been made by the Minister of Health, Welfare and Sport. However, the recall had been imposed on the basis of the Official Controls Regulation – so by the NVWA, not on the basis of a mandate from the Minister but on the basis of its own authority. As the NVWA stated that it took responsibility for the Minister's decision, the court found that there had only been a formal error to which the court did not attach any further consequences.

Under the new Official Controls Regulation, the Minister is the competent authority (Commodities Act Decree implementing regulation on official controls and other official activities). However, there is a possibility to change this by ministerial regulation.

28.3 Food safety requirements

The recall obligations under Article 19 GFL are triggered when the food business operator has reason to believe that 'food safety requirements' have not been met. The arsenic case raises two aspects of the question of what is meant by 'food safety requirements'. These are the relationship to Article 14 GFL and to non-compliance with specific requirements.

In the past, I have sometimes argued that the 'food safety requirements' in Article 19 cover all requirements that have something to do with food safety.[285] I have even gone so far as to argue that this is an important distinction between Article 14 and Article 19. In Article 14, 'unsafe food' is a material/scientific concept. Article 14 is concerned with the likely effect of a food on health, or material acceptability for consumption, as the case may be. Article 14 operates with little or no legal fictions. In particular, it does not state that a food is deemed unsafe if it does not meet certain requirements. The only legal fictions in Article 14 are the legal presumption of unsafety of the entire batch (paragraph 6) and the legal presumption of safety in compliance with standards (paragraphs 7 and 8). In that interpretation Article 19 is *not* linked to material properties of the foodstuff, but to non-standard conformity. The advantage of that reading, I believed, was that there are no complicated material questions to be answered in case of norm violation.

I have been severely criticised for this interpretation by colleagues. It was put to me that – also under general principles of legislative technique – it must be assumed that where the law uses the same term several times, that term has the same meaning each time. My interpretation overlooked the fact that the term 'food safety requirements' is not only used in Article 19 but is also the heading of Article 14. In other words, the reference in Article 19 to 'food safety requirements' is a specific reference to Article 14 not a general reference to requirements regarding food safety. Put another way, Article 19 only applies when Article 14 might apply.

In practice, the two interpretations still mix.[286] The importance of this case-law is that the court explicitly takes a position. The latter interpretation, i.e. that 'food safety requirements' means Article 14, is the one currently in force in the Netherlands.

28.4 Food safety limits

The interpretation that food safety limits do not directly trigger Article 19 lends some urgency to the relationship between them and Article 14.

285 Bernd van der Meulen, The Core of Food Law. A Critical Reflection on the Single Most Important Provision in All of EU Food Law, European Food and Feed Law Review 2012, pp. 117–125.

286 See Chapters 33 and 40 of this book.

Incidentally, the urgency is a limited one. In most cases, the regulation setting limits itself determines the legal consequences in terms of a marketing ban. The possible absence of a recall obligation need not be a major problem as the Official Controls Regulation allows the regulator to impose a recall (or other measure).

The court and the Tribunal concur with the Standing Committee's interpretation.[287] The Standing Committee applies an e-contrario reasoning to paragraphs 7 and 9. Within certain limits, the Committee derives from the legal presumption of safety in the case of norm conformity a legal presumption of unsafety in the case of non-conformity. Legal scholars such as Paul Scholten and Adriaan Pitlo stress that e-contrario reasoning should be applied with the utmost caution. (Pitlo writes in his System of Dutch Private Law: 'The reasoning has not unjustly always been seen in a somewhat bad light. It is somewhat too easy, too cheap and may stem from, and lead to, thought inertia.') The first point of reference in determining the legislator's intent is the words of the law. Reasoning by analogy and e-contrario are, by definition, only applicable to cases not covered by the text of the law. Article 14 focuses on the concept of unsafe food. If the legislator had intended to rely on a legal presumption of unsafety in cases of non-conformity, it would have been very obvious, given the structure of Article 14, for the legislator to have said this in so many words. However, the legislature has instead moved away from the focus on unsafety to introduce a legal presumption of safety linked to norm conformity. In the structure of Article 14, non-conformity is not an isolated fact. The logical approach would be to fall back on the second paragraph and examine whether there is harmfulness to health or unfitness for consumption. In some other Member States, following that route involves a presumption of unfitness in the case of non-conformity.

However, the Dutch court opts for the Committee's e-contrario reasoning. When food safety standards are exceeded, a legal presumption of unsafety applies. This is a rebuttable legal presumption, but the burden of proof is on the person who wants to argue that the product is safe.

28.5 Standard of proof

The Trade and Industry Appeals Tribunal (*CBb*) elaborates on the standard of proof.

> Having regard to Article 1(1) of Regulation (EC) No 178/2002, the purpose of this Regulation is to provide the basis for the assurance of a high level of protection of public health and consumers' interests in relation to food, whilst ensuring the effective func-

---

287 Standing Committee on the Food Chain and Animal Health (SCoFCAH) now Standing Committee on Plants, Animals, Food and Feed (SCoPAFF) Guidelines for the implementation of Articles 11, 12, 14, 17, 18, 19 and 20 of Regulation (EC) No 178/2002 on General Food Law, 26 January 2010. https://food.ec.europa.eu/system/files/2016-10/gfl_req_guidance_rev_8_en.pdf/.

> tioning of the internal market. When assessing whether a food is injurious to health, pursuant to Article 14(4) of this Regulation, consideration shall be given not only to the probable immediate and/or short-term and/or long-term effect which the food has on the health of a person consuming it, but also to the effect on their offspring. Furthermore, the likely cumulative toxic effects as well as the particular physical sensitivities of a specific category of consumers, in case the food is intended for that category of consumers, shall be taken into account in that assessment. The Tribunal will also consider in this light the risk assessments by the NVWA (*Netherlands Food and Consumer Product Safety Authority eds*) and RIVM, on the one hand, and the reports submitted by the appellant on the other, on which the NVWA and RIVM (*National Institute for Public Health and the Environment eds.*) based the contested decision.

The Tribunal notes that it is not in dispute between the parties that acute health effects are not expected to occur as a result of consuming the soft and/or energy drinks containing the contaminated additive. However, given the standard formulated above, the Tribunal looks beyond the risk to be feared from the tangible contamination and also considers long-term exposure.

28.6 Evidence

Elsewhere in this book under the keyword 'recall' (Chapter 33), we have seen that, despite the statutory right to counter-expertise, in the context of the question of whether a recall obligation existed, courts set aside later research as rebuttal evidence.

In this case, the Tribunal does allow a report submitted four years after the fact 'as this report serves to supplement or further elaborate its view that the finished product is safe in all respects.'

This may be due to the fact that the investigation was not aimed at establishing what the factual situation was at the time of the recall, but is about whether the facts as established at the time of the recall constitute a situation of unsafety.

28.7 Guidance

The judges in this case give weight to the Standing Committee's guidance document. It is already almost traditional for the European Commission to issue interpretative documents to accompany new pieces of food legislation. These documents evoke mixed reactions. On the one hand, it is appreciated that the Commission writes down what the law means in a language that mere mortals can understand. On the other hand, there is criticism that if the Commission can write it down in understandable language, it would have been better to do so in the law itself. The interpretative documents are always accompanied by the disclaimer that the European Court of Justice has the final say on the meaning of the law. In practice, however, these documents have great authority.[288]

The Standing Committee's Guidance document is a bit of an oddity. The Committee is primarily a political body with all sorts of national agendas. It is known, for instance, that the Netherlands did its best to have the interpretation included in the Guidance to Article 18 that traceability data should be able to be provided within four hours. However, they did not get the other Member States on board for this interpretation. The example does show that the line between interpreting the law and supplementing it can be wafer-thin. Not for nothing did legal scholars in the Netherlands in the past call such texts 'pseudo-legislation' (a term coined by Van der Hoeven), before Van Kreveld launched the term 'policy rules' in his PhD thesis.

The NVWA lays down many of its interpretations in 'infobladen/informatiebladen' (info sheets). There, too, we occasionally see a certain tendency towards pseudo-legislation rearing its head.

EU-level documents tend to be referred to as 'soft law'. The info sheets are examples of what the Dutch General Administrative Law Act (GALA or in Dutch Awb) calls 'policy rules'. Characteristically, they are based on administrative power, not legislative power. With policy rules, the administrative body cannot therefore impose externally effective obligations. Policy rules apply as a form of self-binding. In other words, by recording interpretations of the law in writing, the administrative body commits itself to actually applying those interpretations. This does, however, include the inherent competence to deviate from its own line of action in distressing cases.

To my knowledge, we do not yet really have a doctrine on the meaning of European Commission interpretative documents for national implementing authorities. From a Union loyalty point of view, there seems to be a case for considering national authorities in Member States (or at least in the Netherlands) to be bound by them as if they were their own policy rules.

Bernd van der Meulen
Director European Institute for Food Law (www.food-law.nl)

---

288 CJEU emphasises the nonbinding nature of soft law in, among others, CJEU 22 September 2016 (C-113/15 ECLI:EU:C:2016:718) Breitsamer and Ulrich and in CJEU 22 June 2023 *Taminco and Arysta LifeScience Great Britain v Commission* ECLI:EU:C:2023:513 – Chapter 20 of this book.

CHAPTER 29

Horse meat – Suspicion of (possibly) unsafe product

Annotated by Barbara Mutsaers

Trade and Industry Appeals Tribunal (CBb) 30 January 2017, ECLI:NL:CBB:2017:6 NJB 2017/388 and JW 2017/25 (*Trustee Selten v Vice-Minister for Economic Affairs*)

Consignments of meat taken into custody. Mixing of beef with horse meat. Meat from one horse contaminated with the drug phenylbutazone supplied. Traceability. Fraud.

Main considerations of the Tribunal:

> Unlike the respondent, the Tribunal is of the opinion that non-compliance with the obligation in Article 18 of Regulation 178/2002 that food must be traceable at all stages of production, processing and distribution does not, in itself, mean that there is therefore also a (suspected) unsafe food as referred to in Article 14(2) of that Regulation. Regulation 178/2002 does not provide any textual or systematic basis for that conclusion. This means that the detention decided on by the respondent in the detention decisions is founded on an incorrect legal basis. The respondent wrongly failed to recognise this in the contested Decision I and maintained that basis, with the result that that decision is inadequately reasoned in so far as it relates to the detention decisions and is contrary to Article 7:12(1) of the Awb (Dutch General Administrative Law Act, eds.).
> The Tribunal further considers that the [...] opinion that the respondent was correct in its detention decisions to officially detain the consignments of meat in question on the basis of Article 54(2)(g), in conjunction with Article 19(1) of Regulation 882/2004, does not mean that the defendant was allowed to continue the detention of the consignments of meat in question at the time of taking the contested Decision I.
> Having regard to the text of Article 19(1) of Regulation 882/2004, in the opinion of the Tribunal, this provision does not provide any authority for the mere taking into official detention of food that does not comply with feed or food law, but it follows imperatively that the competent authority takes such food into official detention and, having heard the food business operator responsible for the consignment, and having verified, where necessary, the existence of adverse effects as referred to in Article 19(1)(c) of that Regulation, takes the measures referred to in that provision under (a).

 | HTTPS://DOI.ORG/10.1163/9789004751699-031

29.1 Relevance

The horse meat affair is an essential inclusion in this book. Although this published ruling of 30 January 2017 has been anonymised, it is known that this case concerns (one of) Selten B.V.s[289], namely Vleesgroothandel Willy Selten B.V. This ruling discusses, on the one hand, the importance of conclusive administration and traceability. On the other hand, it also makes clear that a national authority must comply with the rules once it has taken a batch of foodstuffs into official custody. To top it all, there is also the consideration that, according to the Trade and Industry Appeals Tribunal (hereafter the Tribunal or CBb), the lack of traceability does not in itself mean that there is therefore a (suspicion of an) unsafe food product.

29.2 The case

First, the facts of this case are that Selten purchased more than 300 tonnes of horse meat (horse carcasses) in 2011 and 2012 and subsequently sold and/or stocked in a cold store all of this horse meat, sometimes mixed with beef, as beef.

On 15 February 2013, as part of a criminal investigation, batches of Selten's meat, which had been found in business premises of other companies, were seized. This seizure was lifted with effect from 15 March 2013.

An alert through the so-called Rapid Alert System for Food and Feed (RASFF) from Britain dated February or March 2013 revealed that meat from one horse contaminated with the drug phenylbutazone had been supplied to Selten.

Subsequently, by decision of 15 March 2013, the Netherlands Food and Consumer Product Safety Authority (NVWA), on behalf of the Vice-Minister for Economic Affairs (the respondent), placed the meat stock previously seized in the context of the criminal investigation in official detention[290] pursuant to Article 54(2)(g) and (h) in conjunction with Article 19 Regulation 882/2004.[291]

To this end, the respondent first argued that it was found that the origin, traceability and identification of these batches of meat could not be established. According to the respondent, failure to trace food at all stages of production, processing

289 Vleesgroothandel Willy Selten B.V. and/or Wiljo Import en Export B.V.

290 The context here was consignments found on the business premises of Grolleman Coldstore B.V., Vriesoord B.V. and De Jong Coldstores B.V., as reported in the ruling of the preliminary relief judge of the Tribunal of 20 December 2013, ECLI:NL:CBB:2013:280.

291 Regulation (EG) No 882/2004 of the European Parliament and of the Council of 29 April 2004 on official controls performed to ensure the verification of compliance with feed and food law, animal health and animal welfare rules. This Regulation is no longer in force and was repealed on 13 December 2019 and replaced by Regulation (EU) 2017/625, discussed later in this chapter.

and distribution violates Article 18 Regulation 178/2002 (also known as the General Food Law (GFL)).[292]

Secondly, the respondent based its detention on the fact that sampling of the batches of meat declared as beef showed that horse meat had been used and that the origin of this horse meat was unclear. Since the provenance of the horse meat is unclear and the horse meat cannot be directly traced to (a) particular batch(es) of horse meat, these consignments must be regarded as unsafe within the meaning of Article 14 of Regulation No 178/2002.

Thirdly and finally, the respondent argued that a report from the English authorities dated February or March 2013 and a sampling dated 21 May 2012 showed that Selten processed horse meat from Britain containing the substance phenylbutazone. Residues of this drug are not allowed to be present in food because it poses a risk to human health. Since it cannot be shown in which meat the horse meat containing phenylbutazone was processed, it follows from Article 14(6) of Regulation 178/2002 that the entire batch of meat in which meat from this horse may have been processed must be deemed by operation of law to be unsafe food in accordance with Article 14 of Regulation 178/2002. This entire batch of meat may not be placed on the market. The respondent requested Selten to indicate within 60 days the final destination of the consignments of meat placed under official detention.

By decision of 22 April 2013, the respondent placed the batches of meat found on 15 April 2013 in official custody, applying Article 54(2)(g) and (h) in conjunction with Article 19 of Regulation 882/2004.[293] By decision of 25 July 2013, the respondent supplemented the list of Selten's batches of meat placed in official custody with other batches of meat.

By judgment of the District Court of Oost-Brabant dated 16 April 2013, Selten was declared bankrupt and the appellant, J.A. van der Meer, was appointed receiver. By letter, the appointed receiver (the appellant in these proceedings) requested the respondent to revoke the decisions of 15 March 2013, 22 April 2013 and 25 July 2013 (hereinafter, the 'sequestration decisions'). The respondent rejected this request by decision of 11 December 2013. By decision of 12 December 2014 (hereinafter 'Contested Decision I'), the respondent declared the appellant's objection to the detention decisions unfounded – in so far as they concerned the official detention of foodstuffs.

By the Contested Decision I, the respondent also dismissed as unfounded the appellant's objection to the decision of 11 December 2013, which rejected the appel-

292 Regulation (EC) No 178/2002 of the European Parliament and of the Council of 28 January 2002 laying down the general principles and requirements of food law, establishing the European Food Safety Authority and laying down procedures in matters of food safety.

293 The context here is consignments of meat found on the business premises of Vice Versa Koelvrieshuis B.V., as reported in the ruling of the preliminary relief judge of the Tribunal of 20 December 2013, ECLI:NL:CBB:2013:280.

lant's request to withdraw the official detention of the batches of meat and sale of the stock according to a plan of action proposed by the appellant, and also upheld this decision.

In a decision dated 20 December 2013,[294] the preliminary relief judge of the Tribunal preliminarily held that the respondent had soundly rejected the appellant's request for withdrawal of official detention.

On 30 January 2014, the appellant had a conversation with the deputy inspector general of the NVWA, during which the appellant requested the release of the frozen stock for human consumption. By e-mail message dated 6 February 2014, the appellant provided mass balances for 2011 and 2012, which he said were conclusive. According to the appellant, this makes it possible to establish that all outgoing meat comes from EC-approved slaughterhouses and is therefore EC-worthy.

The respondent rejected the appellant's request for the release of the frozen stock for human consumption by decision of 20 February 2014 (hereinafter 'Contested Decision II'). The respondent argued that no assurance could be obtained as to the correctness and completeness of a mass balance drawn up on the basis of Selten's records, in particular because of proven imperfections in the processing of purchases in its administration system (Reflex). Therefore, even the new mass balance cannot serve as a basis for the traceability of meat placed in official detention, so it cannot be released. If batches of meat are not traceable, they should be classified as unsafe, thus violating Articles 14 and 18 of Regulation 178/2002. Furthermore, the freezer stock may contain meat that contains meat from the horse in which phenylbutazone has been found. All such meat should therefore be deemed to be unsafe food within the meaning of Article 14 of Regulation 178/2002.

The appellant lodged an appeal against Contested decision I and objected to Contested Decision II. In the supplementary appeal, the appellant, referring to the provisions of Section 7:1a(1) of the General Administrative Law Act (Awb), requested the respondent to agree to direct appeal to the Tribunal. The respondent agreed and forwarded the objection to the Tribunal for consideration as an appeal.

29.3 Judgment of the Tribunal

The Tribunal upheld both of the appellant's appeals.

Importance of traceability

In this ruling, the Tribunal first looks at whether the respondent was right to use the power given in Article 54(2)(g), in conjunction with Article 19(1) of Regulation

294 ECLI:NL:CBB:2013:280.

882/2004, to place under official detention all the consignments of meat to which these decisions relate, in the detention decisions of all the consignments of meat concerned. Indeed, Article 54(1) Regulation 882/2004 – which has since been repealed, but the article has been converted into another article of another regulation, about which more below – stipulates that when the competent authority finds a case of non-compliance, it shall take measures to ensure that the operator rectifies the situation. In deciding on those measures, the competent authority shall take into account the nature of the non-compliance and the relevant antecedents of the operator. Paragraph 2(g) stipulates that, where appropriate, such measures may include those provided for in Article 19 of Regulation 882/2004. For example, Article 19(1) refers to the measure, after having heard the food business operator responsible for the consignment, to (a) instruct that the food be destroyed, subjected to special treatment, or sent back from the Community or (b) where the food has already been placed on the market, to monitor or, if necessary, order its recall, withdrawal or destruction before taking any of the above measures; or (c) to verify that foodstuffs do not have adverse effects on human and animal health, either directly or through the environment, during the period or pending the implementation of the measures referred to in points (a) and (b).

The Tribunal inferred from Contested Decision I that the respondent based detention for all detention decisions on three distinguishable grounds. These three grounds have already been mentioned above under the facts. The Tribunal discusses each of them separately in the present judgment. The Tribunal concludes that the respondent was justified in finding that Selten had failed to comply with the traceability requirements laid down in Article 18 in conjunction with Article 3(14) of Regulation 178/2002. In the opinion of the Tribunal, it follows from the criminal suspicion that Selten had falsified its records, namely with the trading and processing of horse meat, as well as from the investigation that took place at the time of the first impounding decision – 15 March 2013 – that at that time there were sufficient doubts about the reliability and accuracy of the records to assume that it was not possible to trace and follow the batches of meat in question through all stages of production, processing and distribution. This despite being required under Article 18 of Regulation 178/2002. The results of the criminal investigation confirm that those doubts were justified and, in the opinion of the Tribunal, provide a sufficient basis for the assumption that Selten's records were so deficient as to make it impossible to trace and track the said batches of meat. In what the appellant has submitted regarding the mass balances drawn up on its instructions, the Tribunal sees no basis for a different opinion. The Netherlands Food and Consumer Product Safety Authority (*NVWA*) has investigated this and ruled that no certainty can be obtained regarding the accuracy and completeness of a mass balance drawn up on the basis of Selten's records. The Tribunal therefore finds that the appellant's documents do not refute the finding of non-compliance by Selten with the traceability requirement.

But lack of traceability does not in itself mean that a (suspected) unsafe food product therefore exists

The second issue addressed in this ruling is Article 14 of Regulation 178/2002, which stipulates that food shall not be placed on the market because it is unsafe. Paragraph 6 of that article states that where an unsafe food is part of a batch or consignment of the same class or description, all the food in that batch or consignment shall be presumed to be unsafe, unless a detailed examination provides no evidence that the rest of the batch or consignment is unsafe.

The respondent has argued in this case that because the origin of the horse meat is unclear and the horse meat cannot be directly traced or traced to a particular batch of horse meat, the entire batch should be considered unsafe. The Tribunal does not entirely follow the respondent in this.

Unlike the respondent, the Tribunal is of the opinion that non-compliance with the obligation in Article 18 of Regulation 178/2002 that food must be traceable at all stages of production, processing and distribution does not, in itself, mean that there is therefore also a (suspicion of an) unsafe food as referred to in Article 14(2) of that Regulation. Regulation 178/2002 does not provide any textual or systematic basis for that conclusion. This means that the detention decided on by the defendant in the detention decisions is based on an incorrect legal basis. The respondent wrongly failed to recognise this in Contested Decision I and maintained that basis, with the result that that decision is inadequately reasoned in so far as it concerns the detention decisions and is contrary to Article 7:12(1) of the Awb.

And it is a bit of an oversimplification to say that all batches of meat are unsafe because of one infected horse

Moreover, the Tribunal does not follow the respondent in its position that Article 14(6) of Regulation 470/2009 entails that all the batches of meat at issue which originate from Selten must be placed in storage, due to the fact that the said sampling showed that Selten processed horse meat from Great Britain containing the substance phenylbutazone.

In view of Article 16(1) of Regulation 470/2009,[295] phenylbutazone may not be administered to food-producing animals within the Community. It is also not in dispute between the parties that residues of this drug should not be present in the horse meat processed by Selten from Great Britain. As appears from Article 1(1) of

295 Regulation (EC) No 470/2009 of the European Parliament and of the Council of 6 May 2009 laying down Community procedures for the establishment of residue limits of pharmacologically active substances in foodstuffs of animal origin, repealing Council Regulation (EEC) No 2377/90 and amending Directive 2001/82/EC of the European Parliament and of the Council and Regulation (EC) No 726/2004 of the European Parliament and of the Council.

Regulation 470/2009 and the preamble to this regulation (inter alia, points 2, 3 and 31), the aim of this regulation is to guarantee food safety and thus protect public health. In the opinion of the Tribunal, therefore, the conclusion is justified that (horse) meat containing phenylbutazone should be deemed an unsafe food within the meaning of Article 14(2) of Regulation 178/2002. In view of the public health interest to be protected by Regulation 470/2009, the Tribunal considers a broad interpretation of the phrase 'batch or consignment of the same class or description' referred to in Article 14(6) of Regulation 178/2002[296] to be justified.

In the opinion of the Tribunal, it is reasonable to use the argumentation that provides the basis for the conclusion that there is an unsafe foodstuff in the sense of Article 14(6) of Regulation 178/2002,[297] to delineate the same batch or consignment that constitutes an unsafe foodstuff in the sense of Article 14 of that Regulation. Without further substantiation, the Tribunal fails to see that in this case the mere fact that Selten processed the meat of one horse contaminated with phenylbutazone can be a sufficient basis for the respondent's assumption that the meat of this horse forms part of *all* the batches of meat placed in detention at the time of the detention decisions. Although it is not immediately clear in which batch(es) of meat this horse's meat ended up, the Tribunal is of the opinion that the respondent wrongfully failed to investigate this further. In this regard, the Tribunal also considers it significant that the appellant submitted a plan of action to the respondent, in which he stated that the horse meat contaminated with phenylbutazone was stored – and, as the Tribunal understands, also processed – by Selten in the period from 23 May 2012 to 23 June 2012, and that this meat was stored at two private limited companies. According to the appellant, it is therefore possible, taking into account a wide safety margin, to separate this meat from other batches of Selten meat in cold stores and freezer stores and not to place it on the market, on the basis of Best Before dates and arrival and departure documents. In doing so, the appellant specifically identified the item numbers, products and weights involved. In the contested decision, the respondent erred in not responding to this detailed proposal by the appellant or assessing it with reasons. The Tribunal reprimands the respondent for this. In preparing Contested Decision I, in so far as it concerns the dismissal of the objections to the detention decisions, the respondent did not gather the necessary knowledge around the relevant facts in connection with the application of Article 14(6) of Regulation 178/2002.[298] In that connection, that decision also lacks adequate reasoning. To that extent, Contested Decision I was taken in breach of Article 3:2 and Article 7:12(1) of the Awb.

In addition, the Tribunal also points out that, despite the fact that the respon-

296 In the ruling Article 14(6) of Regulation (EC)470/2009 is mentioned, but I believe they meant Article 14(6) of Regulation (EC) 178/2002.

297 Ibid. 272.

298 Ibid. 272.

dent, in the detention decisions, rightly placed the consignments of meat in question in official detention pursuant to Article 54(2)(g), in conjunction with Article 19(1) of Regulation 882/2004 – which is no longer in force, about which more below – that does not mean that the respondent was entitled to continue the detention of the consignments of meat in question at the time when it took Contested Decision I. On the contrary. It follows from Article 7:11 of the Awb that the reconsideration in objection is generally based on the factual situation and the law in force at the time the decision on the objection was made. In view of the text of Article 19(1) of Regulation 882/2004, in the opinion of the Tribunal, this provision does not provide the authority to merely take food that does not comply with feed or food law into official custody. Indeed, it follows imperatively that the competent authority taking such food into official detention, having heard the food business operator responsible for the consignment and having verified, where necessary, the existence of adverse effects as referred to in Article 19(1)(c) of that Regulation, shall take the measures referred to in that provision under (a). It is therefore not about taking a measure and then quietly waiting. This has been repeatedly misunderstood by the respondent.

- Contested Decision I was taken approximately 21, 20 and 17 months after the taking of the detention decisions, respectively. In view of this considerable lapse of time, and in light of Article 7:11 of the Awb and Article 19(1) of Regulation 882/2004, the respondent should, in the opinion of the Tribunal, have considered at the time when it took Contested Decision I whether the detention could still be continued at that time without, if necessary, having ascertained whether there were negative consequences as referred to in Article 19(1)(c) of Regulation 882/2004, (one of) the measures to be taken referred to in that provision under (a). The respondent erred in failing to do so. Contested Decision I therefore lacks sound reasoning and was taken in violation of Section 7:12(1) of the Awb in so far as it concerns the seizure decisions. In view of the foregoing, the Tribunal therefore finds that the appeal against Contested Decision I in so far as it relates to the seizure decisions is well founded and that to that extent this decision qualifies for annulment.
- Concerning Contested Decision I, the respondent dismissed as unfounded the objection to the primary decision of 11 December 2013 rejecting the appellant's request to revoke the detention decisions. To this end, the respondent considered that, because there was no certainty as to the accuracy of the records and because undocumented batches of meat were involved, it could not be ruled out that the meat contaminated with phenylbutazone had been mixed with beef and thus constituted foodstuffs whose safety had not been established. The Tribunal is of the opinion that, with this reasoning, the respondent again disregarded the scope of Article 19(1) of Regulation 882/2004 – as referred to above. This means that Contested Decision I, in so far as it relates to the decision of 11 December 2013, also lacks proper reasoning and was taken in violation of Article 7:12(1) of

the Awb. The appeal is therefore well founded to that extent and Contested Decision I is eligible for annulment to that extent.
- Concerning Contested decision II, the respondent rejected the appellant's request for the release of the freezer stock for human consumption. In the view of the Tribunal, the respondent's reasoning for Contested Decision II here too disregarded the aforementioned scope of Article 19(1) of Regulation 882/2004. Contested Decision II was therefore also not properly reasoned and was therefore taken in violation of Article 3:46 of the Awb. The appeal against Contested Decision II is therefore well founded and that decision should be annulled.

The Tribunal therefore saw fit to order the respondent to make a fresh decision on the appellant's objections to the detention decisions and the decision of 11 December 2013, as well as the appellant's application for the release of the freezing stock for human consumption, within 12 weeks of the sending of the ruling. This with due regard to what the Tribunal has considered in this ruling. In so doing, the Tribunal points out that Article 19(1) of Regulation 882/2004 requires the food operator to be heard before such a measure is taken, but that this does not mean that the operator must be given a deciding vote in it. The respondent had given the appellant the opportunity several times – unsuccessfully – to indicate what final destination it believed should be given to the batches of meat. In the opinion of the Tribunal, it was therefore not necessary for the respondent to hear the appellant again on this matter pursuant to Article 19(1) of Regulation 882/2004.

29.4 Commentary

Regulation (EC) 882/2004

I have already mentioned several times that Regulation (EC) 882/2004 is no longer in force and was repealed with effect from 13 December 2019 by Regulation 2017/625.[299] However, that does not diminish the relevance of this ruling. Indeed, Article 54(2) of Regulation 882/2004, referred to in this ruling, has been converted

299 Regulation (EU) 2017/625 of the European Parliament and of the Council of 15 March 2017 on official controls and other official activities performed to ensure the application of food and feed law, rules on animal health and welfare, plant health and plant protection products, amending Regulations (EC) No 999/2001, (EC) No 396/2005, (EC) No 1069/2009, (EC) No 1107/2009, (EU) No 1151/2012, (EU) No 652/2014, (EU) 2016/429 and (EU) 2016/2031 of the European Parliament and of the Council, Council Regulations (EC) No 1/2005 and (EC) No 1099/2009 and Council Directives 98/58/EC, 1999/74/EC, 2007/43/EC, 2008/119/EC and 2008/120/EC, and repealing Regulations (EC) No 854/2004 and (EC) No 882/2004 of the European Parliament and of the Council, Council Directives 89/608/EEC, 89/662/EEC, 90/425/EEC, 91/496/EEC, 96/23/EC, 96/93/EC and 97/78/EC and Council Decision 92/438/EEC.

into Article 138(2) of Regulation 2017/625 and Article 19(1) of Regulation 882/2004, also referred to repeatedly in the judgment, into Article 66 of Regulation (EU) 2017/625.[300]

Traceability

In addition, this ruling is relevant for a number of reasons. Notably, it underlines the importance of good record-keeping and traceability. As a food business operator, having your records in order can prevent a lot of damage. If you can prove that certain batches are safe, it can significantly reduce the size of the batch subject to measures. This is in view of Article 14(6) of Regulation 178/2002. This is not only an issue with the measure of seizure, but also, for example, with recalls and such like.

Companies must be able to track from whom they have received products and to whom they have supplied products. This requirement is based on the 'one step back, one step forward' approach. For food business operators, this means:

- they must have a system that allows them to quickly and accurately identify the direct supplier(s) and direct buyer(s) of their products;
- they should be able to establish a link between 'supplier – product' (which products have been supplied by which suppliers); and
- they should be able to establish a link between 'buyer – product' (which products were supplied to which buyers).[301]

According to an interpretative document of Regulation 178/2002,[302] the name, address of the supplier and the nature of the products supplied by him and the name, address of the customer and the nature of the products supplied to him and date of delivery must be able to be made available to the competent authority quickly

300 Incidentally, the ruling ignores what happens when a competent authority makes use of the authorisation that Article 54 Regulation 882/2004 offered and that Regulation 2017/625 now offers. The rules on this matter are actually the same. Thus, in its decision about the measures to be taken in the case of non-compliance, the competent authority must take into account the nature of the non-compliance, as well as any previous cases of non-compliance by the operator. In addition, the competent authority must provide the operator or their representative with the following: (a) a written notification of its decision regarding the action or measure to be taken and the reasons for that decision, and (b) information on the possibility to appeal that decision and the applicable procedures and deadlines for setting up that appeal. Furthermore, both the now repealed Regulation and Regulation 2017/625 form the basis to charge or let off the responsible operators regarding all costs arising from former Article 54 Regulation 882/2004 and now Article 138 Regulation 2017/625.

301 See also in this context recitals 28 and 29 and Article 18 of Regulation (EC) 178/2002.

302 Standing Committee on the Food Chain and Animal Health (SCoFCAH) now Standing Committee on Plants, Animals, Food and Feed (SCoPAFF) Guidelines for the implementation of Articles 11, 12, 14, 17, 18, 19 and 20 of Regulation (EC) No 178/2002 on General Food Law, 26 January 2010, https://food.ec.europa.eu/system/files/2016-10/gfl_req_guidance_rev_8_en.pdf/.

and accurately in all cases. The Netherlands Food and Consumer Product Safety Authority (*NVWA*) maintains a timeframe of four hours for this purpose.[303] Moreover, without prejudice to sector-specific rules, the regulation does not explicitly require operators to establish a link (so-called internal traceability) between incoming and outgoing products. Thus, the regulation does not require an operator to keep records of which raw materials have been incorporated into which finished product. Needless to say, if an operator does not have this in place or does not have it in sufficient order, any recall or other action could be far-reaching. So this is advisable.[304]

Party

In addition, this ruling shows, that it can indeed be an advantage if you, as a food business operator, come up with your own plan of action and proposals. For example, make a proposal for a study to demonstrate that, with regard to some of the food affected by the measure, there is no evidence that it is unsafe. The national authority cannot simply ignore a proposal/plan; it will have to respond to it. Like the plan of action presented in this case.

Moreover, in earlier summary proceedings between the receiver of Selten and the respondent in 2013 – to be discussed below – the preliminary relief judge of the Tribunal ruled that there is no reason to order the Vice-Minister for Economic Affairs to determine what additional or alternative measures and/or additional investigations are necessary to proceed with the lifting of measures – in this case, release of the batches of meat.[305] It is therefore the operator's own responsibility. Experience shows that removing a suspicion of, or classification as, an unsafe product is not always an easy task. In this context, see also, for example, the (overview) article published at the end of 2019 on countermeasures.[306] It contains several tips.

Ex nunc

In addition, it also clearly follows from this ruling that there is no infinite right of retention for the Netherlands Food and Consumer Product Safety Authority (*NVWA*).

303 *Tweede Kamer, vergaderjaar* 2012/2014, 2246, Aanhangsel van de Handelingen. See also Policy ruling of the Dutch Minister of Public Health, Welfare and Sport and the Minister of Agriculture, Nature and Food Quality of 21 June 2022, 3373903-1030201-VGP, concerning further elaboration of Articles 18, 19 and 20 of Regulation (EC) 178/2002 (*Beleidsregel nadere invulling Verordening (EG) 178/2002*) Staatscourant 2022, 16878. In Belgium a term of 48 hours is applied. See https://favv-afsca.be/nl/themas/controles/interne-controles-door-bedrijven/meldingsplicht/meldingsplicht-en-meldingslimieten.

304 This requirement is therefore also often added to private standards.

305 Recital 12. Prelim. relief judge, Trade and Industry Appeals Tribunal 20 December 2013, ECLI:NL:CBB:2013:280.

306 Mutsaers, 'Contra-expertise, een hoop obstakels', published in Waar & Wet 2019/82.

Thus, on the one hand, this ruling reaffirms that even in cases like this, the starting point for the reconsideration of decisions in objection is that this reconsideration must take place ex nunc: i.e. assessed according to the facts and circumstances and the applicable law at the time of the decision on the objection. After all, it is not only considered whether the interested party's complaint is well founded, but also whether the contested decision can be upheld. Thus, the administrative body does not review but decides again, as if it were a new primary decision. Incidentally, there are exceptions to the principle of ex-nunc review. The review of sanction decisions is, in principle, assessed ex tunc. The tenability of, for example, an order for periodic penalty payments must be assessed on the basis of the facts and circumstances known at the time of its imposition.[307] If the review finds that the breach has since ended and no penalty payments have yet been forfeited, it must be concluded that the authority to take enforcement action has lapsed. In such a case, a reconsideration ex nunc is reasonable.[308] The respondent in this case made light of the situation. She should have been expected to do more.

It is also noted that use of the articles in Regulation 882/2004 (now Regulation 625/2017) also entails obligations. Such as hearing the food business operator.

Inadequate traceability is not in itself unsafe

What is also striking is the Tribunal's opinion that although the traceability was not in order, that does not mean per se that there is a (suspicion of an) unsafe food product. However, it is relevant to read that opinion carefully. Indeed, the wording 'per se' is very important. So you need something more, the Tribunal is saying. In my opinion, this verdict of the Tribunal is fully justified. After all, it is also possible that traceability is not in order but that there is nothing wrong with the product at all and that it is a safe food. Just as it is possible that traceability is in order – one knows where which product comes from and where that product was delivered – but that the product is still unsafe. Traceability is only a tool. For example, the Guidelines for the implementation of Articles 11, 12, 14, 17, 18, 19 and 20 of Regulation 178/2002, Conclusions of the Standing Committee on the Food Chain and Animal Health of 2010,[309] also states that traceability by itself does not ensure that a product is safe. It only helps ensure that food safety problems can be managed. Article 14 of Regulation 178/2002 then states when a product is unsafe. Food is deemed to be unsafe if it is considered: (a) injurious to health; or (b) unfit for human consumption. Whereby

---

307 See for example ABRvS 28 December 1999, AB 2000, 107.

308 See for example ABRvS 12 February 1999, AB 1999, 321.

309 Standing Committee on the Food Chain and Animal Health (SCoFCAH) now Standing Committee on Plants, Animals, Food and Feed (SCoPAFF) Guidelines for the implementation of Articles 11, 12, 14, 17, 18, 19 and 20 of Regulation (EC) No 178/2002 on General Food Law, 26 January 2010 https://food.ec.europa.eu/system/files/2016-10/gfl_req_guidance_rev_8_en.pdf/.

Article 14 indicates when this is the case, as well as (i) what to do if an unsafe food is part of a batch or consignment of the same class or description and (ii) also if there are other reasons to suspect that a food is unsafe (see, for example, Article 14(8) of Regulation 178/2002).

It seems that in previous preliminary injunction proceedings, the Tribunal has tended to hold that lack of traceability triggers the presumption of unsafe food. However, those cases involved additional facts leading to that presumption of unsafety. Moreover, both cases were preliminary rulings and, in preliminary injunction proceedings, one should bear in mind that a judge is reluctant to lift measures or release products, especially if they may be unsafe products.

The preliminary injunction proceedings

On 20 December 2013,[310] the preliminary relief judge of the Tribunal ruled on the applications for preliminary relief in the cases between J.A. van der Meer, receiver in the bankruptcy of Vleesgroothandel Willy Selten B.V. and Wiljo Import en Export B.V. (Selten), the Vice-Minister for Economic Affairs, the respondent. Selten's receiver had objected to the detention decisions and requested that they be revoked. The respondent had indicated that it would defer a decision on possible revocation for investigation purposes and subsequently that the reason for not revoking the official detention decisions was also the basis for suspending decision-making. Selten's receiver then applied to the preliminary relief court for injunctive relief. Pursuant to the provisions of Article 8:81 Awb, if an appeal has been lodged with the Tribunal against a decision or, prior to a possible appeal, an objection has been lodged, a preliminary injunction may be granted at the request of the court if the urgency of the matter, in view of the interests involved, so requires. However, the preliminary relief judge rejected the request. The respondent indicated that the seizure cannot be revoked and in these proceedings put forward the same arguments as mentioned above (no conclusive movement of goods, origin/traceability cannot be demonstrated, phenylbutazone was found in a batch of horse meat, uncertainty of mass balance). The preliminary relief judge ruled at the time that Selten's receiver had failed to successfully challenge the respondent's position. The Tribunal ruled that the respondent was able to take the position that it could not be established with certainty that any batch of meat came entirely from EC-approved slaughterhouses. The safety of the entire stock of meat is therefore not established and the respondent was therefore entitled to declare these batches of meat unsafe within the meaning of Article 14 of Regulation 178/2002. The respondent was able to allow the interest of food safety to prevail in its assessment in this case, as no risk should be taken with it. A risk to food safety is clearly present in this case. The respondent

310 Trade and Industry Appeals Tribunal 20 December 2013, ECLI:NL:CBB:2013:280.

was also entitled to take the view that the seizure of the batches of meat examined at random – which were found to come from EU-approved slaughterhouses – cannot be revoked either. Besides the fact that this was a 'paper' sample and the corresponding batches of meat were not linked to the batches located in the cold store, it could not be ruled out that these batches of meat had also been mixed with undocumented, i.e. unsafe, meat, the preliminary relief judge said. Incidentally, it was virtually certain that horse meat contaminated with phenylbutazone was present in either batch. The requests were therefore rejected.

Another horse meat case concerns the ruling of the preliminary relief judge of the Tribunal of 17 February 2014[311] on the request for preliminary relief in the case between Van Hattem Vlees B.V., as appellant, and the Vice-Minister for Economic Affairs, as respondent. In that case the respondent had imposed an administrative order on the appellant for infringement of Articles 18 and 19 of Regulation 178/2002, in conjunction with Article 6.2(1) of the Animals Act (WD) and Article 2.4(1)(b) of the Animal Products Regulation. The order required the appellant to initiate procedures by a certain date to withdraw from the market all batches of meat it produced and processed from 1 January 2012 until 23 January 2014. The Secretary of State based the order on Article 54(2)(c) of Regulation 882/2004 (now Article 138(2) of Regulation 625/2017) and Article 8.5 of the WD. The Vice-Minister took the view that the appellant had failed to demonstrate traceability of all batches of meat produced by it between 1 January 2012 and 23 January 2014, thereby failing to ensure that the batches of meat in question complied with food safety requirements. The dispute therefore revolved primarily around whether the appellant had indeed failed to demonstrate traceability. In the judge's preliminary opinion, this question should be answered in the affirmative. The preliminary relief judge holds first of all that – in view of Articles 17 and 18 of Regulation 178/2002 – the responsibility for traceability lies with the appellant as operator of the slaughterhouse/meat processing plant. It is therefore incumbent on the appellant, if doubts arise in this regard, to provide the Vice-Minister, by means of its records or otherwise, with such insight into the origin and destination of the meat produced by it as to demonstrate traceability within the meaning of the regulation. The appellant was given the opportunity to do so by the Vice-Minister but, in the latter's opinion, failed to do so. The hearing was even adjourned and the appellant submitted further documents, but even this did not result in clarification of the difference between the horses supplied and those presented for inspection. The Tribunal's conclusion was that the Vice-Minister's assessment regarding the incomplete traceability of the batches of meat produced by the appellant was not considered incorrect. As the origin of the meat was insufficiently clear and there were indications that more meat had left the business than had entered, there was a risk that batches of undocumented,

311 Trade and Industry Appeals Tribunal of 17 January 2014, ECLI:NL:CBB:2014:58.

possibly even untested, meat had been processed. Given this fact, in the preliminary opinion of the Tribunal, the Vice-Minister could reasonably impose the order under administrative coercion. It is a very drastic but not disproportionate measure in the given situation, the Tribunal said, and Van Hattem Vlees B.V.'s request was therefore rejected.

Even after this 2017 ruling, there have been several more rulings dealing with traceability. Take, for example, the Tribunal's ruling of 20 November 2018.[312] Again, this was a case involving an official impoundment of batches of meat because they were insufficiently traceable. The Netherlands Food and Consumer Product Safety Authority (*NVWA*) inspectors had found, among other things, that the stock of meat was often not or only partially labelled with batch numbers. The stock list was also found to contain defects. The inspectors also found that a lot of meat had been removed from the boxes in which it had been delivered by the producer. The blocks were placed on pallets and shrink-wrapped. A pallet tag was then affixed with an EC number from an approved EC company, which is not allowed, the official report said. A Best Before date was also applied. Underlying documentation was not present, making it impossible to ascertain whether the tracing/origin of the meat could still be determined. By the primary decision, which was upheld on appeal, the Minister of Health, Welfare and Sport (respondent), among other things, extended the official detention of the batches of meat. Following the court, the Tribunal finds that the respondent was right to decide to extend the detention of the meat. However, the ruling shows that merely keeping a list of which products you have received from which suppliers and which products you have delivered to customers can be insufficient. On the contrary, the case revealed that it is not sensible to remove and destroy original packaging and identifying marks if you want to prove that traceability has been assured at all stages.

2019

The Selten issue continued even after the ruling on 30 January 2017, and this led to the Tribunal's ruling of 12 February 2019.[313] In April 2017, the respondent decided to lift the official detentions of 15 March 2013 and 22 April 2013. By this decision, the respondent rejected the request for the release of the frozen stock placed in detention for human consumption or for sale to an Asian country. The appellant's objection was rejected, following which it lodged an appeal. Primarily, the respondent took the position that the appellant's objections were inadmissible due to the lack of a legal interest in bringing proceedings, pointing to the decision of 16 April 2015, which granted permission to use the consignments of meat placed in storage as category 2 material for animal feed, which decision had since acquired formal legal

312 Trade and Industry Appeals Tribunal 20 November 2018, ECLI:NL:CBB:2018:571.

313 Trade and Industry Appeals Tribunal 12 February 2019, ECLI:NL:CBB:2019:61.

force.[314] At the time of the judgment of 30 January 2017, the Tribunal was not aware of this decision of 16 April 2015, because it had not been informed about it by the parties. Had the Tribunal been aware of this decision, it would not have been necessary, in the aforementioned judgment, to order the defendant to take a new decision regarding the duration of the official detention and the request for the release of the consignments of meat placed in detention. Indeed, the decision of 16 April 2015 already concerns such a decision. Therefore, the defects found in the judgment of 30 January 2017 in the decisions challenged in those proceedings had already been remedied by the decision of 16 April 2015. Furthermore, it is established that after that decision of 16 April 2015, part of the freezer stock was sold as category 2 material. To that extent, therefore, the appellant is no longer an interested party in the consignments of meat at issue, so that the procedural interest of the appellant in the proceedings concerning those consignments has lapsed. The remaining part of the consignments of meat taken into custody has since been destroyed, so that the appellant also no longer has a procedural interest in adjudicating those remaining consignments. To the extent that the appellant referred to the damage suffered by Selten or the estate as a result of the primary decisions, the Tribunal considers that the decision of 16 April 2015 remedied the defects found in the contested decisions of 20 February 2014 and 12 December 2014. Whether the 16 April 2015 decision was lawful or timely could no longer be assessed, as its formal legal force precludes this. Selten's receiver's appeal was therefore dismissed as unfounded.

This ruling in the horse meat fraud case has mainly contributed to a better understanding of the food law role of traceability. On the one hand, it has been established that defects in or lack of traceability do not per se make a product unsafe or subject to the recall obligation. On the other hand, it can no longer be maintained that no duty of internal traceability at all follows from Article 18 GFL.

Barbara Mutsaers
Lawyer and Industry Leader Food & Life Sciences at AKD law firm, Eindhoven and member of the editorial board of the magazine Waar & Wet (Sdu Publishers)

314 Formal legal force is used to show that there can no longer be a legal procedure against a decision with the administrative judge, therefore rendering it legitimate (i.e. untouchable by legal means).

CHAPTER 30

Heksenkaas (Witch's cheese) – Copyright on the taste of a food product

Annotated by Bram Duivenvoorde

Court of Appeal Arnhem-Leeuwarden 23 May 2017, ECLI:NL:GHARL:2017:6697 (*Levola v Smilde*)
Court of Justice EU 13 November 2018, ECLI:EU:C:2018:899 (*Levola v Smilde*)

Copyright on the taste of a food product? Copyrighted 'work'. Accurate and objective form of expression.

Main considerations:

Court of Appeal:

> 6.3 In its judgment of 16 June 2006 (ECLI:NL:HR:2006:AU8940 Lancôme), the Supreme Court considered the following with regard to the question of whether a fragrance could qualify as a copyright work:
> 3.3.2 In assessing this objection, the starting point is that it does not dispute that an odour – which will hereinafter in each case include an odour combination – may be eligible for copyright protection. That starting point is correct. The description of what is to be understood as a 'work' within the meaning of Article 10 of the Dutch Copyright Act (Aw), in addition to the non-exhaustive enumeration of types of work, is general and does not prevent the inclusion of a smell. As a result, for the question whether a smell is eligible for copyright protection, the decisive factors are whether it is a product susceptible to human perception and whether it has its own original character and bears the creator's personal stamp. It is true that the notion of 'work' of the Copyright Act finds its limits where original character relates only to that which is necessary to obtain a technical effect, but since in the case of a perfume there is no question of a mere technical effect, even the latter requirement does not preclude the granting of copyright protection to the smell of a perfume. The fact that the characteristics of the human sense of smell impose limits on the ability to distinguish odours and that the extent to which individual odours can be distinguished varies from person to person do not alter the foregoing, nor does the fact that the specific nature of odours means that not all of the provisions and limitations in the Copyright Act can apply to them in full, having regard, for example, to the use of a perfume, which cannot be denied to the normal consumer, and which by its nature necessarily entails diffusion of the fragrance.

 | HTTPS://DOI.ORG/10.1163/9789004751699-032

An odour cannot, even for the purposes of copyright, be identified with the substance or substances that produce the odour. That substance or these substances act as – not necessarily exclusive – embodiments of the work that is the fragrance, and they also help to ensure that the fragrance should not – because it is too volatile or unstable – be excluded from copyright protection. In the case of a perfume, therefore, the smell should be distinguished from the mixture of substances from which the perfume is composed and from which, when the mixture is exposed to the open air, the gaseous fragrance substances are released that can be perceived with the olfactory organ.'

[…]

7.2 In these proceedings Levola claimed against Smilde, inter alia, an injunction against the reproduction and distribution of the product *Witte Wievenkaas*. Levola argued that the *Witte Wievenkaas* product is a reproduction of (the taste of) its *Heks'nkaas*, which tatse, according to Levola, is protected by copyright.

By reproducing this taste without permission and distributing it (by selling the '*Witte Wievenkaas*'), Smilde infringed Levola's copyright. Levola brought (party) expert reports into the proceedings from which, in its view, it follows that the (overall impression of) the taste of *Heks'nkaas* is to the same as Smilde's *Witte Wievenkaas*. According to Levola, 'taste' lends itself to copyright protection, defining taste, in summary, as that which is perceived by the senses when consuming a food product, consisting of a combination of basic tastes and the 'mouthfeel' caused by, inter alia, the viscosity and consistency of the product.

[…]

7.8 In view of the above, the Court of Appeal is of the opinion that there is reasonable doubt as to whether and to what extent EU law allows for copyright protection of the flavour of a food product. The answer to this question has (possible) consequences for the obligation of the party invoking such protection. As explained above, in the opinion of the Court of Appeal, a decision of the Court of Justice of the European Union (CJEU) is necessary in this case in order to be able to rule in this matter. The Court of Appeal therefore sees reason to submit the preliminary questions mentioned in paragraph 8 above to the CJEU.

CJEU:

41 […] first, the authorities responsible for ensuring that the exclusive rights inherent in copyright are protected must be able to identify, clearly and precisely, the subject matter so protected. The same is true for individuals, in particular economic operators, who must be able to identify, clearly and precisely, what is the subject matter of protection which third parties, especially competitors, enjoy. Secondly, the need to ensure that there is no element of subjectivity – given that it is detrimental to legal certainty – in the process of identifying the protected subject matter means that the latter must be capable of being expressed in a precise and objective manner.

42 The taste of a food product cannot, however, be pinned down with precision and

objectivity. Unlike, for example, a literary, pictorial, cinematographic or musical work, which is a precise and objective form of expression, the taste of a food product will be identified essentially on the basis of taste sensations and experiences, which are subjective and variable since they depend, inter alia, on factors particular to the person tasting the product concerned, such as age, food preferences and consumption habits, as well as on the environment or context in which the product is consumed.

43 Moreover, it is not possible in the current state of scientific development to achieve by technical means a precise and objective identification of the taste of a food product which enables it to be distinguished from the taste of other products of the same kind.

44 It must therefore be concluded, on the basis of all of the foregoing considerations, that the taste of a food product cannot be classified as a 'work' within the meaning of Directive 2001/29.

30.1 Introduction

Copyright protects all kinds of creations, such as music, literature and paintings. Creations with a practical application can also count on copyright protection. Think, for example, of the design of a car, a shoe or a building. But can copyright protection also apply to the taste of a food product? That question came up in a case between two Dutch companies and was eventually challenged all the way to the European Court of Justice.

30.2 The case

Heksenkaas (also known as: *Heks'nkaas*) is a dip containing cream cheese and fresh herbs. The product was created in 2007 by a Dutch retailer, who concluded a deal with Levola in 2011. The crux of the deal: Levola received the intellectual property rights in exchange for a share of the turnover.

FIGURE 30.1 A tub of *Heksenkaas*.

The product was a hit. And so in 2014, competitor Smilde came out with a similar product: *Witte Wievenkaas*. Levola started proceedings: it felt Smilde was infringing copyright as to the taste of *Heksenkaas*. This was new: never before had copyright on the taste of a product been litigated in the Netherlands.

Trademark proceedings also followed: is the name *Witte Wievenkaas* too similar to the trademark *HEKSENKAAS*?[315] Another interesting issue, but in this case note I will limit myself to the issue of copyright on the taste of the food product.

30.3 Background: copyright on taste

Copyright law is largely harmonised, via, among other instruments, the Copyright Directive (2001/29/EC) and the case-law of the European Court of Justice. The question of whether something is a copyrightable 'work' is also harmonised in European case-law, which shows, among other things, that a copyrightable 'work' must be original, in the sense that it is the creator's own intellectual creation.[316] Exactly what other criteria a 'work' must satisfy was not yet clear on the basis of European law. Similarly, it was not yet clear whether the taste of a foodstuff could be a copyrightable 'work'.

Also in the Netherlands, as mentioned, copyright on taste had not been litigated before. However, the Supreme Court had previously ruled that copyright on the smell of a perfume was possible.[317] But that ruling was seriously questioned: was the Supreme Court's ruling in line with European law? Meanwhile, the French supreme court had also ruled on copyright on smell and had decisively rejected it.[318]

A quick look at Wikipedia also shows that taste is not easy to classify.[319] Primary tastes (sour, salty, sweet, bitter, umami and – controversially – fatty) may still be relatively easy to measure, but the entire taste experience also depends on, for instance, smell and texture. And then there are the special impressions of, for example, pepper ('hot' taste) and menthol ('cold' taste). On top of that, the taste experience often differs from person to person.[320] And the same foods can also produce differences. For example, the taste of a food can vary according to how long and how it is stored. In short: the tasting experience of a food is not easy to fathom.

---

315 For the ruling of the Supreme Court on this issue, see: HR 13 July 2018, ECLI:NL:HR:2018:1215, ECLI:NL:HR:2018:1217 and ECLI:NL:HR:2018:1214 (*Heksenkaas/Witte Wievenkaas*).

316 See among others CJEU 4 October 2011, ECLI:EU:C:2011:631 (*Premier League*).

317 Supreme Court Netherlands 16 June 2006, ECLI:NL:HR:2006:AU8940 (*Lancôme*).

318 Supreme Court France 10 December 2013, ECLI:FR:CCASS:2013:CO01205.

319 See https://en.wikipedia.org/wiki/Taste.

320 See also M. van Tiel, Smaakbescherming in optima forma: over de juridische bescherming van smaak, IER 2018/23.

30.4 The preliminary questions

The Dutch court was therefore not facing an easy task. The District Court of Gelderland initially rejected Levola's claims. On appeal, the Arnhem-Leeuwarden Court of Appeal decided to refer preliminary questions to the European Court of Justice. The questions in brief:

1. Does EU law oppose the protection of the taste of a foodstuff through copyright? For example, is copyright protection possible only in respect of creations that can be perceived by sight and/or hearing? And does the instability of a foodstuff and/or the subjective nature of the tasting experience make copyright protection impossible?
2. If copyright protection of taste of a foodstuff is possible: what are the requirements for such protection? What is protected: the taste or (also) the recipe? And how should courts determine and compare tastes?

Interesting questions! So, what does the EU Court decide?

30.5 Court of Justice: no copyright on taste

The European Court states first that a copyrightable work must be a form of expression of the object of copyright protection. In other words, it must be a concrete form of expression, not an abstract idea or concept. This is no surprise.

More importantly, the European Court requires that the concrete form of expression must be identifiable with sufficient precision and objectivity.[321] This means, first and foremost, that the court must be able to determine in a clear and precise manner what is protected through copyright. This should also be the case for competitors. Indeed, they must be able to determine what they are and are not allowed to do. Secondly, the identification of the work should not be subjective. After all, through a subjective determination of the form of expression, it is not clear to the outside world what exactly is protected and when there is infringement. This would be bad for legal certainty.[322]

Can the taste of a food satisfy this requirement of accurate and objective identification? The European Court ruled as follows:

> 42 The taste of a food product cannot [...] be pinned down with precision and objectivity. Unlike, for example, a literary, pictorial, cinematographic or musical work, which is a precise and objective form of expression, the taste of a food product will be identified essentially on the basis of taste sensations and experiences, which are sub-

---

321 CJEU 13 November 2018, ECLI:EU:C:2018:899 (*Levola v Smilde*), rec. 40–41.
322 Ibid.

> jective and variable since they depend, inter alia, on factors particular to the person tasting the product concerned, such as age, food preferences and consumption habits, as well as on the environment or context in which the product is consumed.
> 43 Moreover, it is not possible in the current state of scientific development to achieve by technical means a precise and objective identification of the taste of a food product which enables it to be distinguished from the taste of other products of the same kind.
> 44 It must therefore be concluded, on the basis of all of the foregoing considerations, that the taste of a food product cannot be classified as a 'work' within the meaning of Directive 2001/29.

According to the European Court, the taste of a food can therefore not be accurately and objectively determined. Taste perception is too subjective and variable for that. Furthermore, according to the Court, with the current technical means it is not even possible to accurately and objectively determine and distinguish the taste of a foodstuff from the taste of other products. Hence, no copyright protection for the taste of *Heksenkaas* nor for the taste of other foodstuffs. This also means that the European Court did not have to address the second question of the Arnhem-Leeuwarden Court of Appeal on, inter alia, the requirements for the protection of taste and the determination and comparison of tastes.

30.6 Disappointment and relief

The ruling is undoubtedly a disappointment for companies that invest a lot of time and money in developing new food products. At the same time, the ruling may be a relief for other professionals in the food sector. Developers, retailers and restaurant owners can continue to experiment to their heart's content. In doing so, they may draw inspiration from others without worrying that their products will be too similar to a competitor's. After all, a taste cannot be monopolised through copyright.

Bram Duivenvoorde
Associate professor at Utrecht University

CHAPTER 31

MPA – Liability in food safety crisis

Annotated by Bernd van der Meulen

The Hague Court of Appeal 19 December 2017 ECLI:NL:GHDHA:2017:3936 (*interlocutory judgment*);
The Hague Court of Appeal 25 September 2018 ECLI:NL:GHDHA:2018:2468 (*final judgment*);
Advocate General 17 April 2020 ECLI:NL:PHR:2020:393 (to overturn)
Dutch Supreme Court 16 October 2020 ECLI:NL:HR:2020:1628 (*affirmed*) (*Rined v Wyeth*)
North Holland District Court 23 April 2014 ECLI:NL:RBNHO:2014:3627 (*first instance*)

The company that disposes of pharmaceutical waste without ensuring that it is properly processed is at fault for this waste entering the food chain. The company that markets residual products as animal feed without ensuring its origin is also at fault.

Main considerations of the Hague Court of Appeal in the interlocutory judgment:

> Wyeth acted unlawfully by transferring the waste sugar water (or having it transferred) without proper investigation to Bioland, which did not have the necessary permits to process the waste. If Wyeth had not offered its MPA-containing sugar water to Bioland, the damage would not have occurred. This sufficiently establishes a causal link. For the time being, it seems plausible that the damage would also not have occurred if Wyeth had complied with its obligations under the EU Waste Shipment Regulation (Regulation (EEC) 259/93 on the supervision and control of shipments of waste within, into and out of the European Community).[323] Without sufficient investigation, Zeeland Voeders – in the persons of [C] and [director appellant] – used a product, of which it (allegedly) knew no more than that it came from Bioland's lemonade industry, with which it had not previously done business, for processing into animal feed. [Director appellant] should have realised [...] that this incurred a risk. It is plausible that, had Zeeland Voeders exercised due care, it would have discovered that this was not waste from the lemonade industry, after which it would not have used the product as a raw material for animal feed. Even then, the injury would not have occurred. The knowledge of [director appellant] should also be imputed to [appellant] and Oostburg Varkens. This leads to (a considerable degree of) own fault.

323 Now replaced by Regulation (EC) No 1013/2006 of the European Parliament and of the Council of 14 June 2006 on shipments of waste (footnote added).

 | HTTPS://DOI.ORG/10.1163/9789004751699-033

For now, however, there is no reason to assume 100% own fault, as the district court did in the contested judgment.

The regulations which Zeeland Voeders failed to comply with are primarily intended to protect the environment and human and animal health and not to indemnify Wyeth against claims for compensation for damage arising from having its sugar water waste, which qualifies as hazardous within the meaning of 91/689/EEC [324] and as amber list within the meaning of the EU Waste Shipment Regulation, transferred to a party (Bioland) which does not have the permits to process that waste, in breach of the provisions of that EU Waste Shipment Regulation. Added to this, Wyeth did not confine itself to the sugar water waste, which could not actually cause any harm because it contained no substances other than water, sugar and dye, but also disposed of the MPA-containing sugar water in the same way. This charge must be borne by Wyeth, because it knew or should have known better than anyone else the exact composition of the waste and realised the risks involved in its improper handling. On the other hand, Zeeland Voeders was at least negligent and gullible. Judgment is reserved on the question of what consequences this should have, pursuant to Article 6:101(1) Dutch Civil Code (BW), for the compensation obligation – in other words, the apportionment of blame.

31.1 Introduction

A large-scale reform of food law was undertaken after the turn of the millennium. The immediate cause was a series of food safety crises. The most important of these, but certainly not the only one, was the BSE crisis because it led to an inquiry in the European Parliament. When the inquiry committee's report was considered, a vote of no confidence was tabled against the European Commission. This could have become the first time that the European Parliament sent the Commission packing. The only reason this did not become the first time the Commission was dispatched was that the Commission donned the hairshirt and promised a complete reform of the European food safety system. This reform took the form of a Green Paper, *The general principles of food law in the European Union*, and the White Paper on Food Safety. However, the realisation that the food safety system was structurally deficient was not only triggered by the BSE crisis. The sense of urgency was reinforced by a succession of food safety scandals and animal health problems. These included the Belgian dioxin crisis, in which waste transformer oil was mixed with chicken feed, as well as the MPA affair. The MPA affair is the focus of our attention in this book because litigation continues over a quarter of a Century after the facts about compensation for the resulting damages. The role of liability law in the food

324 Replaced by Directive 2008/98/EC of the European Parliament and of the Council of 19 November 2008 on waste (footnote added).

sector largely escapes notice, presumably because many issues are resolved through insurance and because damages suffered by consumers are so widespread that it is rarely worthwhile for individual consumers to pursue them.

31.2 Compensation in food law

Where the prohibition to place unsafe food on the market has been breached (Article 14 GFL) or there has otherwise been an infringement on food law requirements, there are in principle three legal grounds on which compensation for damages caused thereby can be claimed.

The person who was supplied with the unsafe food can (1) bring an action out of contract. In that situation, three things must be proven: non-conformity, damages and causation. The person who is not in a direct contractual relationship with the person who put the unsafe food on the market can (2) bring an action in tort. In that case, four things have to be proven: unlawfulness (in this case violation of Article 14 GFL or a more specific provision), damages, causation and imputability (formerly called 'fault'). For consumers, the legislature has created a less stringent regime called (3) product liability. The consumer bringing an action from product liability has to prove the existence of a defective product, damage and causation. The consumer does not have to prove that the producer is at fault. This is therefore referred to as strict liability. Product liability law uses a very broad concept of producer that includes the brand holder and the ingredient supplier. Those who acted solely as vendors are not considered producers. Nevertheless, the former are generally the first point of contact for consumers, but this is then out of contract.

This case does not involve consumers. The parties are at a somewhat greater distance from each other in the chain and do not maintain a direct contractual relationship. The proceedings are therefore conducted in tort.

31.3 MPA

The rulings included above are a few in a web of proceedings against a pharmaceutical company – AHP Manufacturing B.V. – which uses the trade name Wyeth Medica Ireland ('Wyeth'), among others.

It all started at the end of the last century. Wyeth manufactured 'the' (contraceptive) pill in Ireland. The active ingredient in Wyeth's drug was medroxyprogesterone acetate (MPA). MPA is a synthetic substance with hormonal action to prevent conception.

Like many drugs, Wyeth's pill was literally sugar-coated for consumers. A waste product of the coating process was sugar water. Through a waste broker in Ireland, Cara Environmental Technology Ltd., Wyeth shipped this sugar water to Bioland, a processing plant in Belgium. Initially, it was only sugar water allegedly without MPA residue; later shipments certainly did contain MPA. The receiver in Bioland's bankruptcy chose not to destroy the sugar water, but rather to sell it for animal feed. The buyer resold part of it to Rined.

The case came to light when, in May 2002, breeding sows at three pig farms based in the province of North Brabant in the Netherlands suffered fertility problems. After investigation, the meat from the pigs on those farms was found to contain MPA. It subsequently came to light that pigs at (many) more pig farms had been fed sugar water containing MPA, including sugar water from Bioland.

Due to the presence of MPA in the pigs, this hormone also entered the food chain. Whether this means that human consumers were also de facto on the pill, the story does not reveal. Since the use of hormones is banned in animal husbandry, residue is subject to a zero limit. The pigs that had eaten the contaminated feed were deemed unfit for human consumption by the government and were therefore culled.

Some of the parties involved in this entanglement include

1. Wyeth, from whom the contaminated sugar water originated (some 800 tons).
2. Cara, the waste broker (went bankrupt)
3. Bioland, to whom the waste water was shipped (went bankrupt)
4. Several businesses kept anonymous in the court rulings who bought sugar water from Bioland and sold it to other businesses
5. Rined, a feed business to whom sugar water was sold
6. Zeeland voeders, another feed business to whom sugar water was sold
7. Oostburg Varkens, one of the pig farms that received contaminated pig feed

Rined has credited invoices for the MPA-contaminated pig feed it supplied and had to rent silos to store that feed. It also paid compensation to pig farmers who had held it liable for losses suffered as a result of supplying MPA-contaminated pig feed. These were undoubtedly claims for compensation out of contract.

Rined and some of the pig farms sued Weyth and Cara for damages based on tort. After Cara's bankruptcy procedures continued against Weyth.

31.4 Legal battle

Ultimately, the court ruled that Wyeth had acted negligently by disposing of its waste without ensuring that the receiving company was entitled and able to dispose of it properly and without notifying the authorities. It was therefore held liable for the losses suffered by Rined and by the pig farm. Both Rined and the pig farm had also acted negligently by not verifying the origin of the sugar water it used to supply to its customers and to feed their pigs respectively. Therefore, they had to bear part of the damages themselves. On 25 September 2018, the court ruled on the merits, stating that the damages should be divided. 30% should be borne by Wyeth and 70% by Rined. On 16 October 2020, the Supreme Court of the Kingdom of the Netherlands upheld the court's rulings. At the time of writing, the parties are still discussing the amount of total damages suffered (to be divided 30:70).

Also with regard to the pig farm the courts ruled that their own fault contributed to the damage to such an extent that it had to bear part of the damages itself.

The own fault on the part of Rined and the prig farm consisted of not checking the origin of the sugar water. Moreover they knew or should have known that Bioland was not GMP+ certified.[325] Also the pig farm that itself was not GMP+ certified had been closed by authorities for a longer period than certified competitors and it received a lower compensation from authorities for culled animals. The district court ruled that damages related to not being certified were caused by the pig farm itself, not by Wyeth. The court of appeal shared liability between Wyet and the pig farm.

GMP+ is a private standard.[326] GMP+ certification is not a legal obligation. In literature private standards are usually referred to as 'voluntary'. In a previous publication[327] I pointed out that businesses may economically depend on powerful customers to such an extent that in practical terms the need to be certified can hardly be called 'voluntary'.

Now tort law adds another layer to the *de facto* binding force of private standards when the presence or absence of private certification influences the extent to which a customer may trust its supplier and the extent to which the customer must exercise due diligence, on pain of 'own fault' in liability.

To an outsider in tort law it seems a bit counterintuitive that the unreliability of the supplier is protected under the law against its gullible customer. The fact that the party that put the contaminated product into circulation escapes an important part of the liability for the consequences, is in stark contrast with the case law of the EU courts who raise food safety above all else and leave the financial risks to the businesses even when they did nothing wrong such as Malagutti, Lidl and Ute Reindel.[328]

Bernd van der Meulen
Van der Meulen Food Legal Services (www.BerndvanderMeulen.eu)

---

325 On GMP+ see https://www.gmpplus.org/.

326 On this issue see also chapter 4, section 4.4. in this book.

327 See On private standards in food law, see: Bernd van der Meulen (ed.) Private food law. Governing food chains through contract law, self-regulation, private standards, audits and certification schemes, European Institute for Food Law series, Brill 2011 (open access) https://brill.com/edcollbook-oa/title/68768.

328 See chapters 1 and 4 in this book.

CHAPTER 32

XTC – Unfit for consumption for humans and animals

Annotated by Bernd van der Meulen

Trade and Industry Appeals Tribunal (*CBb*) 11 November 2018 ECLI:NL:CBB:2018:490 (*x v Minister of LNV*)

Not harmful but unsafe. Maize that has become contaminated with drug waste via fertilisation is an unsafe feed because food produced by the animals could contain traces of MDMA (XTC), which makes it unfit for human consumption.

Main considerations:

> In the opinion of the Trade and Industry Appeals Tribunal (CBb), consumers should be able to assume under normal circumstances that the maize used as food does not contain MDMA. MDMA is prohibited under Section 2 of the Opium Act from being brought into or outside the Netherlands, being present, manufactured or processed. Assuming the normal use of maize as food, the respondent was correct to take the view that the appellant's maize contaminated with MDMA was unacceptable and therefore unfit for human consumption. The respondent was therefore correct to classify the appellant's maize as unsafe food in accordance with Article 14(1) of Regulation 178/2002.
>
> Under Article 15(1) and (2), second indent, of Regulation 178/2002, unsafe feed may not be placed on the market if it is considered to 'make the food derived from food-producing animals unsafe for human consumption'. Furthermore, under Article 4 of Regulation 183/2005 read in conjunction with Article 57 of the Dutch Animal Products regulation, an animal keeper is not allowed to administer a feed containing a chemical contaminant.
>
> In the opinion of the Tribunal, the respondent was correct in its position that if MDMA-contaminated maize were fed to food-producing animals, it would make the food produced from those animals unsafe for human consumption. The MDMA from the contaminated maize can pass through the food-producing animals into the food produced from those animals. Those foods must then be considered unfit for human consumption due to the presence of MDMA. In view of this, the respondent was right to classify the appellant's maize as unsafe animal feed. Because of the presence of MDMA in the appellant's maize, the respondent was also justified in taking the view that that maize should be classified as chemically contaminated animal feed.
>
> In contrast to what the appellant seems to assume, the respondent did not contend that appellant's claim that its maize was considered harmful to public and animal health.

 | HTTPS://DOI.ORG/10.1163/9789004751699-034

As food safety was at issue at the time the primary decisions were taken, the respondent was right, in the Tribunal's opinion, to refrain from hearing the appellant.

32.1 Introduction

The appellant in this case owns a farm. Apparently, the farm is facing contamination from drug waste. The cause is not clear from the judgment.[329] In any case, 3,4-methylenedioxymethamphetamine (MDMA), better known as XTC, ended up in the manure cellars. The spreading of some of this manure caused the MDMA to enter the soil and along that path into maize grown by the appellant. The Minister of Agriculture, Nature and Food Quality imposed an order under administrative coercion on the appellant to have the maize in question destroyed or designated for purposes other than as food and feed to animals. The legal debate seems to revolve around Article 14(2)(b) and (5) of Regulation 178/2002 (GFL). Under this provision, a food is unsafe if it is unfit for consumption. The respondent and the Trade and Industry Appeals Tribunal (*CBb*) applied the same principle to animal feed. Because food safety is at stake – but not public health, according to the Trade and Industry Appeals Tribunal (*CBb*) – the Tribunal (*CBb*) considers it justified that the duty to be heard under Article 4:8 of the Dutch General Administrative Law Act was not observed.

32.2 Unsafe food

Article 14 GFL prohibits the marketing of unsafe food. According to the second paragraph, 'unsafe' means both injurious to health and unfit for human consumption. The concept of 'unfit for human consumption' is difficult to grasp. In the past, I have recommended[330] to understand this concept as a refutable legal presumption of injuriousness. Since the Berger judgment of 2013 (see Chapter 10 of this book), it has been clarified that even food that is established as not harmful can be unsafe in a legal sense. In this case, the CBb presented in so many words to the appellant: 'Contrary to what the appellant seems to assume, the respondent did not contend that the appellant's maize was considered harmful to public and animal health.'

329 On 20 July 2023 an article appears in the newspaper *de Volkskrant* (*Economie*, p. 19) about possible biogas fraud. A residue of biogas production is 'digestate'. Digestate is used in farming as a fertiliser. If illegal (chemical) waste is dumped in the fermentation process, this can get into the food chain via the digestate. It is plausible that this sort of biogas fraud forms the background to this case.

330 B.M.J. van der Meulen, 'The Core of Food Law. A Critical Reflection on the Single Most Important Provision in All of EU Food Law', European Food and Feed Law Review 2012, pp. 117–125.

The Tribunal (*CBb*) considers this against the appellant's argument that the highest value measured in the maize in question would be 17 ug/kg and that in another case it would have been considered that, at levels between 1 and 60 ug/kg measured there, there is no danger to animal health and public health.

It is not evident that the legislature indeed intended to bring within the scope of the concept of unsafe a category of foods that are known to be non-harmful.

Rather, the explanatory memorandum to the proposed regulation[331] seems to want to address evidentiary problems that may arise in contaminated food:

> Food unfit for human consumption or contaminated food is also considered to be unsafe in this Regulation. Food, for example, putrid food, may or may not be potentially injurious to health but it is unacceptable for human consumption and *may be* injurious to health. *It may be almost impossible to prove injury or probable injury to health with such food*, so this separate factor is proposed in relation to safety within this Regulation.' Similarly, food which is contaminated with, for example, insect parts, or beef which may be contaminated with the hairs from cattle, *may not be injurious* to health but it is still not reasonable to expect this to be used for human consumption, and *proving that it may be potentially injurious to health should not be required to declare that it is unsafe*. (emphasis added).

As mentioned, however, the Tribunal's unqualified equating of unfit with unsafe puts it in good company with the EU Court of Justice. This ruling provides a nice illustration of where that can lead in practice.

The Minister and the Trade and Industry Appeals Tribunal (*CBb*) declare the maize not only unfit for consumption within the meaning of Article 14 GFL, but also unsafe within the meaning of Article 15 GFL. The first two paragraphs of Article 15 GFL read as follows:

> 1. Feed shall not be placed on the market or fed to any food-producing animal if it is unsafe.
> 2. Feed shall be deemed to be unsafe for its intended use if it is considered to:
> - have an adverse effect on human or animal health;
> - make the food derived from food-producing animals unsafe for human consumption.

It appears undisputed in this case that there are no adverse effects on animal or human health. The Tribunal bases the conclusion that there is an unsafe animal feed on this consideration: 'MDMA from the contaminated maize will be able to pass

331 Proposal for a Regulation of the European Parliament and of the Council laying down the general principles and requirements of food law, establishing the European Food Authority, and laying down procedures in matters of food COM(2000) 716.

through food-producing animals into the food produced from those animals.' What I miss are considerations as to whether the MDMA will be digested or excreted by the animals. Apparently, the appellant has not submitted anything in this regard.

What remains is that products that could potentially contain traces of MDMA are unacceptable to consumers. Why unacceptable? Probably because the idea of drug waste is abhorrent.

This has weakened the concept of 'unsafe food'. An exploratory empirical study[332] in a number of Member States shows that it is indeed weak. Anything that does not please the competent authorities can be labelled as such. This ranges from grain that has been stored outside for lack of sufficient shed space, to products coming from an unregistered food company, to colour or taste defects, to horse meat sold as beef. It makes the food safety standard highly subjective. Which does not help legal certainty.

It is therefore commendable that the CBb does not limit itself to a marginal assessment of whether the respondent could reasonably take the position that the animal feed or foodstuff is unsafe, but takes an explicit position itself. Such a method of complete review would seem to allow for a substantive debate.

32.3 Importance for legal practice

The classification of a food as unsafe under Article 14 GFL has potentially far-reaching consequences. In practice, it is not always easy to determine whether a food is harmful to health. But that criterion is clear. However, if food that is not harmful to health can also be classified as unsafe because it is unfit for human consumption, this introduces great uncertainty when no clear criteria exist as to when a food then qualifies as unfit. This gives the regulator a lot of room to assess on a case-by-case basis whether a product is unacceptable for human consumption. This can then be discussed in court.

Bernd van der Meulen
Food Legal Consultant

332 Daniela Polinski and Bernd van der Meulen, 'Unfit for Human Consumption. The elusive element in the EU food safety concept of Article 14 GFL', European Institute for Food Law Working Paper Series 2019/01 https://food-law.nl/Working-papers.

CHAPTER 33

Recall – Administrative pressure

Annotated by Bernd van der Meulen

Trade and Industry Appeals Tribunal 23 August 2012 ECLI:NL:CBB:2012:BX6798
Rotterdam District Court 24 October 2013 ECLI:NL:RBROT:2013:8226
Trade and Industry Appeals Tribunal 24 November 2015 ECLI:NL:CBB:2015:397
The Hague District Court (civil) 9 March 2016 ECLI:NL:RBDHA:2016:2245
Trade and Industry Appeals Tribunal 23 October 2018 ECLI:NL:CBB:2018:546

Recall, decision, the recall obligation under Article 19 GFL applies by operation of law. When the regulator notifies a food company of the existence of this obligation, it is generally not an appealable decision.

Main consideration Trade and Industry Appeals Tribunal (CBb) 23 August 2012:

> The CBb is of the opinion that merely drawing the appellant's attention to an obligation arising directly from Article 19(1) of Regulation (EC) No 178/2002 [GFL] does not entail a change in the appellant's legal position and is not aimed at legal effect. The NVWA's letter of 10 August 2012 therefore does not constitute a decision within the meaning of Article 1:3, first paragraph, of the Awb (Dutch General Administrative Law Act (GALA)– Algemene wet bestuursrecht (Awb)).

Main considerations Rotterdam District Court (*rechtbank: Rb*) 24 October 2013:

> The report received by the claimant on 1 August 2011 from Bureau de Wit, which was engaged by [buyer], in which that agency found, based on ISO standard 6579, that the batch of chilli powder was not contaminated with salmonella, cannot play a role in the question of whether the obligation referred to in Article 19(1) [GFL] arose for the claimant on 27 July 2011. Results of counter-expertise will generally not be known until after the obligation referred to in Article 19(1) has arisen.
> In view of the foregoing and since it is not in dispute between the parties that the appellant failed to carry out a recall, there was a breach of Section 2(10) of the Commodities Act Decree on Preparation and Handling of Foodstuffs in conjunction with Section 19(1) [GFL] and the minister was entitled to impose an administrative fine on the appellant.

 | HTTPS://DOI.ORG/10.1163/9789004751699-035

Main consideration Trade and Industry Appeals Tribunal (*College van Beroep voor het bedrijfsleven: CBb*) 24 November 2015:

> In so far as the appellant put forward that – in addition to the negative result for the presence of salmonella in the consignment of chilli powder in the 'certificate of weight & quality' of Geo-Chem Laboratories Pvt. Ltd. dated 1 July 2010 and the certificate of analysis of NofaLab B.V. of 5 July 2010 – it had at its disposal the investigation report of Bureau de Wit of 28 July 2011, issued on behalf of [name 3], which reported the absence of salmonella as the result of the analysis of four samples left for counter-expertise, the CBb is of the opinion that the appellant could not draw the conclusion from this that the immediate initiation of proceedings could be waived. With that counter-expertise, the incorrectness of the results of the investigation carried out by the laboratory of the Netherlands Food and Consumer Product Safety Authority (*NVWA*)was not established, so it cannot be said that there were therefore no (longer) reasons to assume that the chilli powder did not comply with food safety regulations.

The Hague District Court (civil), 9 March 2016:

> The Court agrees with the State that the NVWA did not address a 'recall order' to Abattoir Amsterdam on 20 November. Abattoir Amsterdam's assertion that there was such an order is undermined in the first place by the contents of the […] e-mail and the […] letter from the NVWA. On behalf of Abattoir Amsterdam, even after being explicitly asked about it at the hearing, no concrete facts about a (verbal) communication by employees of the NVWA were put forward that could lead to a different opinion. The fact that, on behalf of the NVWA, it was pointed out several times that Abattoir Amsterdam had to take measures does not constitute such a communication. Like the peremptory notices contained in the letter of 5 December 2014 […] under the heading 'consequences of non-implementation of obligation', these notices cannot reasonably be understood in any other way than that the NVWA is thereby rightly drawing Abattoir Amsterdam's attention to Abattoir Amsterdam's recall obligation and the consequences of non-compliance. The recall obligation follows from Article 19(1) [GFL] and, therefore, no order by the NVWA is necessary for the recall obligation to arise, nor a reliance by the NVWA on Article 7 [GFL]. Furthermore, particularly with regard to previously mentioned mandatory notices, it is important to note that it is not in dispute that the Netherlands Food and Consumer Product Safety Authority (*NVWA*) must take firm action at all stages of production, processing and distribution against food business operators who fail to comply with their recall obligations under Section 19(1) [GFL]. This also follows from Article 17(2) [GFL].
> It is not in dispute that the present case does not involve a government act. If the NVWA's notification of 20 November 2014 and 5 December 2014 should already be regarded as a government act relevant for this assessment – after all, Abattoir Amsterdam's recall obligation is not the (direct) consequence of this notification, but follows

from Article 19(1) of the Regulation – Abattoir Amsterdam's argument stumbles on the fact that (the alleged extent of) the damage was foreseeable and that the alleged damage is therefore part of its normal business risk. The Court agrees with the State that in the trade in animal products there is not only an objectively foreseeable risk of diseases (cf. HR 20 June 2003, AB 2004, 84), but also a similar risk of the presence of (prohibited) substances that are harmful to health, such as prohibited muscle enhancers. The fact that the latter risk is foreseeable and rests first and foremost on the business owner is confirmed by the fact that Article 19(1) [GFL] places the primary responsibility on the business owner for (preventive) measures with a view to the threat to food safety posed by, inter alia, these prohibited substances. Furthermore, this is confirmed by the statement made on behalf of Abattoir Amsterdam that it had experienced recalls before, and by the fact that Abattoir Amsterdam, as it stated at the hearing when asked, could have insured itself against the damage at issue – which it itself estimated at approximately EUR 100,000 – but did not do so because of the level of the premiums. Finally, Abattoir Amsterdam did not put forward anything on the basis of which it should be held in this particular case that, despite the foregoing, there were disproportionate adverse consequences. The mere assertion, correct in itself, that Abattoir Amsterdam did nothing wrong cannot lead to such a finding.

Trade and Industry Appeals Tribunal (*CBb*) 23 October 2018:

According to settled CBb case-law, as evidenced by its judgments of 23 August 2012 (ECLI:NL:CBB:2012:BX6798) and 24 November 2015 (ECLI:NL:CBB:2015:397), an obligation to recall follows automatically from Article 19 [GFL]. Only failure to comply with this obligation can lead to the adoption of enforcement decisions. That is then open to objection and appeal and, in that context, the question can be raised as to whether a recall obligation existed.

At the hearing, the appellant argued that the 30 June 2017 e-mail should be seen as imposing an exceptional burden. The Trade and Industry Appeals Tribunal (*CBb*) does not agree with the appellant in this. At the time the e-mail of 30 June 2017 was sent, unlike the situation at issue in the Rotterdam District Court ruling of 15 July 2016 cited by the appellant, there was no failure to comply with the obligation following from Article 19 [GFL]. Indeed, the responsibility for complying with that obligation at that time lay with the appellant as operator. This means that the respondent did not have the power to take an enforcement decision at that time. In addition, the 30 June 2017 e-mail makes no mention of possible sanctions in the event of non-compliance with the recall obligation.

However, the Tribunal (*CBb*) considers that the 30 June 2017 e-mail, given its wording, suggests a power to impose a recall obligation that does not exist, and the appellant was thus misled. The CBb recommends that, where appropriate – unlike the e-mail of 30 June 2017 – the respondent explain to operators that, in view of Article 19 [GFL], they have their own responsibility if food does not meet food safety requirements, and that the obligation to impose a recall may follow from this article.

33.1 Decision and administrative law

In 1932, the first administrative law handbook was published in the Netherlands. That book included a contribution by C.W. van der Pot entitled 'The forms of administration'. This could easily be the most influential publication in the history of Dutch administrative law. According to Van der Pot, the essence of the executive's work is taking decisions with legal consequences in individual cases. He thus laid the foundation for an administrative law that connects to the legal act of the administrative body. This focus on the legal act in setting standards and in legal protection has a downside: namely, that factual acts of administrative bodies – even though they may be very far reaching, especially in the context of enforcement – largely escape standard-setting and legal protection.[333] Thus, the structure of the Dutch administrative law system puts a premium on it for administrative bodies when they manage to evade the 'decision' as defined in GALA (*Awb*) as a form of their governance. Again and again, we (therefore) see 'informal' forms of governance developing.[334]

The rulings discussed in this chapter variously raise the question of whether or not pressure from the NVWA on food companies to initiate recalls comes under the legal responsibility of the government.

I use the term 'administrative pressure' in this annotation to distinguish it from 'administrative enforcement'. The latter is the legal act of the administrative body creating an obligation for the addressee to act. Pressure is the factual action by which the regulator for a food business fills in or allows the conditions of application to be filled in so that the legal standard becomes applicable. I will start by discussing this administrative pressure in general, then the connection first with the potential liability for damages, and thereafter with questions of jurisdiction. Following that, I will discuss some other aspects of the rulings and of recall more generally: namely, the right to counter-expertise, the obligations arising from the first paragraph of Article 19 and the position of the consumer in the recall. I conclude with a brief reflection.

333 I once tried to make a contribution to the development of an administrative legal theory concerning physical action, but we have to admit that it is still largely uncharted territory. B.M.J. van der Meulen, De normering van het feitelijk handelen der overheid, Nederlands Tijdschrift voor Bestuursrecht 1991, p. 267–278.

334 I wrote an article about this in 2008 for the Liber Amicorum Twan Tak: B.M.J. van der Meulen, Vormen van besturen, in: M. Heldeweg e.a. Publiekrecht in de Leer; Liber Amicorum Twan Tak, Wolf Legal Publishers 2008, p. 169–190. On the issue of avoiding appealable decisions in food law, see also: Vicente Rodríguez Fuentes, Food alerts and RASFF, in Altinay Urazbaeva, et al. The functional field of food law. Reconciling the market and human rights, Brill 2019.

33.2 Administrative pressure

It is now established case-law that when the NVWA provides a food company with information regarding the possible unsafety of a foodstuff, this triggers the obligation of Article 19 GFL without a decision to that effect.

If the operator considers that there is no Article 19 situation, there are in principle two paths to legal protection. The operator can wait for an administrative fine and challenge it (as in Trade and Industry Appeals Tribunal (*CBb*) 24 November 2015) or the operator can turn to the civil court as a general residual judge (as in The Hague District Court (civil) 9 March 2016). The aforementioned examples also immediately show that these avenues are unlikely to be successful. A third avenue is conceivable, but I have not yet encountered it in case-law. If the court were to consider that it would be disproportionately onerous to resort to a fine, then the court could by way of exception regard the notification that there is an Article 19 situation as an appealable administrative legal judgment.[335]

The civil chamber of the Hague district court seems to be caught in the administrative law web. The consideration that the recall obligation follows from Article 19(1) GFL, and that no additional action by the Netherlands Food and Consumer Product Safety Authority (*NVWA*) is necessary for the recall obligation to arise, is in line with administrative-law case-law. However, the further substantiation that the court comes up with suggests that in practice, an enforcement decision is the issue more often than not. After all, the District Court considered that, particularly with regard to compelling notices, it is important that it is not in dispute that the NVWA should take vigorous action at all stages of production, processing and distribution against food business operators who fail to comply with their recall obligations under Article 19(1) GFL and that this also follows from Article 17(2) GFL. Article 17(2) GFL concerns the enforcement responsibility of Member States. However, forceful action requires the exercise of powers, not merely informal communication.

33.3 Damages

The civil judgment reveals another interest that is bound up with the presence or absence of a decision. That is the presence or absence of a connecting factor for government liability in tort if the assessment of the administrative body were found to be incorrect, or from administrative compensation (lawful act) if the burden on

335 On the subject of the administrative legal judgment, see the contribution by Oswald Jansen in Chapter 37 of this book. This in-depth analysis shows that anyone placing their hope in this will soon be disappointed. See also his short introduction to Dutch food law at the beginning of this section of this book.

the company were to be disproportionate. The food company implementing the Article 19 obligation does so entirely at its own expense and risk.

33.4 Lack of competence?

The ruling of the Trade and Industry Appeals Tribunal (*CBb*) of 23 October 2018 deals with a case in which the CBb had to rule on an e-mail which, according to them, 'in view of its text, suggests a power to impose a recall obligation'. The CBb ruled that there had not yet been a violation of the Article 19 obligation and that NVWA therefore had no power of enforcement. The CBb ruled that the e-mail did not create any legal effect.

I consider the Tribunal's line of reasoning to be both systematically and factually incorrect. Only when an administrative body is manifestly incompetent, in an absolute sense, to take a decision of the type alleged, can it result in the absence of a decision. As a rule, however, a lack of competence is grounds for annulment. Absolute lack of competence does not exist in this case. The regulator is incompetent only to the extent that, according to the Trade and Industry Appeals Tribunal (*CBb*), the conditions of application for exercising the power, which in principle belongs to the regulator, have not yet been met, because, according to the CBb, there is no violation. If the application condition for the exercise of the power to make a decision is not met, the court must annul that decision rather than concluding that there was no decision at all.

The CBb appears to believe that the regulator cannot impose an obligation that already exists on another basis (in this case, Article 19 GFL). However, the Official Controls Regulation, now Regulation (EU) 2017/625 (hereinafter Controls Regulation),[336] expressly grants the regulatory body the power to impose a recall (Article 138). While it is conceivable that this power could be exercised in situations where Article 19 GFL does not apply, this will be the exception rather than the rule.

Finally, in my view, the CBb is wrong in assuming that enforcement powers only exist in cases of non-compliance with an Article 19 obligation. Article 19 applies when there is reason to believe that a food does not comply with food safety requirements. Is that not then a suspicion of non-compliance with Article 14 GFL? But if so, then the power denied by the CBb does in fact exist. The condition of application of Article 138 is that there is a case of non-compliance with a requirement covered by the Controls Regulation. According to Article 2, the Controls Regulation covers almost the entire agri-food domain.

---

336 At the time of the judgments Regulation (EC) 882/2004 still applied. The regulations are substantively the same on the relevant matters for this discussion. For the convenience of the reader, I refer to the current provisions.

33.5 Counter-expertise

The Controls Regulation is primarily a piece of instrumental legislation. It directly grants coercive powers to competent authorities in Member States. However, it does not provide rights for aggrieved parties ('complainants') and only marginally protects whistleblowers and rights of defence.[337] One of the few defence rights that the Controls Regulation does provide for is the right to a second expert opinion (Article 35). The courts leave that right in shreds. The Rotterdam District Court is very laconic in considering that counter-expertise results will generally only be known after the obligation referred to in Article 19(1) has arisen. More substantive, and therefore more worrying, is the Tribunal's consideration that the counter-expertise did not establish the incorrectness of the results of the investigation carried out by NVWA's laboratory, so that it cannot be said that there were therefore no (more) reasons to assume that the product did not comply with the food safety regulations. This consideration goes to the heart of the right to counter-expertise. This right only has meaning if the outcome of the second investigation between the regulator and the food company is decisive. It is difficult to see how counter-expertise could ever prove the incorrectness of the investigation conducted by the regulator. If that is the criterion, then the right to counter-expertise has become meaningless.

In Trade and Industry Appeals Tribunal (*CBb*) 3 July 2018 ECLI:NL:CBB:2018:326[338], the Tribunal did assign significance to a counter-expertise. Perhaps the difference lies in the fact that in that case there was no difference of opinion as to whether there was contamination (a circumstance that gives reason to suspect within the meaning of Article 19) but as to whether the contamination found constituted a legally relevant risk (as had been disputed by the company from the outset).

33.6 Article 19 obligations

The above rulings conclude relatively quickly an obligation to initiate proceedings to recall products. Perhaps, in the circumstances, this is in fact justified. However, it is not the compelling conclusion from applicability of Article 19 in all cases. Recall is an ultimum remedium.

The English text describes the legal consequences as follows: 'it shall immediately initiate procedures to withdraw the food in question from the market where

337 See my contribution to this subject: Enforcement of EU agri-food law. Regulation (EU) 2017/625 on official controls and other official activities performed to ensure the application of food and feed law, rules on animal health and welfare, plant health and plant protection products, ERA Forum 2018 p. 623–641. https://doi.org/10.1007/s12027-018-0532-5.

338 See chapter 28 of this book.

the food has left the immediate control of that initial food business operator and inform the competent authorities thereof. Where the product may have reached the consumer, the operator shall effectively and accurately inform the consumers of the reason for its withdrawal, and if necessary, recall from consumers products already supplied to them when other measures are not sufficient to achieve a high level of health protection.'

It is clear that a consumer recall is required only when necessary. This already follows from the words 'if necessary'. But to what does the phrase – 'when other measures are not sufficient to achieve a high level of health protection' – refer? The Dutch text suggests by a different sentence order that this refers only to the recall. In the Dutch version, the phrase is nothing more than a repetition or clarification of the words 'if necessary'.

The German and French texts correspond to the English text. In those texts, the phrase grammatically refers to the entire preceding sentence. The Dutch sentence order grammatically, unlike the English (German and French), cuts off the possibility of sufficing with other measures from the public information. This does not seem logical. Why should the public be informed when it is not necessary for the protection of the public? Rather, it seems logical to me to relate the possibility to suffice with other measures, if this can achieve a high level of health protection, to the whole provision of the legal effects of Article 19 of Regulation 178/2002. The duty of the operator is then to ensure consumer safety, and if this cannot be achieved by other means, to withdraw the product from customers and, if necessary, also recall it from consumers. For this interpretation, see EU Food Law Handbook edition 2020, p. 325. Arguments of proportionality favour this interpretation.

An overly literal reading of Article 19 Regulation 178/2002 would lead to absurd results. If performing logistical movements[339] were mandatory in all cases, rendering or processing on the premises of the purchaser would be prohibited. Since the Food Information Regulation (in Article 24(1)) designates products after their use-by date as unsafe by operation of law (within the meaning of Article 14 of Regulation 178/2002!), companies would be obliged to recall any remaining products from their customers on every day on which such a date expires. In practice, it is the customer who disposes of products past their expiry date. It cannot reasonably be assumed that the legislator intended to prohibit this practice and intended absurd consequences. It therefore makes more sense to read Article 19 of Regulation 178/2002 as allowing the operator to fulfil its obligations by solving the food safety problem. It is only when the operator is unable to do so by means of its own choice that the use of the means expressly mentioned in Article 19 is mandatory.

339 In other words physically taking back the product.

33.7 Consumer and recall

In 2016, Mars launched an international recall after bits of plastic were found in its Snickers bars. Consumers were urged to send affected products to a reply address in Belgium. I had a bag of mini Snickers that could be stuffed into a letterbox with a bit of good will and aggression. Sometime later, I received a note from Mars thanking me for participating in the recall, and a bag of M&Ms as a gift. Intrigued as to why I received a gift and not a replacement product, I contacted Mars customer service. I presented the question and, in particular, also asked how Mars saw its relationship with the affected consumer. It turned out to be a difficult question. I was put through several times and they promised to call me back. The person who called me back apologised: Mars could not answer this question because it concerned company confidential information. This incident made me realise that, as a rule, there is no legal relationship between a producer who recalls a consumer's product and the consumer whose product is recalled (unless it is a matter relating to tort). Damage to the product itself is not covered by product liability and the contractual relationship is with the retailer not the manufacturer.

Recall is a food safety measure, not a measure to protect consumers' economic interests. Against that background, it is understandable that the legislature did not concern itself with consumer rights in the recall. The result, however, is that the consumer has an economic incentive not to follow the route devised by the legislature by responding to the recall by the manufacturer, but instead to exercise the rights under contract by seeking reimbursement of the purchase price or a replacement product from the retailer. The financial and logistical problem then falls to the retailer. From this perspective, the statutory scheme could be called suboptimal.

33.8 Reflection

I observe that lawyers are increasingly advising their clients to exercise restraint around administrative pressure in situations where the clients themselves believe that there is no public health risk. The reason is that by complying with the regulator's wishes, they give up their legal protection and accept all the risks.

This is precisely why the informal route is popular with the government. It can create a recall situation by making factual announcements without running any legal or financial risk itself in the matter.

From the case-law discussed here, the fine does not seem to offer a very promising legal route for the food company either, but at least it is one.

There's still something not quite right about the regulator's intervention in the responsibilities of the food company. If the regulator considers that a recall should take place, the regulator has the power under Article 138 of the Controls Regulation to make that happen. It may well provide a legal avenue for food companies to file an application (i.e. request as an interested party to take a decision (Article 1:3 paragraph 4 and Article 4:1 Awb) to exercise the power under Article 138 of the Controls Regulation in such a case. If the application is granted, this is an appealable decision. However, legal protection is also available if it is rejected or not decided on time (Article 6:2 Awb).

Bernd van der Meulen
Director European Institute for Food Law (www.food-law.nl)

CHAPTER 34

Organic barley grass powder – Powers of regulator Skal

Annotated by Silvia Gawronski and Dagmar Linstra

Trade and Industry Appeals Tribunal 9 October 2018, ECLI:NL:CBB:2018:534
Trade and Industry Appeals Tribunal 13 April 2021, ECLI:NL:CBB:2021:387

Main considerations:

Trade and Industry Appeals Tribunal (CBb) 2018:

> The Tribunal is of the opinion that the respondent correctly decided to (maintain) the blockade and to decertify the batch at issue. After all, it is not disputed that tebufenozide has been found in this batch and that this product does not appear on the list of plant protection products allowed for organic production pursuant to Annex II to Regulation 889/2008. Therefore, this batch does not meet the requirements of the Regulation and the Implementing Regulation regarding organic production. The appellant's arguments against the respondent's considerations concerning Ecocert's investigation into Shenghui and the subsequent decertification of that company and of the entire 2015 harvest do not require discussion. The presence of the unauthorised substance in the blocked and decertified batch is a sufficient basis for the decisions at issue.

Trade and Industry Appeals Tribunal (CBb) 2021:

> The Tribunal notes first of all that Article 30(1) and first subparagraph of Regulation 834/2007 is the basis for the respondent's authority to decertify a batch of agricultural products and the respondent referred to this in the decertification decision. That provision has its own assessment framework and the use of that power has the effect that the consignments of sunflower seeds may not be marketed with reference to the organic production method. Contrary to the opinion of the parties, Article 91 of Regulation 889/2008 is not (also) the basis for that authority. That provision has a different framework of assessment compared to Article 30(1) and first subparagraph of Regulation 834/2007 and provides for the possibility of taking temporary conservation measures in anticipation of the application of Article 30 of Regulation 834/2007. This follows from the scope of the power conferred on the Commission in Article 38(c) of Regulation 834/2007, in conjunction with the place of Article 91 in Title IV of Regulation 889/2008 and the text of this provision. Article 38(c) of Regulation 834/2007 empowers the Commission to adopt provisions to implement the control system laid down in Title V of this Regulation, which includes Article 30. Article 91 is included

 | HTTPS://DOI.ORG/10.1163/9789004751699-036

in Title IV of Regulation 889/2008 which deals with controls and, given the text, includes measures in case of suspected infringements and suspected irregularities. Therefore, in the opinion of the Tribunal, in order to answer the question whether the respondent is authorised under Article 30 of Regulation 834/2007 to decertify the consignments of sunflower seeds, it is irrelevant whether the respondent, when taking measures as referred to in Article 91 of Regulation 889/2008, followed the procedure described in that article, as well as whether the substantive criteria set out in that article were met. As regards the relationship between Article 30(1) and first subparagraph of Regulation 834/2007 and Article 91 of Regulation 889/2008, in view of what has been considered above, there is no doubt that requires a preliminary question to be put to the Court of Justice of the European Union . The Tribunal will therefore not proceed to do so.

Article 30(1) of Regulation 834/2007 requires the respondent, where an irregularity is found with regard to compliance with the requirements of this regulation, to ensure that the labelling and advertising of the entire batch or production run affected by the irregularity does not contain any reference to the organic production method, provided that it is proportionate to the importance of the requirement infringed and to the nature and specific circumstances of the irregular activities. This is done by the respondent's decertification decision.

34.1 Introduction

In case of irregularities in the organic production process of a product, the organic certificate may be revoked or even an entire enterprise may be decertified. Subsequently, the products can no longer be marketed as organic. Such a sanction can have far-reaching (financial) consequences for the parties involved: after all, consumers are willing to pay more for organic products than for non-organic products.

The reason for our contribution is a ruling from 2018. In that matter, the Trade and Industry Appeals Tribunal (CBb) addressed the question of whether there was a notification within the meaning of Article 91(1) of Implementing Regulation 889/2008 laying down detailed rules for the implementation of Council Regulation 834/2007 on organic production and the labelling of organic products with regard to organic production, labelling and control[340] ('889/2008'). Both regulations have since been amended by Regulation 848/2018, but this does not affect the value of the case-law. Given that the relevant case-law deals with the formerbio-regulations and that there are currently still cases pending in which these regulations play a role, we use references to the old articles in this article. The CBb ruled in the affir-

340 309 ELI: http://data.europa.eu/eli/reg/2007/834/2022-01-01.

mative. Therefore, Skal could base its authority to decertify on Article 91 Regulation 889/2008.

We address the potential legal impact of this ruling. In the Netherlands the Stichting Skal Biocontrole (hereinafter 'Skal') was designated as the regulatory and enforcement authority under Article 27(4)(a) of Regulation 834/2007 on organic production and labelling of organic products[341] ('Regulation 834/2007'), see Article 15 of the Agricultural Quality Decree 2007. Skal derived its authority to (de)certify from two sections of law. These are Article 30 Regulation 834/2007 and Article 91 Regulation 889/2008. How do those articles intersect? What are the specific conditions under which Article 30 of Regulation 834/2007 applies, versus Article 91 of Regulation 889/2008? Is their interplay unambiguous, or would preliminary questions to the Court of Justice be beneficial for clarity?

The Tribunal's view on this topic partly follows from subsequent case-law. We will first discuss the general framework of supervision and enforcement by Skal, before turning to an issue from 2018. We will then discuss in more detail the assessment frameworks of the two jurisdictional bases (Article 30 Regulation 834/2007 and Article 91 Regulation 889/2008) in more detail. We will address possible views on the interrelationship of those articles, including the view the CBb expressed in a more recent matter from 2021. Finally, we will highlight why, in our view, clarity on the jurisdictional basis is important, addressing the two main differences of the assessment frameworks, specifically in the areas of (i) evidence, and (ii) proportionality.

34.2 Supervision and enforcement by Skal in the context of organic production processes

For products to be marketed as 'organic', they must adhere to specific laws and regulations. Non-compliance necessitates the removal of the 'organic' designation, requiring the products to be sold as conventional.

Within the European Community, (the supervision of) certification of organic production processes was centrally regulated by Regulation 834/2007 – now Regulation 848/2018. Its primary objective is to establish a framework for the sustainable development of organic production, ensuring the effective functioning of the internal market, fair competition, consumer confidence and safeguarding consumer interests. Notably, article 27 Regulation 834/2007 included the 'Control system', mandating that Member States implement such a control system and designate one or more competent authorities responsible for conducting controls.

More specific requirements for organic production, labelling and control of rel-

---

341 310 ELI: http://data.europa.eu/eli/reg/2008/889/2022-01-01.

evant products (which were specified in Article 1 Regulation 834/2008), are laid down in Regulation 889/2008. These can pertain to diverse areas, such as the specific rules for seaweed production and organic wine production.

In the Netherlands, those organic seaweed and wine producers fall under the supervision of Skal. European regulations regarding the organic production process are embedded through the Agricultural Quality Act and the Agricultural Quality Decree 2007. In these, the Ministry of Agriculture, Fisheries, Food Security and Nature is designated as the 'competent authority' (see Article 17(b) Agricultural Quality Decree 2007). Enforcement and control powers are assigned to Skal as the 'controlling authority' (see Article 15 Agricultural Quality Decree 2007). Skal is a so-called independent administrative body – a private-law foundation with a statutory task *and* powers in accordance with the EU Regulation, the Agricultural Quality Act and Section 5 ('Enforcement') of the Dutch General Administrative Law Act (Awb).

For example, under Article 28(1) of Regulation 834/2007, Skal had to be informed by an operator who produces, processes, stores or imports products from a third country before products may be marketed as organic products or conversion products. But companies also have an 'obligation to notify' Skal under Article 91(1) of Regulation 889/2008: operators who consider or suspect that a product produced, prepared or imported by them, or a product supplied to them by another operator, is not in compliance with organic production rules, shall initiate procedures to remove from that product any indication relating to the organic production method from the product or to segregate and identify the product. They may start processing or packaging or marketing the product only after the suspicion has been resolved, unless the product is marketed without indications referring to the organic production method. In case of such suspicion, operators must immediately inform the control authorities or control bodies. In general, it is also Skal that carries out the controls and – if necessary – takes enforcement action in case of violations of the rules on the organic production process.

34.3 Notifying Skal of contaminated barley grass powder: a notification within the meaning of Art 91 Regulation 889/2008

In the present case, the company had purchased a batch of barley grass powder from China.[342] In doing so, the company contracted a laboratory to conduct sampling and analysis of the batch. The sample taken, representing 0.042 mg/kg of the batch, showed the presence of the unauthorised insecticide tebufenozide. This substance is not listed in Annex II of Regulation 889/2008 as an authorised plant protection product for organic production methods. The company informed Skal of

342 CBb 9 October 2018, ECLI:NL:CBB:2018:534.

these findings. The company also stated it would not market the product as organic until further notice.

At some point, the company changed its mind and intended to market the product as an organic product after all. After being apprised of this by the company, Skal decided to block the product (or keep the existing blok in place) and to decertify the batch as an organic product. The company disputed this decision, leading the case to be brought before the CBb.

The Trade and Industry Appeals Tribunal (*CBb*) ruled that the decision to impose (and then maintain) the blockade was justified, as was the decertification of the batch of barley grass powder, Simply because the substance found was not authorised. Therefore, the batch did not meet the requirements of Regulation 834/2007 and Regulation 889/2008 regarding organic production methods. In the view of the Tribunal (*CBb*) a 'clean' retention sample from another batch did not change that, in the. Nor did the fact that it was a dried product.

The legitimacy of Skal's enforcement action was also part of the debate. The CBb concluded that the company's notification about pesticide found in the barley grass powder batch could be regarded as a notification within the meaning of Article 91 of Regulation 889/2008.[343] That notification empowered Skal to make use of its enforcement powers under Article 91 of Regulation 889/2008.[344]

In short, on the basis of the case annotated here, the CBb has confirmed that Skal may be authorised to block and decertify a batch pursuant to Article 91 Regulation 889/2008 following a notification by the Company. This raises the question whether Skal's also has such authority following its own inspection. Or does Article 30 Regulation 834/2007 then govern the situation and would that make a difference?

34.4 Mutual relationship of Art 30 Regulation 834/2007 and Art 91 Regulation 889/2008: different interpretations

The Trade and Industry Appeals Tribunal (*CBb*) issued a judgment in 2021 on the interrelationship of Article 30 Regulation 834/2007 and Article 91 Regulation 889/2008. The CBb finds that Article 91 Regulation 889/2008 does not need to be

343 See recital 3.4.: 'Finally, the appellant's argument that she did not intend to make a notification as referred to in Article 91, paragraph 1, of the Regulation with her e-mail message of 20 November 2015 to the defendant regarding the results of the Eurofins Labco investigation cannot be followed. After all, the absence of such intention is in no way apparent from that message. To the extent that the appellant states that she only wanted to ask a question about a drying factor of the product in and with that message, the Tribunal notes that such a question is not apparent from the message either.'

344 See recital 3.4 '4. The foregoing means that, in the opinion of the Tribunal, the defendant in this case was authorised to block and decertify batch D-150629 and could reasonably have exercised that authority.'

followed by Skal before Skal can decertify on the basis of Article 30 Regulation 834/2007. According to the CBb, it is irrelevant whether Skal followed the procedure of Article 91 of Regulation 889/2008 when taking measures.[345] In doing so, the CBb found it unnecessary to put preliminary questions to the Court of Justice, because '[*a*]*s regards the relationship between Article 30(1) and first subparagraph of Regulation 834/2007 and 91 of Regulation 889/2008, there is, in view of what has been considered above, no doubt that requires a preliminary question to be put to the Court of Justice of the European Union (Court).*'[346]

In our view, the comment that there is 'no doubt' about the interrelationship may have been phrased too strongly. Especially given the Tribunal (*CBb*) already discussed that Skal could decertify under Article 91 Regulation 889/2008 *because* the company's notification had to be seen as a notification within the meaning of Article 91 Regulation 889/2008. Could that consideration not, in fact, mean that if a company does not inform Skal of its own observation of a possible infringement or irregularity, Skal is therefore not entitled to the basis of jurisdiction of Article 91 Regulation 889/2008 in that situation? At the same time, in our view, it is not inconceivable that Article 91 Regulation 889/2008 is an elaboration of Article 30 Regulation 834/2007. After all, Article 91 Regulation 889/2008 is included in Regulation 889/2008, also called the 'Implementing Regulation' (*what's in a name*?). However, that leaves the question what the basis of for the authority to decertify is in a case where Skal becomes aware of a (possible) infringement or irregularity after its own inspection. In any case, we believe this provides ample food for thought.

Some doubt about the interrelationship between Article 30 of Regulation 734/2007 and Article 91 of Regulation 889/2008 does exist, in our view. Even if only because the mutual relationship of the relevant articles can be interpreted in different ways. Below, we provide a (non-exhaustive) list of possible views:

1. The jurisdictional bases coexist;
2. Article 91 Regulation 889/2008 is an elaboration of Article 30 Regulation 834/2008;
3. The basis of jurisdiction depends on how Skal became aware of the infringement/irregularity;
   - Article 91 Regulation 889/2008 only applies when a notification has been made and Article 30 always applies;
   - Article 91 Regulation 889/2008 only applies when a notification has been made and Article 30 only applies when Skal has become aware of a possible infringement/irregularity based on an inspection.

345 CBb 13 April 2021, ECLI:NL:CBB:2021:387, recital 5.2.
346 CBb 13 April 2021, ECLI:NL:CBB:2021:387, recital 5.2.

34.5 Clarity for the sake of consumer confidence (and legal certainty for businesses)

One might argue that it makes no difference to a company whether it faces decertified on the basis of Article 91(2) of Regulation 889/2008 or Article 30 of Regulation 834/2008. After all, decertified is decertified: the legal consequences are the same and the damage to companies is no less costly. However, this distichtion is more than mere legalistic nuance. In fact, we identify two key differences between the review framework of Article 91(2) of Regulation 889/2008 and Article 30 of Regulation 834/2008: namely, the burden of proof and the role of proportionality. Therefore, clarification by the Court of Justice of the European Union would, in our view, be higly adviseable and certainly not trivial.

Burden of proof

Article 91(2) of Regulation 889/2008 mandates an initial 'substantiated suspicion'. The precise definition of 'substantiated suspicion' remains somewhat unclear, but it unequivocally necessitates more than mere suspicion, and certainly more than rumour or hearsay. For instance, a positive test result for a banned substance could constitute such a (substantiated) suspicion. However, in the system of Article 91(2) Regulation 889/2008, this does not suffice for decertification. After all, the substantiated suspicion must still be investigated within a timeframe determined by Skal. Should the suspicion be confirmed at the end of this period, then, and only then, can definitive decertification proceed. In cases in which the suspicion is unconfirmed (either disproved or lingering doubt remains), the batch must be released.

In contrast, Article 30 Regulation 834/2007 differs, requiring for an irregularity to be *found*. Skal subsequently carries the burden of proving this finding. Crucially, administrative law does not demand 'physical, scientifically indisputable evidence'; rather, the finding of an irregularity must be sufficiently plausible. Practically, this leads to significant discussion: what is the approach with one positive versus ten negative samples? And how should a situation be handled where a positive sample is found, but the substance could not reasonably have entered the product (for example, if it is not used in the product's conventional cultivation)?

Proportionality

Another key distinction is the role of 'proportionality'. Indeed, the attentive reader will note that Article 30 Regulation 834/2007 grants Skal the authority to decertify in instances of observed violations of laws and regulations, but only when such a measure is 'proportionate to the relevance of the requirement' that has been violated and to the nature and specific circumstances of the irregular activities. Consumer confidence in organic products is a significant factor in this regard. Nonetheless, this cannot be a *decisive* argument, as it would render this subordinate clause in the European regulation meaningless. The point is, if Skal were to allow a product to market after a violation was found, consumers would end up with a product that is not fully compliant. Instead, the focus ought to be on whether the violation's severity, duration, extent, and nature are such that they warrant decertification.

At the same time, it bears mentioning that, beyond this specific proportionality test, all decisions made by a Dutch administrative body must satisfy the proportionality test pursuant to Article 3:4 Dutch General Administrative Law Act (Awb). This counts for an Article 30 decertification, but thus also for an Article 91(2) decertification in which no 'European proportionality test' is applied. Under the Dutch proportionality test, the administrative body is mandated to balance all interests involved in the decision, ensuring that any adverse consequences for parties are not disproportionate to the decision's intended objectives. In our opinion, it could therefore be unreasonable, for example, to decertify substantial batches of organic product merely due to a minor administrative omission violating European laws and regulations – such as the absence of a transport document or a single positive sample, especially when no other evidence suggests it is a non-organic product and multiple clean counter-samples exist. Indeed, this often causes millions in damages to the parties involved, when there is no food safety issue but 'only' a quality issue.

34.6 Conclusion

Clarifying the underlying principles, in our opinion, is crucial for two reasons: it will provide legal certainty to the organic market and, moreover, secure the integrity of the organic production process itself, as intended by the pertinent regulations. Specifically, as outlined in the policy framework: '*the policy framework should pursue the objective of ensuring fair competition and a proper functioning of the internal market in organic products, and of maintaining and justifying consumer confidence in products labelled as organic. It should further aim at providing conditions under which this sector can progress in line with production and market developments*' attention should not only be given to consumer confidence in organic products, but also to fostering the progress of organic product production.

For nurturing consumer confidence, ensuring the legal certainty of companies and advancing organic production processes, the transparancy and mutual trust between these actors is crucial. This extends beyond just to openness displayed by companies and Skal to consumers, adressing questions like, 'Is this carton of milk truly organic?' In our view, consumer trust is also served by openness between the companies and Skal. After all, when considering the reserve. Consumer confidence could actually be jeopardized if companies refrain from reporting irregularities to Skal, fearing a rigid policy or enforcement action based on an unclear or incorrect basis for jurisdiction. The organic sector and the consumer would arguably not benefit from a situation where organic products become more expensive and scarce because fewer parties dare to 'go organic'.

Uncertainty about the basis of jurisdiction is, in our view at least, unsustainable for the organic sector and should therefore be avoided, particularly in light of the European Union's Green Deal ambitions for the organic sector.

Silvia Gawronski
Partner International Trade & Commodities at Van Traa law firm, Rotterdam

Dagmar Linstra
Associate Insurance & Liability at Van Traa law firm, Rotterdam.

CHAPTER 35

Refrigeration in the slaughterhouse – Legal unity in food law

Annotated by Silvia Gawronski and Mirjam von Meijenfeldt

Trade and Industry Appeals Tribunal 6 February 2018, ECLI:NL:CBB:2018:10, NJB 2018/428, AB 2018/84 cf. R. Ortlep, JW 2018/18;
The Hague Court of Appeal (criminal division) 13 March 2013, ECLI:NL:GHDHA:2013:BZ9420;
Court of Justice of the European Union 2 May 2019, C-98/18, ECLI:EU:C:2019:355;
Trade and Industry Appeals Tribunal 12 May 2020, ECLI:NL:CBB:2020:333, NJB 2020/1334

Establishment term. Fine for slaughterhouse for loading meat with temperature above 7°C. Can the meat be cooled in a refrigerated lorry or should this take place in the slaughterhouse itself? Reference for a preliminary ruling concerning the interpretation of Regulation 853/2004.

Main considerations:

The Hague Court of Appeal:

> […] that the regulation does not impose any requirements on the design and operation of the premises at which the meat is located during the stage of storage other than (continuous) refrigeration and adequate ventilation. Furthermore, the regulation does not require that, after the post-mortem inspection during the stage of storage, the meat must be a maximum of 7°C at all times until it can be transported. Only at the time the meat can be transported (and during that transport) should the temperature not exceed 7 degrees Celsius.

Trade and Industry Appeals Tribunal (CBb):

> 1.1. The appellant operates a food business which processes and markets veal, meat products and by-products. […] The slaughter process at the appellant can be summarised as follows. […] The carcasses and carcass parts (hereinafter: the meat) are then cooled on slaughter hooks in the cold store and/or refrigerated lorry. The cold store is an extension of the room where the slaughtering and butchering of the animals takes place and connects seamlessly to that room. As an extension of the cold store, there is a refrigerated lorry which, the appellant said, connects to the cold store

 | HTTPS://DOI.ORG/10.1163/9789004751699-037

> via the dock of the area designated for slaughter and butchering. The meat is loaded from the cold store to a refrigerated lorry. During loading, the temperature of 10 representative risk parts is measured for each refrigerated lorry. In the refrigerated lorry, the meat is stored and, according to the appellant, further cooled until the prescribed temperature of 7°C is reached throughout the meat. [...] According to the appellant, after the prescribed temperature has been reached throughout the meat via a continuous drop in temperature, the transport documents are drawn up and issued. The meat is then transported in the refrigerated lorry. [...]
> 7.2. What is in dispute, however, is whether the appellant's refrigeration process, in which the required temperature of 7°C for the meat is achieved not in the slaughterhouse but in a refrigerated lorry, complies with the provisions of Annex III, Section I, Chapter VII, introductory sentences and points 1 and 3 of Regulation (EC) No 853/2004.
> 7.3. The Tribunal, in an earlier case between the same parties, by decision of 20 March 2017 (ECLI:NL:CBB:2017:69) answered this question in the negative [...].
> 7.4. [...] The Tribunal is, for the time being, of the opinion that there is no reason to go back on its considerations in the judgment of 29 March 2017, but now recognises that the interpretation of the provisions of Union law relevant here is not so evident that there can reasonably be no room for doubt. In these circumstances, the Tribunal [...] considers it appropriate to refer to the Court of Justice, on the basis of Article 267 of the Treaty on the Functioning of the European Union, the preliminary question formulated below.

Court of Justice of the European Union:

> 48. [...] Annex III, Section I, Chapter VII, points 1 and 3, to Regulation (EC) No 853/2004 of the European Parliament and of the Council of 29 April 2004 laying down specific hygiene rules for food of animal origin must be interpreted as meaning that the chilling of meat after slaughter must be carried out in the slaughterhouse itself until the meat has reached a temperature throughout of not more than 7 °C before any loading of the meat into a refrigerated truck.

35.1 Introduction

Regulation (EC) No 853/2004 *laying down specific hygiene rules for food of animal origin* contains detailed rules, including with regard to the slaughter process. Several judges – in criminal, civil and administrative courts – have addressed the question of whether or not meat may be chilled in a refrigerated lorry on the slaughterhouse premises instead of inside the slaughterhouse building itself. The applicable provision, Annex III, Section I, Chapter VII, introductory paragraph and points 1 and 3, reads as follows in so far as relevant:

> Food business operators must ensure that the storage and transport of meat of domestic ungulates takes place in accordance with the following requirements.
> 1. (a) Unless other specific provisions provide otherwise, post-mortem inspection must be followed immediately by chilling in the slaughterhouse to ensure a temperature throughout the meat of not more than 3 °C for offal and 7°C for other meat along a chilling curve that ensures a continuous decrease of the temperature. However, meat may be cut and boned during chilling in accordance with Chapter V, point 4;
> […]
> 3. Meat must attain the temperature specified in point 1 before transport, and remain at that temperature during transport. […]

The judges arrive at different answers. The main questions are what is to be understood by 'chilling in the slaughterhouse' (sub 1) and by 'transport' (sub 3). In other words, can chilling in a refrigerated lorry on the slaughterhouse premises qualify as chilling 'in the slaughterhouse' and/or does chilling in a refrigerated lorry on the slaughterhouse premises qualify as 'transport'?

The The Hague economic police judge (magistrate in economic criminal law matters) ruled that cooling in a refrigerated van was not allowed and constituted an offence. On appeal, as cited above, the economic criminal law division of the Court of Appeal came to a different opinion (ECLI:NL:GHDHA:2013:BZ9420). The civil chamber of the The Hague District Court could not agree to chilling in a refrigerated van (ECLI:NL:RBSGR:2011:BU1554) nor could the The Hague Civil Court on appeal (ECLI:NL:GHDHA:2013:4678). While the Administrative Court in Rotterdam saw room for such an interpretation (ECLI:NL:RBROT:2015:6266), the Trade and Industry Appeal Tribunal (the 'Tribunal' or CBb) did not (ECLI:NL:CBB:2017:69). As such, the Tribunal ruled that the correct interpretation cannot reasonably be in doubt. In the above ruling, the Tribunal (ECLI:NL:CBB:2018:10) did in the end ask preliminary questions. Ultimately, it sees its initial judgment upheld. This case proves once again that the road to legal unity in food law can be a long one.

35.2 The Trade and Industry Appeal Tribunal stands by its earlier judgment… for now

The interlocutory judgment at the centre of this case has many interesting legal aspects. We will limit ourselves to discussing the preliminary question, which the Court of Justice of the European Union (the 'Court') has since answered on 2 May 2019 (C-98/18, ECLI:EU:C:2019:355). In this interlocutory ruling, the Tribunal (CBb) sees no reason in principle to deviate from its previously adopted line that transport begins with the loading of the meat into the refrigerated lorry, regardless of whether, how, and for how long a refrigerated lorry is on the premises of the establishment.

Annex III, Section I 'Meat of domestic ungulates', Chapter VII 'Storage and transport', introductory wording and points 1 and 3 of Regulation 853/2004 lay down requirements for the storage and transport of meat. Under the aforementioned provision, meat must be chilled immediately after post-mortem inspection at the slaughterhouse to ensure a temperature of 7°C via a continuous drop in temperature – anywhere in the meat. Meat must reach the temperature of 7°C before it can be transported and must maintain that temperature during transport.

The definitions of Regulation 178/2002 and Regulation 852/2004 as well as Annex I of Regulation 583/2004 have been declared applicable.[347] But none of these regulations further defines the terms 'transport' or 'loading'. It is not incomprehensible, therefore, that the Tribunal (CBb) relies on the term 'slaughterhouse', which is defined. Presented schematically, the following definition ladder then follows. The definition ladder is presented schematically in Table 35.1 below.

In the Tribunal's opinion, the refrigerated lorry is not a unit of a food business and therefore not an establishment as referred to in Article 2(1)(c) of Regulation 852/2004. As The Hague Court of Appeal also ruled in its judgment of 2 July 2013,[348] according to the Tribunal (CBb), a refrigerated lorry is by its nature intended for the transport of meat and not for the slaughter and butchering of meat. The fact that cold stores can be part of a slaughterhouse does not make a refrigerated lorry also such a part. Nor can a refrigerated lorry be regarded as belonging to the 'unit' of the food business, because a refrigerated lorry is not permanently intended

347 Article 2 of Regulation (EC) 853/2004.
348 The Hague Court of Appeal 2 July 2013, ECLI:NL:GHDHA:2013:4678.

TABLE 35.1

Term	Definition	Location
Slaughterhouse	'an *establishment* used for slaughtering and dressing animals, the meat of which is intended for human consumption'	Regulation (EC) No 853/2004, Annex I, Article 1(1.16).
Establishment	'any unit of a *food business*'	Regulation (EC) No 852/2004, Article 2(1)(c).
Food business	'any undertaking, whether for profit or not and whether public or private, carrying out any of the activities related to any stage of production, processing and distribution of food'	Regulation (EC) No 178/2002, Article 3(2).

for the use of the food business but for transporting the meat to the customer.[349] Furthermore, the CBb considers that it is clear from the preamble to Regulation 853/2004 that its main purpose is to ensure a high level of consumer protection with regard to food safety. In the light of this, there would be no place for an interpretation of terms that could detract from the guarantees the regulation is intended to provide.

In other words, temperature requirements must be met before the food leaves (the establishment of) the slaughterhouse and enters the refrigerated lorry.

As a result, for the time being, the appellant has not complied with the condition laid down in Annex III, Section I 'meat of ungulates kept as domestic animals', Chapter VII 'storage and transport', introductory wording and point 3 of Regulation 853/2004, according to which the meat must have reached the temperature of not more than 7 °C referred to in point 1 of this provision before it can be transported.

However, the appellant interprets the provision differently. According to the appellant, it does not follow from this provision that the meat must have reached a temperature of 7°C before it can be *moved* to a refrigerated lorry if it is further chilled there (but not yet *transported*). According to the appellant, the process of chilling freshly slaughtered meat to a temperature below 7°C in a parked refrigerated lorry in the slaughterhouse car park is in accordance with the aforementioned regulations. Especially since the refrigerated lorry is connected to the production line and is essential for the actual chilling process.

35.3 … The Tribunal (CBb) had its doubts

The Tribunal (CBb) acknowledges in the interim ruling that its own interpretation of the regulation is not so straightforward, i.e. there could be other interpretations. The Tribunal (CBb) does not elaborate on its doubt. Perhaps it lies in the (changed) facts. In this case, what is certain is that the meat, at the time of loading into the refrigerated lorry, is not yet intended to be transported, as it is first kept on the slaughterhouse premises for quite some time. Loading, in the narrow sense of the word, could then not yet be considered *transport* if the meat is not yet intended to be transported to the customer. Or perhaps the Tribunal (CBb) recognised that under some circumstances a refrigerated lorry could still be part of the establishment within the meaning of the regulation, and wondered where that boundary lay.

349 The Hague Court of Appeal 2 July 2013, ECLI:NL:GHDHA:2013:4678.

35.4 A parallel with the concept of 'establishment'

What struck us about this interlocutory ruling and doubt is the view of the concept of 'establishment'. If we draw a parallel with the concept of 'establishment' under the Dutch Environmental Management Act, it is defensible that the refrigerated lorry does in fact belong to the establishment. Pursuant to Article 1.1(1) of the Dutch Environmental Management Act (Wm), the term 'establishment' in this act and the provisions based on it is understood to mean: 'any activity undertaken by humans as a business or on a scale as if it were a business, which tends to be carried out within a certain boundary'.[350] We assume that the slaughterhouse meets these three establishment criteria. Furthermore, pursuant to Article 1.1, subsection 4, of the Environmental Management Act, elsewhere in this Act and the provisions based on it, the term 'establishment' means 'an establishment belonging to a category designated pursuant to subsection 3. For this purpose, the installations belonging to the same undertaking or institution which have technical, organisational or functional links between them and are located in close proximity to each other shall be considered as one establishment. [...]' If there are two of the three bonds, it can be said to be one establishment. If the appellant has control over both the site, the building as well as the refrigerated lorry at the time of refrigeration and the refrigerated lorry is connected to facilities such as electricity present there – it could be argued that the refrigerated lorry satisfies two of the three bonds making the refrigerated lorry part of this establishment at that time. Although the case does not provide a definitive answer to the existence of these bonds, nor to the technology behind the refrigerated lorry so that it can be considered an installation, we consider it very possible that a parallel can be drawn with the environmental regulations so that there can be said to be one establishment.

35.5 Unintended consequences of broad interpretation by the Tribunal (CBb)?

We also noticed that no attention is paid to the other consequences of the broad interpretation of the term 'transport'; i.e. including loading within the establishment. In any case, we thought it would be advisable to consider this carefully: after all, is the refrigerated lorry part of the (slaughterhouse) establishment if it goes all the way into the slaughterhouse?[351] And what about the situation where a company

350 (1) business or on the scale of a business; (2) within a certain boundary; (3) is usually carried out. For the sake of brevity, we assume these three criteria are known.

351 Some of these aspects are discussed in a ruling of the Tribunal of 2020 (ECLI:NL:CBB:2020:808). The case concerned a refrigerated lorry that was connected by open doors to the dock shelter of the slaughterhouse. As yet, the Tribunal (CBb) does not seem to be affected by this line of thinking.

chooses to do the refrigeration in its own or rented (mobile) refrigeration unit on the premises instead of in the slaughterhouse itself, because the refrigeration unit is broken, or when it is economically more advantageous at production peaks to put a refrigerated lorry on the premises instead of expanding the refrigeration facilities? And if these situations lead to a different conclusion: Could you arrange things by contract, e.g. by having the refrigeration on the slaughterhouse's own premises done by the slaughterhouse staff at the slaughterhouse's own expense and risk?

Consideration should also be given to the spill-over effect to other establishment issues: for example, is a mobile waste shredder on the premises of an electronics manufacturer part (or a unit) of the facility? And does it then matter whether it arrives at the site and stays for a few hours, whether it is placed in the factory hall or whether it is rented by the company? Indeed, if it is part of the establishment, then the establishment licence may need to be extended.

Finally, we noted that the Tribunal's position also excludes technological innovation to reach a hygienic, safe but also cheaper and/or more efficient solution for chilling meat, such as in a connected refrigerated lorry. After all, the Tribunal concludes and interprets the concept of safe, but the ruling offers no connecting factors to infer that it has been scientifically or technically substantiated that the technology of the refrigerated van does indeed compromise the level of consumer protection.

35.6 Preliminary question answered

The preliminary question has now been answered by the Court. The Court infers from the wording of Annex III, Section I, Chapter VII, points 1 and 3 in Regulation 853/2004 that refrigeration must indeed take place in the slaughterhouse itself. Although the Court does not clarify this, it is clear from the considerations that it does not consider the refrigerated lorry to be part of (the establishment of) the slaughterhouse, at least not in this case. It considers – in short – that a refrigerated lorry is intended for transport, not for refrigeration. Nor is the function of a refrigerated lorry aimed at chilling meat; the cooling systems in a refrigerated lorry are designed to maintain low temperatures during transport. In addition, slaughterhouses facilitate optimal chilling under optimal conditions, which is not the case with chilling that can take place in a refrigerated lorry (using a rule of thumb to determine temperature). Although the means of transporting meat itself must meet a number of requirements, they are exempt from the authorisation procedure to be applied by the competent authorities and are not covered by the slaughterhouse operator's authorisation. Furthermore, allowing refrigeration in a refrigerated lorry on the premises of a slaughterhouse to 7°C would not be in line with the main purpose of hygiene rules, which aim to ensure a high level of consumer protection with regard to food safety. In this context, the European legislator made food business operators responsible for ensuring food safety and (thereby) obliged them to slaughter animals in slaughterhouses that comply with technical requirements on construction, layout and equipment (Annex III to Regulation 853/2004).

Although rather conservatively framed, sufficient room (for innovation) can be derived from this mode of reasoning for, for example, a refrigerated lorry that is aimed at cooling meat, is also designed as such, contains sound measuring instruments and is formally included in an approval procedure. The Tribunal (CBb) has more or less made the Court of Appeal's conclusion and considerations its own by upholding its consideration in 7.3. We hope this means that the path to innovative solutions – where the refrigerated lorry does form part of the approval procedure and thus of the establishment – has deliberately remained open and has not been unnecessarily 'cut off'.

Silvia Gawronski
Partner International Trade & Commodities at Van Traa law firm, Rotterdam

Mirjam von Meijenfeldt
Legal counsel renewable energies at VARO Energy

CHAPTER 36

Alpro – Protected terms for dairy

Annotated by Bernd van der Meulen

Den Bosch Court of Appeal, 19 December 2017, ECLI:NL:GHSHE:2017:5731 (*NZO – Alpro*); confirmed by HR 30 August 2019 ECLI:NL:HR:2019:1293
Court of Justice of the European Union 14 June 2017, ECLI:EU:C:2017:458 (*Tofu-Town*)
Court of Justice of the European Union 4 October 2024, ECLI:EU:C:2024:826 (*Beyond Meat*)

The protection of dairy terms in the Common Market Organisation Regulation (Regulation (EU) No 1308/2013) is limited to the designation of products. However, other uses in relation to non-dairy products are permitted.

Main considerations:

Den Bosch Court of Appeal in NZO – Alpro:

> The use of the reserved dairy designations
> 3.6.1. In its assessment, the court makes the following points. In its aforementioned judgment of 14 June 2017 (TofuTown), the CJEU ruled, with regard to the interpretation of Article 78(2) and Part III of Annex VII, that those provisions must be interpreted as precluding the use of the designation 'milk' and the designations reserved by that regulation exclusively for milk products, in marketing or advertising, a purely vegetable product, even if those designations are supplemented by clarifying or descriptive additions which point to the vegetable origin of the product concerned. According to the CJEU, the exception set out in Part III(5) in the second paragraph applies only to those products listed in the Commission decision of 20 December 2010 (Decision 2010/791/EU), that being the decision establishing the list of products referred to in (now) Annex VII, Part III(6), second paragraph. There is no dispute between the parties that this exception does not apply to [Alpro]'s products in this case.
> 3.6.2. Subject to that interpretation by the CJEU, [Alpro] may not use the designation 'milk' and the other reserved dairy designations (which include the designations 'yoghurt' and 'cream') as the (sales) designation of its soy products, nor may it use those reserved dairy designations to designate its products in any way. In doing so, it makes no difference whether it uses the reserved designations on the packaging of its products or otherwise in the marketing or presentation of those products.

 | HTTPS://DOI.ORG/10.1163/9789004751699-038

3.6.3. Contrary to NZO's argument, in the opinion of the court of appeal this interpretation does not imply that [Alpro] is not allowed to use the designations 'milk', 'yoghurt' and 'cream' (or any other reserved dairy designation) in any way whatsoever on its packaging or otherwise when presenting its products. Indeed, such use is only prohibited as a designation or indication of its soy products. According to the CJEU in its judgment of 14 June 2017 (TofuTown), if the reserved dairy designations are used by [Alpro] in any other way, such use is not – as such and without further ado – contrary to the provisions of Regulation (EU) No 1308/2013.

CJEU in TofuTown:

Article 78(2) and Annex VII, Part III, to Regulation (EU) No 1308/2013 [...] establishing a common organisation of the markets in agricultural products [...] must be interpreted as precluding the term 'milk' and the designations reserved by that regulation exclusively for milk products from being used to designate a purely plant based product in marketing or advertising, even if those terms are expanded upon by clarifying or descriptive terms indicating the plant origin of the product at issue, unless that product is listed in Annex I to Commission Decision 2010/791/EU of 20 December 2010 listing the products referred to in the second subparagraph of point III(1) of Annex XII to Council Regulation (EC) No 1234/2007.

CJEU in Beyond Meat

Articles 7 and 17 of Regulation (EU) No 1169/2011 (...), read in conjunction with Article 2(2)(o) and (p) and Article 9(1)(a) thereof, must be interpreted as meaning that they specifically harmonise, within the meaning of Article 38(1) of that regulation, the protection of consumers from the risk of being misled by the use of names, other than legal names, consisting of terms from the butchery, charcuterie and fish sectors to describe, market or promote foods containing vegetable proteins instead of proteins of animal origin, including in their entirety, and thereby preclude a Member State from adopting national measures regulating or prohibiting the use of such names.
Article 38(1) of Regulation No 1169/2011 must be interpreted as meaning that the specific harmonisation referred to in point 1 of this operative part does not prevent a Member State from laying down administrative penalties for non-compliance with the requirements and prohibitions resulting from that regulation and from national measures that are in conformity with that regulation. However, that specific harmonisation precludes a Member State from adopting a national measure setting the proportions of vegetable proteins below which the use of names, other than legal names, consisting of terms from the butchery and charcuterie sectors to describe, market or promote foods containing vegetable proteins is still authorised.

36.1 'What's in a name? That which we call a cheese, by any other name would smell as sweet'

The Common Market Organisation Regulation (Regulation (EU) No 1308/2013) grants special status to a number of terms traditionally used for dairy products (hereafter also referred to as dairy terms). A legal battle has ignited in several EU Member States between the dairy sector, on the one hand, and companies that use dairy terms to position vegetable products, on the other. In Belgium, for example, Brussels Court of Appeal ruled on 10 March 2015 (Registry No. 2014/AR/1274) that Alpro may not use a number of expressions such as 'yoghurt variation' and 'plant-based variation on yoghurt' on its Facebook page, website, in advertisements or on its packaging. In a German case, the CJEU ruled (CJEU 14 June 2017, case C-422/16 *TofuTown*) that such dairy terms may not be used as *names* of non-dairy products in any case.

Alpro in particular is also in the spotlight in the Netherlands. The brand positions soy- or almond-based products as an alternative to yoghurt, among other things. A question that continued to occupy minds is to what extent dairy terms may be used in such positioning.

The judgment of Den Bosch Court of Appeal is between the *Nederlandse Zuivelorganisatie* (NZO – Dutch Dairy Association) as plaintiff at first instance and on appeal and Alpro. The parties are engaged in a long-running dispute over the use of dairy terms. In May 2012, the Breda District Court granted summary judgment between these parties, ordering Alpro, among other things, to cease using the term 'yoghurt' in certain publications and advertisements. Thereafter, this discussion continued in proceedings on the merits before the Court of Zeeland-West Brabant (Breda). There, NZO was largely ruled against. Most of Alpro's expressions were deemed admissible under the applicable legislation. NZO appealed against this ruling to Den Bosch Court of Appeal. The Court of Appeal delivered its judgment on 19 December 2017. Thereupon, NZO appealed to the Supreme Court. The Supreme Court upheld the judgment of the Court of Appeal.

36.2 Poetic licence

What freedom do food companies have in the choice of words when communicating about their products? To what extent may they use terms to which the legislator has granted special protection in the context of products that do not belong to the protected category?

Food law has specific rules for the naming of food products. Market regulation law adds another layer to that for a number of products including dairy products. The above ruling essentially revolves around how big that layer is.

In this annotation, I turn first to general food law rules on naming and then to the

Common Market Organisation Regulation and its application by Den Bosch Court of Appeal in the above case.

36.3 Naming in food law

Regulation (EU) No 1169/2011 on Food Information to Consumers (the FIC Regulation) focuses specifically on the interests of consumers. According to Article 1(1), the Regulation aims to provide the basis for ensuring a high level of consumer protection with respect to food information. According to Article 3(1), the provision of food information aims to ensure a high level of protection of consumers' health and interests by providing final consumers with a basis for making informed choices and the safe consumption of food.

Articles 9 and 17 of the Food Information Regulation require a food to bear a name. The name in this food legal sense is to be distinguished from what in general speech is called a 'name'. In common parlance, for example, we knew products with names like 'Mars', 'Raider' and 'Golden Wonder'. Such a name is generally protected as a trademark, but sometimes not (e.g. because it names properties of the commodity, such as BioMild). In the literature, such name is also called 'fancy name'. The function of the name within the meaning of the Regulation is to make clear to consumers exactly what the product is, including consumers who are not familiar with the product. And that is why Mars has as its name within the meaning of the Food Information Regulation: 'Milk chocolate with airy filling (32%) and caramel (27%)'. Older Dutch legislation, such as the former *Warenwetbesluit Etikettering van levensmiddelen*, and the *Warenwetbesluit Zuivel* to be discussed below, uses two words synonymously for designation: '*aanduiding*' and '*benaming*'. The Court of Appeal of Den Bosch uses both words interchangeably. In addition, the Court also uses words such as 'name', 'product name' and 'generic name'. It seems that the Court does not distinguish the concepts of 'name' and 'designation' but, with the different words it uses, includes both at the same time.

Except in literal quotations, in this contribution I use the words 'name' in accordance with common parlance for the commercial designation of a product; 'designation' for the labelling designation of a product subdivided into 'legal name', 'customary name' and 'descriptive name'; 'term' or 'protected term' for those words to which legislation has conferred special status.

Article 17 of the Food Information Regulation requires that if there is a legal name, it must be used. This legal name can be found in product-specific EU or Commodity Act regulations. A legal name may not be used for a product that does not comply with these regulations.

An example of product-specific regulations is the Dutch Commodities Act Decree on Dairy (Stb. 2016, 61). This Commodities Act Decree states that designations such as 'cheese' and 'yoghurt' may only be used for products that meet composition requirements detailed in the Decree. Pursuant to Article 2(8) of the Dutch Commodities Act Decree on Dairy, '[it] is prohibited to use the designations specified – or similar to them – by or under this Decree to market other goods than those to which such designations are reserved by or under this Decree.' The term 'dairy' is defined in the decree as follows: 'dairy products, i.e. products derived exclusively from raw milk, on the understanding that substances necessary for their manufacture may be added, provided that these substances are not used to replace, in whole or in part, any of the milk's constituents' (Article 1(m) (i)). 'Raw milk' is then defined as: 'milk secreted by the mammary gland of one or more cows, ewes, goats or buffaloes, which has not been heated (…)' (Article 1(h)).

For products for which there is no legal name, the customary name should be used. An example of this is the *kroket*. In the Dutch Government's view, a legal name may become a customary name when the former is abolished. In that case, it remains the case that it may not be used for another product. For example, in the explanatory notes to the Decree of 12 November 2004 repealing and amending various Commodities Act Decrees (STB[352] 2004, 599), the Government writes, 'Until the entry into force of this Decree, the designations coffee and tea were compulsorily prescribed designations (…). From the entry into force of this Decree, these have become generally accepted designations (…). The marketing of a commodity with, for example, the self-chosen designation *coffee*, when that commodity does not consist of the properly cleaned and roasted seed cores of the coffee bean (species of the genus Coffea), is therefore prohibited even after the entry into force of this Decree.'

If a customary name is also missing or not used, a descriptive name should be used. In the Mars example given above, 'Milk chocolate with airy filling (32%) and caramel (27%)' is thus a descriptive name in which the legal name has been used for the ingredient 'milk chocolate'.

36.4 CMO and TofuTown

Regulation (EU) 1308/2013, the Common Market Organisation (or CMO) Regulation, contains, among many other things, protective regimes for the use of terms for specific food categories, including, among others, dairy terms. The question now is whether this regime adds something to the general food law protection of legal names (including dairy names in the Commodities Act Decree for Dairy), and if so what this 'something' is. This question falls into 2 sub-questions: 1. Is it permissible to use dairy terms as designations for non-dairy products? 2. Is it permissible to use dairy terms for non-dairy products *other than as a designation*?

352 Bulletin of Acts and Decrees.

The Court of Justice of the EU (CJEU) answered the first question in the negative in the TofuTown case (CJEU 14 June 2017, case C-422/16). This also applies in circumstances where further qualifications eliminate the risk of misleading of consumers. The European Commission's Notice on questions and answers on the application of Regulation (EC) No 1169/2011 of the European Parliament and of the Council on the provision of food information to consumers (Official Journal C 196/1, 8/6/2018, para. 2.1.1) also follows this line of argument. This states, for example, that it is prohibited to use the name 'imitation cheese', as the name 'cheese' is reserved exclusively for milk products. It is up to the food business operator to find a suitable name for this substitute product in accordance with the rules on food naming.

The second question, 'Is it permissible to use dairy terms for non-dairy products other than as a designation?', was essentially addressed in the case that gave rise to the judgment printed above.

The CMO Regulation differs from the general food law protection regime in three ways. First, the scope of the CMO Regulation's objectives is broader than that of the Food Information Regulation. Second, the CMO Regulation provides protection to terms independent of product specifications (other than that it concerns milk). Thirdly, the prohibition on using the legal name for other products is more broad. These three aspects taken together provide food for thought as to whether the prohibition on using the protected terms might extend beyond use as a designation.

The purpose of protecting terms in the CMO Regulation is not only to protect consumers from being misled but also to protect the market position of producers of products falling within these protected categories. As the preceding Common Market Organisation Regulation 1234/2007 puts it in recital 51, 'Several legal instruments have been put in place to regulate the marketing and designation of milk, milk products and fats. They pursue the objective of improving the position of milk and milk products on the market on the one hand and ensuring a fair competition between spreadable fats of milk and non-milk origin on the other, both to the benefit of producers and consumers.'

The TofuTown case concerned designations for purely vegetable products containing terms such as 'butter', 'cheese' and 'cream'. The CJEU concluded that the relevant provisions of the CMO Regulation preclude the use, in marketing or advertising, of the designation 'milk' and the designations reserved exclusively for dairy products by that Regulation to designate a purely vegetable product, even where those designations are supplemented by clarifying or descriptive supplements indicating the vegetable origin of the product concerned, unless there is a statutory exception.

Commission Decision 2010/791/EU lists such exceptions in the different EU languages. In Dutch, for example, exceptions apply to peanut butter, coconut milk and *leverkaas* (processed meat containing liver referred to as 'cheese').

The CMO Regulation states in Article 78(2): 'The definitions, designations or sales descriptions provided for in Annex VII may be used in the Union only for the marketing of a product which conforms to the corresponding requirements laid down in that Annex'. The question that remains open after TofuTown is whether this provision refers only to use of the 'definitions, designations or sales description' as a 'name' for a product, or also to other use for the purpose of marketing a product. The preamble to Annex VII, to which this provision refers, equates one of the three terms with the term name: 'For the purposes of this Annex, the "sale description" means the name under which a foodstuff is sold, [...], within the meaning of Article 17 of Regulation (EU) No 1169/2011.' Since the CMO Regulation does not suffice with this single term, but juxtaposes it with the terms 'definition' and 'designation', it is obvious to assume that the three words refer to a broader category than the single word, although that in itself says little about the use made of those words. The third part of the annex concerns dairy terms. The second paragraph of this section of the Annex states, among other things, 'The following shall be reserved exclusively for milk products. a) the following names used at all stages of marketing: following designations, used at all stages of marketing: i) whey, ii) cream, iii) butter, iv) buttermilk, v) butter oil, vi) caseins, vii) anhydrous milk fat (AMF), viii) cheese, ix) yogurt, (...)'. The wording in the Dutch language version is a bit unfortunate. However, comparison with other language versions clearly shows that the phrase which in Dutch reads [in translation]'For dairy products only the following names may be used ...', should say 'Only for dairy products may the following names be used ...'. There is no misunderstanding about this in TofuTown and in the above case. The sixth paragraph reads, 'In respect of a product other than those described in points 1, 2 and 3 of this Part, no label, commercial document, publicity material or any form of advertising [...] or any form of presentation may be used which claims, implies or suggests that the product is a dairy product.' The Den Bosch court applies the sixth paragraph standard (that it may not be indicated, implied or suggested) on a case-by-case basis. The mere use of dairy terms is not considered by the Court to be sufficient to fall under the prohibition of the sixth paragraph. Probably most important for our question is the fifth paragraph: 'The designations referred to in paragraphs 1, 2 and 3 may not be used for products other than those referred to in those paragraphs.'

36.5 Alpro case

In Article 78(2) and the fifth paragraph of Part III of Annex VII, the Court of Appeal of Den Bosch implicitly reads that it is (only) not allowed to use the protected terms as designations (but it is possible in other ways). The only argument that the Court puts forward for that reading is the CJEU's TofuTown judgment, applying the rule of that judgment a contrario.

The Den Bosch Court of Appeal sets out its reading of TofuTown in considerations 3.6.1, 3.6.2 and 3.6.3.

The use of the reserved dairy designations
3.6.1. In its assessment, the court makes the following points. In its aforementioned judgment of 14 June 2017 (TofuTown), the CJEU ruled, with regard to the interpretation of Article 78(2) and Part III of Annex VII, that those provisions must be interpreted as precluding the use of the designation 'milk' and the designations reserved by that regulation exclusively for milk products *to designate*, in marketing or advertising, a purely vegetable product, even if those *designations* are supplemented by clarifying or descriptive additions which point to the vegetable origin of the product concerned. According to the CJEU, the exception set out in Part III(5) in the second paragraph applies only to those products listed in the Commission decision of 20 December 2010 (Decision 2010/791/EU), being the decision establishing the list of products referred to in (now) Annex VII, Part III(6), second paragraph. There is no dispute between the parties that this exception does not apply to [Alpro]'s products in this case.
3.6.2. Subject to that interpretation by the CJEU, [Alpro] may not use the designation "milk" and the other reserved dairy designations (which include the designations 'yoghurt' and 'cream') as the (*sales*) *designation* of its soy products, nor may it use those reserved dairy designations *to designate* its products in any way. In doing so, it makes no difference whether it uses the reserved designations on the packaging of its products or otherwise in the marketing or presentation of those products.
3.6.3. Contrary to NZO's argument, in the opinion of the court of appeal this interpretation does not imply that [Alpro] is not allowed to use the designations "milk", "yoghurt" and "cream" (or any other reserved dairy designation) in any way whatsoever on its packaging or otherwise when presenting its products. Indeed, such use is only prohibited as a *designation* or *indication* of its soy products. According to the CJEU in its judgment of 14 June 2017 (TofuTown), if the reserved dairy designations are used by [Alpro] in any other way, such use is not – as such and without further ado – contrary to the provisions of Regulation (EU) No 1308/2013.' (italics in original).

36.6 Criticism of the ruling of the Den Bosch court

In TofuTown, according to recital 20, the CJEU sought to answer the following question: 'whether Article 78(2) of, and Part III of Annex VII to, Regulation No. 1308/2013 are to be interpreted as precluding the use, in marketing or advertising, of the designation "milk" and the designations reserved exclusively for milk products by that regulation to designate a purely vegetable product, even where those designations are supplemented by clarifying or descriptive supplements indicating the vegetable origin of the product concerned.'

TofuTown concerned the use of dairy terms in the naming of non-dairy products. According to the CJEU, this is not allowed. The Court of Appeal of Den Bosch seems to infer a contrario that the use of dairy terms other than as a designation is therefore allowed, provided that at least possible confusion is overcome by additional information. It is questionable whether TofuTown can be read in this way. By reasoning in this way, the Court of Appeal of Den Bosch promotes the formulation of the question and the delimitation of the facts in TofuTown (that concerned the name) to a judgment of law (the prohibition *only* concerns the name). In doing so, the Court of Appeal of Den Bosch missed the fact that TofuTown did not express an opinion at all on the question before it in its own case. This makes application of an (already hazardous) a contrario reasoning, even more risky.

The Den Bosch Court of Appeal puts the TofuTown judgment under extreme stress, especially in paragraph 3.6.3. The consideration 'After all, such use is only not permitted as a designation or indication of its soy products' seems to be based on nothing but the italics added by the Court of Appeal of Den Bosch itself. The Court makes it even more bonkers in the final sentence of that paragraph of law: 'If the reserved dairy designations are used by [Alpro] in a different way, according to the CJEU in its judgment of 14 June 2017 (Tofu Town), such use is not – as such and without further ado – contrary to the provisions of Regulation (EU) No 1308/2013.' Here, the Court nevertheless suggests that the CJEU would have ruled in terms such as 'as such and without further ado' or at least on a different mode of use. However, there is no consideration in the TofuTown judgment (at least, I have not been able to find it) from which it could be inferred that the CJEU considered that a method of use other than as a designation 'is not – as such and without more – contrary to the provisions of Regulation (EU) No 1308/2013.' The Court of Appeal of Den Bosch is thus putting words into the CJEU's mouth that it did not in fact utter.

The Court of Appeal of Den Bosch gives no other argument for the view that the use of dairy terms other than as a designation is permitted (provided, incidentally, that it is not indicated, implied or suggested that the product in question is a dairy product). In the light of the above, it should not be surprising that NZO filed an appeal in cassation. However, the Supreme Court upheld the judgment of the Court of Appeal of Den Bosch.

36.7 State of play

With the Supreme Court (*Hoge Raad: HR*) ruling, the legal situation is now (in the Netherlands) as follows. The meaning of a number of dairy terms is laid down by law. These words may not be used in commercial messages in a different meaning. Provided this does not mislead the consumer and the legal meaning of the words is respected, the protected dairy words may be used to say something about non-dairy products. For example, Alpro is allowed to communicate 'you will find us in the dairy section of the supermarket'.

36.8 Lobby: Amendment 171

Where dairy organisations failed to achieve the intended result in court, they resorted to lobbying. This resulted in an amendment from the European Parliament to amend the CMO. This amendment reads as follows:

> (32a) In Part III of Annex VII, point 5 is replaced by the following:
> 5. The designations referred to in points 1, 2 and 3 may not be used for any product other than those referred to in that point.
> Those designations shall also be protected from:
> (a) any direct or indirect commercial use of the designation;
> (i) for comparable products or products presented as capable of being substituted not complying with the corresponding definition;
> (ii) in so far as such use exploits the reputation associated with the designation;
> (b) any misuse, imitation or evocation, even if the composition or true nature of the product or service is indicated or accompanied by an expression such as "style", "type", "method", "as produced in", "imitation", "flavour", "substitute", "like" or similar;
> (c) any other commercial indication or practice likely to mislead the consumer as to the product's true nature or composition.
> However, this provision shall not apply to the designation of products the exact nature of which is clear from traditional usage and/or when the designations are clearly used to describe a characteristic quality of the product.

The amendment is prefaced as a codification of TofuTown. The wording is taken from Article 13 of Regulation 1151/2012 on quality schemes for agricultural products and foodstuffs. On closer inspection of the text, the conclusion seems inescapable that the amendment goes far beyond TofuTown. It concerns not only protection against the use of dairy words to designate non-dairy products – i.e. legal protection of the meaning of these words – but also protection against the use of these words in their protected meaning to communicate about other products both when this would be misleading for consumers and when it is not.

Amendment 171 evoked a powerful counter-lobby. Consequently, positions within the European Parliament shifted and the parliament itself withdrew the amendment. A similar amendment for meat terms had already failed at the proposal stage.

36.9 Meat names

As mentioned in the previous section an amendment aimed at protecting meat terms in a similar way as dairy terms was withdrawn. However, lobbying efforts on this issue continued. This led to legislation in some Member States limiting the use of meat terms. Such legislation from France was struck down by the CJEU in its ruling of 4 October 2024 in the Beyond Meat case quoted above. The Court of Justice made it abundantly clear that the FIC leaves not room for national legislation in this regard.

In response to this both the European Commission and the European Parliament took initiatives to regulated the use of meat terms.

On the 14th of May 2025 in Parliament an amendment (nr. 645) was tabled to a proposed Regulation Strengthening of the position of farmers in the food supply chain (2024/0319(COD)).[353] This amendment is as follows:

> (8a) Regulation (EU) No 1308/2013 – Annex VIII – Part IIa 'Meat, meat products and meat preparations'
> 1. For the purposes of this part of this Annex, 'meat' means edible parts of the animals referred to in points 1.2 to 1.8 of Annex I to Regulation (EC) No 853/2004, including blood. The meat related terms and names that fall under Article 17 of Regulation (EU) No 1169/2011 and are currently used for meat and meat cuts shall be reserved exclusively for the edible parts of the animals.
> 2. 'Meat preparations' means fresh meat, including meat that has been reduced to fragments, which has had foodstuffs, seasonings or additives added to it, or which has undergone processes that do not alter the internal muscle fibre structure of the meat enough for the characteristics of fresh meat to be eliminated.
> 3. 'Meat products' means processed products resulting from the processing of meat or from the further processing of such processed products, so that the cut surface shows that the product no longer has the characteristics of fresh meat. Names that fall under Article 17 of Regulation (EU) No 1169/2011 that are currently used for meat products and meat preparations shall be reserved exclusively for products containing meat. These names include, for example:- Steak- Escalope- Sausage- Burger- Hamburger- Egg yolk- Egg white
> 4. The poultry products and cuts defined in Regulation (EU) No 543/2008 laying down detailed rules for the application of Council Regulation (EU) No 1234/2007 as regards the marketing standards for poultrymeat shall be reserved exclusively for the edible parts of animals and products containing poultrymeat.
> 5. The above-mentioned names shall not be used for any product other than the products referred to and shall exclude cell-cultured products.

On the 8th of Octobre 2025 Parliament voted in favour of this amendment.

On the 16th of July also the European Commission proposed a regulation of meat names. This proposal lies hidden in a document bearing the following title: "Proposal for a Regulation of the European Parliament and of The Council amending Regulation (EU) No 1308/2013 as regards the school fruit, vegetables and milk scheme

353 https://www.europarl.europa.eu/meetdocs/2024_2029/plmrep/COMMITTEES/AGRI/AM/2025/09-08/1321108EN.pdf.

('EU school scheme'), sectoral interventions, the creation of a protein sector, requirements for hemp, the possibility for marketing standards for cheese, protein crops and meat, application of additional import duties, rules on the availability of supplies in time of emergencies and severe crisis and securities" (COM(2025)553).

The text of the relevant part of the Commission's proposal to amend Annex VII of the CMO Regulation is as follows:

> *'PART Ia*
>
> ***Meat and meat products designations***
>
> 1. "Meat" means exclusively the edible parts of an animal.
>
> 2. For the purposes of this Part, "meat products" means products derived exclusively from meat, on the understanding that substances necessary for their manufacture may be added provided that those substances are not used for the purpose of replacing, in whole or in part, any meat constituent.
>
> 3. The following names shall be reserved for products derived exclusively from meat at all stages of marketing:
>
> (s) Shoulder;
> (t) Shank;
> (u) Chop;
> (v) Wing;
> (w) Breast;
> (x) Thigh;
> (y) Brisket;
> (z) Ribeye;
> (aa) T-bone;
> (bb) Rump;
> (cc) Bacon.
>
> 4. The term 'meat' and the designations listed in point 3 may also be used in association with a word or words to designate composite products of which no part takes or is intended to take the place of any meat constituent and of which meat is an essential part either in terms of quantity or for characterisation of the product.

Parliament's amendment seems more radical than the Commission's proposal. At the time of writing this annotation, both are still subject to an ongoing legislative process.

36.10 Finally

It is sometimes said that language is alive in the sense that words, meanings and forms of expression are in constant development. The CMO regulation puts this development on hold in terms of the meaning that may be given to certain words.

The CJEU has confirmed this. The Court of Appeal of Den Bosch – with the blessing of the Supreme Court – has refused to halt this development as regards the use to be made of these words.

The status quo with regard to dairy terms seems to be that they may not be used as name for plant based alternatives, but that it is otherwise allowed to use them in text describing use or properties of such products as long as consumers are not misled into believing that these are actually dairy products.

The status quo with regard to meat terms is that they can freely be used in the naming of plant based alternatives as long as consumers are not mislead. Member States have little options to regulate on this. However, no less than two separate legislative initiatives are in the pipeline aiming to grant exclusively to meat terms.

Bernd van der Meulen
Professor, EU food law University of Copenhagen

CHAPTER 37

Bucklings[354] – Guidance and the concept of administrative decision (*besluit*): the administrative legal opinion

Annotated by Oswald Jansen

Trade and Industry Appeals Tribunal (CBb) 15 November 2022, ECLI:NL:CBB: 2022:754
Rotterdam District Court 19 October 2021, ECLI:NL:RBROT:2021:10274;
Rotterdam District Court 23 March 2022, ECLI:NL:RBROT:2022:2051, and
Rotterdam District Court 27 February 2023, ECLI:NL:RBROT:2023:1584.

Administrative decision (besluit), art. 1:3(1) General Administrative Law Axt (GALA) (Algemene wet bestuursrecht:Awb)), administrative legal opinion (bestuurlijk rechtsoordeel)

Main considerations of the Trade and Industry Appeals Tribunal (CBb):

> In the opinion of the Tribunal, there is also no reason to equate the respondent's rejection of the appellant's application with an appealable decision, as far as the possibility of challenging it in court is concerned, because this rejection would be an administrative legal opinion (*bestuurlijk rechtsoordeel*). An administrative legal opinion is an opinion of an administrative authority on the applicability of public law provisions, the application of which falls within the competence of the administrative authority. According to established case-law of the Tribunal [...], the making of an independent and definitive legal opinion about the applicability of a statutory provision in a given situation in respect of which an administrative body has the authority can, in very special cases, be equated with a decision as referred to in Article 1:3, first paragraph, of GALA, as far as the possibility of appealing against this in court is concerned. There is justification for this only in cases in which the legal opinion cannot be deemed to anticipate a decision on the application of the statutory regulation which may be expected or provoked by the person concerned and that can be reviewed by an administrative court, without there being a disproportionately burdensome route to court for the person concerned. To the extent that the respondent's rejection of the appellant's request in this case could already be regarded as an independent and definitively intended legal opinion as referred to in the above-mentioned case-law, the Tribunal is of the opinion that the appellant has not made it plausible that provok-

354 Young, intact male goats, usually under one year old.

 | HTTPS://DOI.ORG/10.1163/9789004751699-039

ing an enforcement decision would be disproportionately burdensome for her. The possible consequences mentioned by the appellant that triggering an enforcement decision could have, namely that the appellant would risk an administrative fine from the respondent, reputation and image damage could occur and the appellant would be obliged to recall the products it put on the market, with all the (economic) consequences that would entail, are insufficient for this. The mere fact that an administrative fine may be imposed on the appellant in the event of a breach of the applicable regulations does not make the provocation of an enforcement decision unreasonably onerous. Indeed, this applies to any administrative fine that can be imposed on a legal person. This is therefore not a special circumstance on the basis of which the administrative legal opinion at issue here should be equated with an appealable decision. The same applies to damage to reputation and image resulting from an administrative fine. Any administrative enforcement decision that becomes publicly known may result in damage to the reputation and image of the party concerned. The appellant has not argued why this should be a special circumstance in her case. The same applies to the disclosure of documents under the Open Government Act (*Wet open overheid*). Again, the appellant did not substantiate why that would be a special circumstance in her specific case [...]. While the respondent has stated that, if the appellant violates the relevant regulations, it will remind the appellant of its obligation to inform the buyers of its products that they have received category 2 material instead of category 3 material, and that the material in question must be withdrawn from the market, the Tribunal does not consider this to be disproportionately onerous for the appellant in the case of a single unapproved buckling.

Main considerations of the Rotterdam District Court:

The claimant has not made it plausible that provoking an enforcement decision would be disproportionately burdensome for her. The potential consequences that provoking an enforcement decision could have, namely that the appellant would risk an administrative fine and damage to its reputation and image, are insufficient for that. The mere fact that an administrative fine may be imposed on the appellant in the event of violation of the applicable regulations does not make provoking an enforcement decision unreasonably onerous [...]. The fact that a turnover-related fine could be imposed on the claimant does not change this. Indeed, this applies to any administrative fine that could be imposed on a legal entity on the basis of this type of offence. This is therefore not a special circumstance on the basis of which the administrative opinion at issue here should be equated with an appealable decision. What the claimant [...] has argued about possible increases in fines on the basis of repeat offences and about the NVWA's[355] intervention policy does not lead to a different assessment. Contrary to the judgment cited by her, the claimant was in a position to directly pursue legal remedies against any sanctioning action by the respondent so that she is not prevented from effectively challenging the lawfulness of the establishment of the offence.

355 NVWA: Nederlandse Voedsel- en Warenautoriteit, the Netherlands Food and Consumer Product Safety Authority.

> There is therefore no situation in which, by provoking an enforcement decision, the plaintiff exposes herself to increasingly onerous sanctions or interventions without being able to effectively challenge the individual breaches leading to this aggravation.'

37.1 The case in the ruling of the Trade and Industry Appeals Tribunal (CBb)

On 21 February 2019, an abattoir specialising in the slaughter of bucklings applied for permission to slaughter bucklings according to a protocol which says that they would be marketed as category 3 material for processing into animal feed by an approved processor without post-mortem inspection. At issue is the interpretation of Article 10 of the Animal by-products Regulation 1069/2009 which defines what constitutes category 3 material. The company believes that without post-mortem inspection, the buckling carcasses should automatically be classified as category 3 material and can be processed into dog food. However, the Netherlands Food and Consumer Product Safety Authority (*NVWA*) considers that the regulation follows the inspection system for meat intended for human consumption and that it therefore requires not only an ante-mortem inspection, but also a post-mortem inspection in order for the slaughtered goat goats to be classified as category 3 material.

The request of 21 February 2019 was rejected by the the Netherlands Food and Consumer Product Safety Authority (*NVWA*) and the objection filed against it was declared unfounded. The Tribunal (*CBb*) concluded that the objection should not be declared unfounded but inadmissible. The NVWA's rejection was not a decision or an administrative legal opinion equivalent to a decision.

37.2 The case in the ruling of the District Court

On 4 December 2020 a producer of ingredients for bakeries/confectioners reported to the Netherlands Food and Consumer Product Safety Authority (*NVWA*) that it had received from a supplier a batch of sesame seeds that contained an excessively high ETO (ethylene oxide) content. The validity of the analysis was doubted by the supplier and the producer, and a re-analysis was carried out (this was also reported to the the Netherlands Food and Consumer Product Safety Authority (*NVWA*) by the supplier). The second analysis showed that the sample was negative. This was also reported to the NVWA, with the manufacturer indicating that no recall would therefore be initiated. In an e-mail dated 23 December 2020, the NVWA subsequently indicated (referring to the analysis results that were in doubt) that there had been contamination in the batch, and because it was not homogeneously distributed, there was insufficient guarantee for the Netherlands Food and Consumer Product Safety Authority (*NVWA*) that the entire batch would meet the ETO standard. In response, the producer reported that it disagreed but proceeded with the ordered recall, albeit under protest, and appealed the decision (the NVWA's e-mail message).

The court assumed that the e-mail message was neither a decision nor an administrative legal opinion that should be equated with it. Indeed, the obligation to recall arises directly from Article 19(1) of Regulation 178/2002.

37.3 Discussion

These judgments focus on a topic of administrative law scholarship that has great importance for practice: the concept of decision (*besluit*). This legal issue is one of the topics in administrative law that makes it hard to win the hearts of the average law student for this discipline. Many of the students who, partly as a result of this, graduate in another area of law discover the importance and interesting nature of administrative law only after their studies and in practice, for example that of food law. I think the two rulings discussed here underline the practical importance of the discipline and of the concept of decision.

The concept of decision[356]

Whether a given act is a decision or not determines the scope of the Dutch General Administrative Law Act (*Algemene wet bestuursrecht: Awb*) and access to the administrative courts. The cases discussed here deal with the latter issue: legal protection and reviewability before the administrative courts. It should be noted, incidentally, that the linking of access to the legal protection provided by the administrative courts to the concept of a decision has been increasingly questioned in legal scholarship in recent years.[357] The desire to make legal protection also about other

356 See on this Chris Backes and Mariolina Eliantonio (eds.), Cases and Materials and Text on Judicial Review of administrative action, Oxford: Hart Publishing 2019, p. 115 ff.

357 See L. van den Berge, 'Macht en tegenmacht in de netwerksamenleving', in: T. Barkhuysen e.a. (eds.), 25 jaar Awb. In eenheid en verscheidenheid, Deventer: Kluwer 2019, p. 3-12; B.J. Schueler, 'Van besluit naar geschil' and R.J.N. Schlössels, 'Het besluit voorbij ... maar via welke route?', in: T. Barkhuysen e.a. (eds.), 25 jaar Awb. In eenheid en verscheidenheid, Deventer: Kluwer 2019, pp. 69–80; L. van den Berge, 'Bestuursrecht in de netwerksamenleving', RMThemis 2018/4, pp. 124–136; M. Scheltema, 'Advies integrale geschilbeslechting in het sociaal domein', NTB 2018/3, pp. 4–7; J.A.F. Peters, 'In de ban van het besluit', A.T. Marseille, 'Weg van het besluit en het bestuur in het sociaal domein: gevolgen voor de rechtsbescherming', B. Assink & A.M.M.M. Bots, 'Het besluit voorbij, of toch niet?', in: R.J.N. Schlössels e.a. (eds.), In het nu... Over toekomstig bestuursrecht, Deventer: Wolters Kluwer 2018, pp. 27–43, pp. 45–60 and pp. 61–77; F.J. van Ommeren & P.J. Huisman, 'Van besluit naar rechtsbetrekking: een groeimodel', in: F.J. van Ommeren e.a., Het besluit voorbij (VAR-reeks 150), Den Haag: BJu 2013, pp. 6–102; J.C.A. de Poorter & K.J. de Graaf, Doel en functie van de bestuursrechtspraak: een blik op de toekomst, Den Haag: Raad van State 2011; Ch.W. Backes & A.M.L. Jansen, 'De wederkerige rechtsbetrekking als panacee voor de gebreken van de "besluiten-Awb"?', in: T. Barkhuysen e.a. (eds.), Bestuursrecht harmoniseren: 15 jaar Awb, Den Haag: BJu 2010, pp. 75–96. Also see Chapter 33 in this book by Bernd van der Meulen.

aspects of the administrative-law legal relationship between administrative bodies and citizens or companies than decisions, and to this end to use a strategic concept of decision, sometimes makes us lose sight of the importance of the dogmatics of the concepts of 'legal act' and 'decision'.[358]

Access to the administrative courts

The general idea of Dutch administrative procedural law is that when the government changes the legal position of a citizen or company, the latter should have the opportunity to challenge it in court. The premise here is that the court comes within reach only after another organ of government, an administrative body, has interpreted the law, determined its policy and made a decision within the given options that changes the legal world (the world of law) of a citizen or company. To be more precise: a decision involves a public law legal act, and that act is aimed at changing the world of law. For the enthusiasts among you, please refer to the textbooks.

Acts equated with a decision

Besides the decisions that are actually public law acts, there are decisions or situations that are equated with a decision by the legislator or the administrative courts. Examples of acts equated with a decision by the Dutch legislator of GALA are the rejection of an application (Article 1:3, second paragraph, GALA) and the failure to decide on time (Article 6:2 GALA).[359] In doing so, the legislator determines the legal consequences of the equivalence (equated to a decision only in terms of legal protection or also in terms of more and other legal consequences).

Administrative law opinion (bestuursrechtelijk oordeel)

The cases discussed here focus on an example of an act of an administrative body equated to a decision by the administrative court: the administrative law opinion.[360] This involves a non-statutory decision which declares what the law or legal situation is (which must be distinguished from a statutory decision declaring the

358 See Rolf Ortlep, 'Meldingsstelsel Leeuwarden', AB Klassiek 2022/40, with many references to legal scholarship and case-law.

359 See on this Chris Backes and Mariolina Eliantonio (eds.), Cases and Materials and Text on Judicial Review of administrative action, Oxford: Hart Publishing 2019, p. 174 ff and Oswald Jansen, Silence of the administration: General Comments and Dutch Law, in: Jean-Bernard Auby (ed.), Droit compare de la procédure administrative – Comparative Law of Administrative Procedure, Brussels: Bruylant 2016, p. 623–646.

360 See a recent article on this subject: S. van Kaam, Bestuurlijk rechtsoordeel, tien jaar jurisprudentie, Praktisch bestuursrecht 2023/42.

law or legal situation).[361] Moreover, the non-statutory decision declaring what the law or legal situation is, the administrative legal opinion, involves a strategic concept of decision at the boundary between actual act and legal act.[362] In essence, it involves the provision of information by a competent administrative body. The non-statutory administrative legal opinion is therefore not a legal act because the legal consequences derive from the law about which the opinion informs.

Examples of opportunities to obtain legal opinions from market authorities

Of course, it is always possible to ask an administrative body whether a particular situation, business process or proposed commercial activity falls within the scope of a particular legal regulation. If that is not expressly provided for by law (as, for example the prior consultation (*Voorafgaande raadpleging* (*VR*) under Article 36 GDPR (and the Police Data Act (*Wet politiegegevens*) and the Judicial and Criminal Records Act (*Wet justitiële en strafvorderlijke gegevens*)) or as a policy that has been made public, that administrative body is not obliged to provide a substantiated substantive response to it. A number of market authorities explicitly offer market participants the opportunity to request an administrative legal opinion. An example of this is the informal view (*informele zienswijze*) that can be given by the Netherlands Authority for Consumers and Markets (ACM) on request. The ACM has published a Working Method (*Werkwijze*) on this (Government Gazette (*Staatscourant*) 2019, 11177) (see also the Working Method (*Werkwijze*) in merger cases (Government Gazette (*Staatscourant*) 2021, 43083)). It sets out a number of criteria for issuing an informal view and a number of requirements for a request. A (small) number of views have been published on ACM's website. The latest informal view I was able to find there is that on the cooperation between hospitals specialised in complex cancer care dated 15 July 2016. Informal views can also be requested from the Dutch Healthcare Authority (NZA) (concerning the application of the care-specific concentration test). The Department of Waterways and Public Works (*Rijkswaterstaat*) of the Ministry of Infrastructure and Water Management (*Ministerie van Infrastructuur en Waterstaat*) has issued legal opinions on whether material is or is not to be considered waste. Due to the large number of requests submitted, there is not enough capacity to deal with new requests. Again, a (large) number of legal opin-

361 See Frank van Ommeren, Wetsinterpretatie: het bestuurlijk oordeel als rechtsvaststellend besluit, JBplus 2000, p. 113-124 as well as the opinions of Advocate General Widdershoven of 12 November 2014, ECLI:NL:RVS:2014:4116, 3.5 and 3.6, 24 January 2018, ECLI:NL:RVS:2018:249, 3.6 and 16 January 2019, ECLI:NL:RVS:2019:86, 3.5. See also Rolf Ortlep, Meldingsstelsel Leeuwarden, AB Klassiek 2022/40.

362 Advocate General Widdershoven in his opinion of 12 November 2014 (ECLI:NL:RVS:2014:4116). See also Van Wijk/Konijnenbelt & Van Male, Hoofdstukken van bestuursrecht, 16e druk, Kluwer: Deventer 2014, p. 163 t/m 166 and Bestuursrecht in de sociale rechtsstaat Band 1 (HSB) 2017/175.

ions have been published on the website.[363] The Dutch Media Authority (*Commissariaat voor de Media*) also issues administrative legal opinions that are published on its website. The last opinion I could find was the administrative legal opinion at the request of KPN on targeted advertising. I mentioned above the possibility (or obligation) to request a prior consultation (VR) from the Dutch DPA (*Autoriteit Persoonsgegevens*). This is mandatory if a data protection impact assessment (DPIA) shows that a proposed processing of personal data has a high privacy risk. On DPA's website, I found the opinion dated 6 August 2022 on the prior consultation about the COVID19 app. On the websites of the Financial Markets Authority (*AFM*) and the Dutch Central Bank (*De Nederlandsche Bank*), I did not find any administrative law opinions or information on the application for such an opinion, and the same applies to the Netherlands Food and Consumer Product Safety Authority's (NVWA) website. Of course, this does not mean – according to case-law – that in practice administrative legal opinions are not or cannot be issued on request.

The need for guidance on request which is as binding as possible

The examples described above involve requests in which the competent market authority is asked for more than just information or general guidance on the interpretation of legal requirements. My impression is that offering the possibility of requesting administrative legal opinions or rulings from a market authority is about obtaining the most binding ruling or opinion possible which offers as much guidance as possible for the company concerned and the market involved. This possibility therefore manoeuvres in administrative law between the concepts of administrative act and administrative decision, the principle of legal certainty and the principle of legitimate expectations (as well as the principle of equality). It mostly involves so-called declaratory decisions which are not legal acts, because the legal consequences do not flow from that decision but from the statutory regulation or public law on which the opinion is sought.

Standard case-law

Back to standard administrative law case-law. In the standard ruling of the Administrative Jurisdiction Division of the Dutch Council of State (*Afdeling bestuursrechtspraak Raad van State* (*ABRvS*)) of 18 August 2004 (Duiven),[364] this administrative court considered:

> Where an administrative body, separate from an application for a permit or a request for enforcement as indicated above, gives an opinion, upon a request to that effect,

363 See https://www.afvalcirculair.nl/onderwerpen/afval/toetsing-afval
364 ABRvS 18 August 2004, ECLI:NL:RVS:2004:AQ7004; AB 2005/106

> as to whether the performance of a certain act requires a permit, this opinion cannot generally be regarded as a public law act within the meaning of Article 1:3(1) of GALA. An exception to the above may be made if applying for a permit is disproportionately onerous, or if requesting the taking of enforcement measures is disproportionately onerous due to the occurrence of irreversible consequences or otherwise.[365]

With this ruling, the Administrative Jurisdiction Division (*ABRvS*) attempted to systematise the doctrine, and the possibility of legal protection against an administrative legal opinion was significantly reduced.[366]

The Trade and Industry Appeals Tribunal (CBb)[367] has similar case-law and so has the other administrative Court of Appeal in the Netherlands, the Central Appeals Council (*Centrale Raad van Beroep* (*CRvB*)).[368]

Definition of administrative law opinion

What is the definition of an administrative law opinion in that case-law? I have borrowed the following description from the opinion of Advocate General (AG) Widdershoven of 24 January 2018:[369]

> An administrative legal opinion is an independent and definitively intended opinion of an administrative body on the applicability of a legal regulation on the application of which that body has powers. Legal opinions relate, for example, to the authorisation requirement of an act or the compatibility of the act with statutory regulations. Sometimes they can be regarded as legally binding decisions and for that reason are appealable.
> An administrative legal opinion can only be appealed if it would be disproportionately onerous for the person concerned to bring the dispute on the interpretation of legal rules before the administrative court via an appeal procedure on an actual decision.

365 See also ABRvS 8 July 2009, ECLI:NL:RVS:2009:BJ1862; ABRvS 28 September 2016, ECLI:NL:RVS:2016:2554; ABRvS 24 April 2019, ECLI:NL:RVS:2019:1356; ABRvS 18 December 2019, ECLI:NL:RVS:2019:4242; ABRvS 12 February 2020, ECLI:NL:RVS:2020:453; ABRvS 2 December 2020, ECLI:NL:RVS:2020:2857; ABRvS 29 September 2021, ECLI:NL:RVS:2021:2170; ABRvS 8 December 2021, ECLI:NL:RVS:2021:2753; ABRvS 14 September 2022, ECLI:NL:RVS:2022:2691; ABRvS 2 November 2022, ECLI:NL:RVS:2022:3163; ABRvS 23 August 2023, ECLI:NL:RVS:2023:3222.

366 See Schlössels/Zijlstra, Bestuursrecht in de sociale rechtsstaat 2017/178.

367 CBb 21 July 1998, AB 1998/437; CBb 13 August 2002, ECLI:NL:CBB:2002:AE6745; CBb 1 July 2014, ECLI:NL:CBB:2014:256; CBb 27 May 2015, ECLI:NL:CBB:2015:175; CBb 17 January 2018, ECLI:NL:CBB:2018:5; CBb 10 September 2019, ECLI:NL:CBB:2019:408; CBb 15 December 2020, ECLI:NL:CBB:2020:983.

368 CRvB 24 November 2009, ECLI:NL 18:5:CRVB:2009:BK4988 and CRvB 29 August 2014, ECLI:NL:CRVB:2014:2889; CRvB 9 May 2017, ECLI:NL:CRVB:2017:2087; CRvB 22 September 2020, ECLI:NL:CRVB:2020:2264; CRvB 15 June 2021, ECLI:NL:CRVB:2021:1421.

369 ECLI:NL:RVS:2018:249

> Only in very exceptional situations an administrative law judgment will be equated with a decision. The general line in case-law is that filing an application for a permit or awaiting a decision on enforcement cannot be considered a disproportionately onerous route.

Alternative route to legal protection disproportionately onerous

The case-law discussed here ultimately revolves around situations where the administrative court assumes that the alternative route is disproportionately onerous. Only then does case-law assume that the interests of an effective remedy prevail. Legal scholarship mostly makes do with the justified observation that case-law on this issue is casuistic with a brief review of the case-law and the observation that the case-law is not generous in allowing appeals against administrative law judgments.[370]

Konijnenbelt wrote:

> This jurisprudence is rather casuistic. Searching and groping, jurisprudence on this subject always seems to ask itself whether the need for legal protection in a particular case should be considered so urgent that the administrative legal judgment should be regarded as a decision; here we see the [...] strategic decision concept emerging in optima forma.[371]

Complicated case-law

The case-law involved is complex, as evidenced by the fact that it took the Rotterdam District Court three judgments to determine, in the same case, whether waiting for an administrative fine is disproportionately onerous for the interested party and therefore, that the administrative law opinion should be equated with a decision. In the judgment of 19 October 2021, the appeal was declared unfounded without a hearing. This is possible if the final judgment (appeal manifestly unfounded) is beyond reasonable doubt. The producer in question appealed against this, and in its ruling of 22 March 2023 – the ruling of the Trade and Industry Appeals Tribunal (CBb), also discussed here, had already been delivered by then – the opposing judge of the Rotterdam District Court ruled that waiting for the decision to impose a penalty in light of the Netherlands Food and Consumer Product Safety Authority (*NVWA*)'s intervention policy could be disproportionately burdensome. This did not leave beyond reasonable doubt that the appeal would be unfounded. Ultimately, the Rotterdam District Court ruled in this case on 27 February 2023 that waiting

370 Schlössels/Zijlstra, Bestuursrecht in de sociale rechtsstaat 2017/178.

371 Van Wijk/Konijnenbelt & Van Male, Hoofdstukken van bestuursrecht, 16e druk, Kluwer: Deventer 2014, p. 164).

for a decision to impose an administrative fine was not disproportionately onerous. And, according to the court, it does not matter that this could be a turnover-related fine. 'Indeed, this applies to any administrative fine that can be imposed on a legal entity based on this type of offence. This is therefore not a special circumstance on the basis of which the administrative law judgment at issue here should be equated with an appealable decision.' This judgment comes some two years after the inadmissibility of the objection was appealed.

Case-law too strict for businesses

As I wrote above, the case-law does show great restraint in granting access to the administrative courts against an administrative law opinion. Although waiting for an administrative fine is in very rare cases considered disproportionately onerous,[372] I cannot help thinking that this case-law is far too strict for companies, and it is too quickly assumed by case-law that provoking a (substantial) administrative fine is not disproportionately onerous for the company concerned.

Need for statutory regulation of the right to petition an interpretation judgment from an economic administrative law court

As I have written before on this matter,[373] the government regulating entrepreneurial activities should provide clarity and legal certainty in doing so. Like every citizen, entrepreneurs want to know where they stand. It befits a service-minded government to provide information upon request and inform entrepreneurs about the content of its regulation and how it applies it. There I suggest including in the rules on administrative procedure the possibility of an application procedure to obtain an interpretative opinion from the economic administrative law court. 'In an environment where there is a risk of severe sanctions and major prudential consequences of government action, it is desirable to obtain a relatively quick ruling on a legal interpretation issue without the need to provoke a sanction, for example, to do so. This possibility should be situated in administrative law. My proposal is to link up with the petition procedure in administrative procedural law on compensation (Art. 8:88 GALA (*Awb*)). Thus, a procedure could be set up to establish a judgment of law, which would be initiated with an application to the administrative court. Similar to the regulation of the petition procedure, a prior opinion of the administrative body – the market authority – should be required. Given the importance of

372 See for example recently ABRvS 2 November 2022 ECLI:NL:RVS:2022:3163; AB 2023/9 annotated by A.C. Hendriks. This case was focused on a natural person (a medical doctor).

373 Oswald Jansen, Aanpassingen boetebevoegdheden voor een steviger handhavingsregime. Naar een stroomlijningswet bestuursrechtelijke handhaving NVWA, Waar & Wet 2023/72.

transparency, such an opinion of the market authority should be known to other market participants and the general public.'[374]

Here, I echo the views of Zijlstra and Schlössels,[375] among others. They write: 'The complicated jurisprudence, whereby an administrative legal opinion is sometimes considered to be a decision and sometimes not, could be ended by regulating by law that in case of sufficient interest a declaratory decision can be requested from the competent administrative body. Another – perhaps even better – option would be to introduce the possibility of directly requesting the administrative court to issue a declaratory order in relation to the application of administrative law regulations if there is sufficient interest. Because this is ultimately what it is all about.'[376]

Prof. dr. Oswald Jansen
Partner and attorney at Libertas Lawyers in Rotterdam and professor of European and Comparative Administrative Law at Balekesir University, Türkiye. Former professor of European Administrative Law and Public Administration at Maastricht University and adjunct professor American University Washington College of Law (AUWCL).

374 Oswald Jansen, Aanpassingen boetebevoegdheden voor een steviger handhavingsregime. Naar een stroomlijningswet bestuursrechtelijke handhaving NVWA, Waar & Wet 2023/72.
375 Bestuursrecht in de sociale rechtsstaat Band 1 2017/178.
376 Bestuursrecht in de sociale rechtstaat Band 1 2017/1789.

CHAPTER 38

Harmful shakers – Prison sentence for sale instead of destruction after recall

Annotated by Oswald Jansen

Overijssel District Court 16 March 2023 ECLI:NL:RBOVE:2023:956; ECLI:NL:RBOVE:2023:957; ECLI:NL:RBOVE:2023:958; ECLI:NL:RBOVE:2023:959; ECLI:NL:RBOVE:2023:962; ECLI:NL:RBOVE:2023:965 (JW 2023/42; Waar & Wet 2023/70).

Criminal law, fine, imprisonment

Main considerations:

> Defendant was guilty of delivering goods which he knew were harmful to life or health and concealed this harmful nature from others. [...] The defendant's actions ensured that the batch of harmful shakers was not destroyed, but entered the market. Food safety is a great good and of eminent importance for public health, which must be handled with great care, especially by those directly or indirectly engaged in food waste disposal. The defendant seriously disregarded this interest, thereby creating health risks. Consumers should be able to trust that food offered for sale is safe. By his conduct, the defendant has grossly damaged confidence in food safety and seriously endangered public health in general. The court charges the defendant with having allowed monetary gain to outweigh public health.

38.1 The case

On 25 March 2020, glass shakers containing herbs and spices were recalled by Verstegen Spices & Sauces due to glass contamination. The following day, the Netherlands Food and Consumer Product Safety Authority (*NVWA*) published a public warning. Glass particles could come loose from the top edge under the cap. On 16 April 2021, the NVWA warned about the use of some glass shakers of the Verstegen brand. An investigation by the NVWA's Intelligence and Investigation Service had revealed that although these had been taken off the market by the manufacturer, they had not been destroyed but reintroduced to the market without its knowl-

 | HTTPS://DOI.ORG/10.1163/9789004751699-040

edge.[377] This warning also received press coverage. The case came to light on 27 March 2021 due to an anonymous tip-off from someone who had bought six shakers from a premises where shop cabinets could be rented to offer products for sale. This anonymous whistleblower also reported that he had cut his finger and tongue when using them. Following this, the investigation 'Charleston' was launched which led to the criminal case against six suspects, including a legal entity. The legal entity was engaged in the collection of organic residues and waste materials released particularly in the food and beverage industry, catering industry and institutions. It was also engaged in recycling waste from the food industry. The company in question was also previously engaged by Verstegen.

On 31 March 2021, the Netherlands Food and Consumer Product Safety Authority (*NVWA*) informed Verstegen, and Verstegen apparently also went to investigate on its own. On 1 April 2021, one of his employees found that Verstegen's shaker jars were being offered online via marktplaats.nl, and Verstegen then engaged the private investigation agency International Security Partners. Using a pseudo-buying campaign, this agency bought 544 jars that were delivered on 6 April 2021. These jars were indeed found to have been withdrawn from the market with the 25 March 2020 recall. Incidentally, the investigation also found in a warehouse a number of bottles of sauce on which three destruction declarations had been (falsely) signed.

On 27 March 2023 the Overijssel District Court handed down the following sentences for the two offences of Article 174 Dutch Criminal Code (DCC) (selling goods knowing that they are harmful to life or health and concealing that harmful nature) and 225 DCC (forgery and using a false document: the false declaration of destruction). The legal entity was fined €45,000, of which €15,000 was suspended with a probationary period of three years. The de facto manager of this legal entity was given a prison sentence of 18 months, of which six months were suspended with a probationary period of three years. On top of that, he was given 240 hours of community service. The trader who provided a shed in which the herb jars were stored and delivered the harmful jars to another person despite being sent a screenshot of the NVWA's public warning was given a prison sentence of 180 days, 162 of which were suspended with a probation period of three years. And he was also given 240 hours of community service on top of that. The trader who had bought 2,400 jars, and resold some of these jars to another trader and furthermore not only offered a number of jars for sale in his shop, but also – after being informed by the Netherlands Food and Consumer Product Safety Authority (*NVWA*) about the harmfulness of the jars – went to the market three more times to offer these jars for sale there, was given 120 days' imprisonment, 105 of which were suspended with a probation period of three years, and community service of 100 hours. The trader who offered some of the harmful jars in the rented shop cupboard, a few others via

377 NVWA-IOD treft mogelijk schadelijke potjes met kruiden aan; gebruik ze niet.' News item of 16 April 2021. Article can found in the NVWA Site Archive: https://nvwa.sitearchief.nl/#archive

a business on marktplaats.nl (the day after he was informed by the NVWA about the harmfulness of the jars) and sold another number of jars to the pseudo-buyer of the private investigation agency International Security Partners (ISP) hired by Verstegen, was given a prison sentence of 150 days, 138 of which were suspended with a probation period of three years, and community service of 200 hours.

38.2 Discussion

This ruling by the criminal division of the Overijssel District Court is included because it illustrates the operation of criminal enforcement in food law. Not every food business operator has a clear idea of the possible applicability of criminal law. The de facto manager in this case (unlike the legal entity) received a hefty sentence. The case involves conviction for two common offences. Besides forgery (Article 225 DCC), there is the deliberate sale of harmful goods (selling, delivering or distributing goods 'knowing that they are harmful to life or health, and concealing that harmful character') (Article 174(1) DCC). The maximum penalty here is a temporary prison sentence of 15 years or a fifth-category fine (€103,000).[378] For legal persons, incidentally, under Article 23(7) of the Criminal Code, this easily becomes the sixth category (€1,030,000 or 10% of the company's worldwide turnover).

Article 174(1) DCC reads as follows:

> He who sells, offers for sale, delivers or distributes goods, knowing that they are harmful to life or health, and concealing that harmful character, shall be punished by imprisonment for a term not exceeding 15 years or a fine of the fifth category.

This common offence has existed for more than 135 years: it was introduced on 1 September 1886 with the new Penal Code, and has been amended only once since then.[379] On 1 May 1984, the fine categories were added with the Act on Classification of Fine Categories (Stb. 1984,81). It was based on Article 318 Code Pénal. The prima facie remarkable component of concealment of harmfulness was added to the text by the legislator because it saw no reason for assuming a criminal offence when the seller and buyer were both aware of the dangerous nature of the goods delivered.[380]

The common criminal law has a number of other and more serious criminal offences, such as those if the offence results in death (Article 174(2) DCC: temporary

378 Every two years the amounts of the six categories of fines are adjusted to an inflation index. The most recent time was 1 January 2024.

379 See on this offence H.J. Gerrits, Verkoop van schadelijke waren. De 'toepassing' van artikel 174 Wetboek van Strafrecht, TBSenH 2019, p. 20–23.

380 Noyon-Langemeijer-Remmelink (NLR), Strafrecht, Article 174 Sr, aant. 1. See for the parliamentary history at that time Smidt II, p. 157 et seq.

imprisonment for thirty years[381] or life imprisonment). This highest penalty threat also applies if the offence is committed with a terrorist intent (see sections 176a and 83a of the Criminal Code). In one of the cases discussed here, the public prosecutor asked for the additional penalty of a professional ban (see section 176(1) DCC). She claimed a professional ban for a period of five years, prohibiting the person concerned from carrying out any (administrative) activities in waste disposal in any way. However, the criminal court considered that a substantial suspended portion of the sentence was sufficient to prevent the defendant from committing similar offences again.

Perhaps in doing so, the court also glanced at the many non-criminal law consequences of a conviction for section 174 Sr. For example, this antecedent explicitly plays a role in numerous reliability tests[382] and in public procurement procedures.[383]

It is not very often that the offence under Article 174 DCC is prosecuted in a case on food law. The closest and a well-known recent Dutch example is the conviction for this offence of the legal entities Chickfriend and Chickclean and their de facto principals in the Fipronil affair (Overijssel District Court 12 April 2021, ECLI:NL:RBOVE:2021:1508; JM 2021/85). In two judgments of 4 August 2020 (ECLI:NL:RBROT:2020:7093; ECLI:NL:RBROT:2020:7094), the Rotterdam District Court acquitted the accused of the intentional variant (174 DCC) and convicted him of the guilt variant (Article 175) of the offence discussed here. It concerned aflatoxin in peanuts. In two judgments of 23 November 2020 (ECLI:NL:RBOBR:2020:5777 and ECLI:NL:RBOBR:2020:5778), the East Brabant District Court convicted defendants of the offence of Article 174 DCC for (in short) transporting meat unrefrigerated. On 4 May 2016 (ECLI:NL:RBROT:2016:3399), the Rotterdam District Court convicted defendants under 174 DCC for fraud involving baby milk powder. The cheaper milk powder containing cow's milk protein (and lactose) was sold with falsified labels as the more expensive lactose-free milk powder specially intended for babies with a lactose allergy.

Outside food law, other well-known cases can be highlighted, such as the dump-

---

381 The maximum of twenty years was increased to thirty years on 1 February 2006 with the Wet verhoging maxima (the Act to increase maxima).

382 See, for example, the policy rule (Beleidsregel 06-01 betrouwbaarheid personen ex Wet toezicht accountantsorganisaties) and the *Besluit toezicht accountantsorganisaties*, (Appendix C (*Appendix C*) that relates to Article 13 of) the Besluit Gedragstoezicht financiële ondernemingen Wft, (de Appendix that relates to) the Besluit Markttoegang financiële ondernemingen Wft, (Appendix A that relates to Article 6 of) the Besluit prudentiële regels Wft, (Appendix 1 that relates to Article 19 of) the Besluit toegelaten instellingen volkshuisvesting 2015, (Appendix that relates to Article 5 of) the Besluit toezicht trustkantoren 2018, (that relates to Article 32 of) the Besluit uitvoering Pensioenwet en Wet verplichte beroepspensioenregeling.

383 See the regulation on declaration of conduct procurements in Article 4.7(1)(b), Procurement Act 2012 and Article 8(1), Procurement Decree.

ing of chemical waste at the Coupépolder landfill in Alphen aan den Rijn by a waste broker (Kemp) that attracted attention. This was partly because the court ruled that the waste materials in question were not 'goods' as referred to in Article 174 DCC because they had no intended use and therefore no longer had any economic value in trade and had to be destroyed. This judgment of the Court of Appeal was upheld by the Supreme Court (HR 16 March 1993, ECLI:NL:HR:1993:AD1846; NJ 1993,718). On 23 November 2011 (ECLI:NL:GHAMS:2011:BU9237), the Amsterdam Court of Appeal incidentally held that the waste materials transferred by the Probo Koala on behalf of Trafigura were 'goods', and thus the conduct was held to fall under the offence of 174 DCC.

Violation of food law is not exclusively a criminal offence. A number of offences have been classified as economic offences and misdemeanours under Article 1(3) in conjunction with Article 2(3) Act on Economic Offences (*Wet op de economische delicten:Wed*), and very many offences as economic offences and misdemeanours under Article 1(4) in conjunction with Article 2(4) Act on Economic Offences.

And finally, in food law, punishment is regulated not only in criminal law, but also in administrative law. Article 32a of the *Warenwet* (the Dutch Act on the safety of goods and foodstuffs) regulates the power to impose an administrative fine for a wide range of offences. While this power to impose fines is not as old as the criminal offence in the ruling discussed here, it has existed since 1 February 2001, more than two decades. In food law, punishment within administrative law is here to stay. That does not alter the fact, as I have written elsewhere,[384] that there is room for improvement in this penalty regime.

Prof. dr. Oswald Jansen
Partner and attorney at Libertas Lawyers in Rotterdam and professor of European and Comparative Administrative Law at Balekesir University, Türkiye. Former professor of European Administrative Law and Public Administration at Maastricht University and adjunct professor American University Washington College of Law (AUWCL).

384 Oswald Jansen, Aanpassingen boetebevoegdheden voor een steviger handhavingsregime. Naar een stroomlijningswet bestuursrechtelijke handhaving NVWA, Waar & Wet 2023/72, p. 4–9.

CHAPTER 39

Food supplements – not medicines after all?

Annotated by Irene Verheijen

East Brabant District Court 25 March 2022, ECLI:NL:RBOBR:2022:1109, JGR 2023/3, annotated by K. van Lessen Kloeke, IER 2022/46, annotated by S. Arayess.

Delineation of product categories. If a product clearly falls exclusively within the definition of other product categories, the Medicinal Products Directive does not apply. Application criterion not relevant when assessing product category. Van Bennekom judgment does not lead to a different judgment. FIC Regulation applicable. Gnw – Geneesmiddelenwet (Dutch Medicines Act)

Main considerations:

> 6. The court first notes that Article 1, sixth paragraph, of the Gnw is the implementation of Article 2 of Directive 2001/83/EC of the European Parliament and of the Council of 6 November 2001 on the Community code relating to medicinal products for human use (OJ 2001 L 311; as amended, inter alia, by Directive 2004/27/EC; hereinafter: Medicinal Products Directive). Article 2 of this Directive concerns an amendment to the same article by Directive 2004/27/EC (of the European Parliament and of the Council of 31 March 2004 amending Directive 2001/83/EC on the Community code relating to medicinal products for human use). Point 7 of the preamble to Directive 2004/27/EC states, inter alia, the following:
> '[…] In order to take account both of the emergence of new therapies and of the growing number of so-called 'borderline' products between the medicinal product sector and other sectors, the definition of 'medicinal product' should be modified so as to avoid any doubt as to the applicable legislation when a product, whilst fully falling within the definition of a medicinal product, may also fall within the definition of other regulated products. This definition should specify the type of action that the medicinal product may exert on physiological functions. This enumeration of actions will also make it possible to cover medicinal products such as gene therapy, radiopharmaceutical products as well as certain medicinal products for topical use. Also, in view of the characteristics of pharmaceutical legislation, provision should be made for such legislation to apply. With the same objective of clarifying situations, where a given product comes under the definition of a medicinal product but could also fall within the definition of other regulated products, it is necessary, in case of doubt and in order to ensure legal certainty, to state explicitly which provisions have to be complied with. **Where a product comes clearly under the definition of other prod-**

 | HTTPS://DOI.ORG/10.1163/9789004751699-041

uct categories, in particular food, **food supplements**, medical devices, biocides or cosmetics, **this Directive should not apply**. […]' (passages in bold by the court)

7. The court deduces from this entry that the Medicinal Products Directive does not apply if a product is clearly exclusively a food supplement. Since, according to established case-law, national law should, as far as possible, be interpreted in accordance with the directive, the foregoing implies that if the product in question is clearly exclusively a food supplement, Article 1(6) Gnw and, consequently, the Gnw does not apply to the product.

8. The court notes that the Minister expressly acknowledged during the hearing that the products traded by the claimant should be regarded as food supplements and not also as medicinal products. In view of this and what has the Gnw does not apply to these products and that, consequently, the Gnw does not apply to these products.

9. Recital 8 states that the question of whether the presentation criterion is fulfilled (is a product presented or offered on the website as a medicine or not?) is irrelevant when assessing whether these products fall under the definition of a medicinal product within the meaning of the Gnw. In this regard, the Minister argued that if a food supplement does not in itself qualify as a medicinal product, this does not alter the fact that it may nevertheless qualify as a medicinal product in view of its presentation, and that this should be assessed under the Gnw. The court does not follow the Minister in this argument. Indeed, as there is no doubt that the claimant's products are food supplements, the presentation criterion, which would be relevant if the Gnw did apply, is left out of consideration.

10. The Minister further argued that if a product were to fall under both the definition of medicinal product and the definition of food supplement, it follows from Union law that the Medicinal Products Directive takes precedence over the Gnw. In support of the position that the claimant's products, as presented on its website at the time of the inspection, should be considered medicinal products on the basis of the presentation criterion, the Minister referred to the judgment of the Court of Justice of the European Union (CJEU) in the Van Bennekom case of 30 November 1983, case number C-227/82, ECLI:NL:XX:1983:BF5706. In this judgment, the CJEU ruled that a product is not only deemed to be 'presented as having therapeutic or prophylactic properties' within the meaning of Directive No 65/65 (of the Council of 26 January 1965 on the approximation of provisions laid down by law, regulation or administrative action relating to proprietary medicinal products (OJ 1965, p. 369)), when it is expressly 'indicated' or 'recommended' as such – possibly on the label, in the package leaflet or also by oral presentation – but also when, implicitly but no less emphatically, it gives the impression to the averagely well-informed consumer, by the manner of presentation, that it has an effect as defined in the first part of the Community definition.

11. The court ruled that this judgment of the CJEU cannot be attributed the meaning that the minister wants to see attributed to it. After all, this judgment relates to Directive no. 65/65 and not to the – currently applicable – Medicinal Products Directive. As considered above under recital 7, the Medicinal Products Directive does not apply to

a product that is clearly exclusively a food supplement. Moreover, on 25 October 2011, the European Parliament and the Council of the European Union published Regulation No 1169/2011 on food information to consumers (FIC Regulation). Recital 8 of the FIC Regulation reads as follows:

> The general labelling requirements are complemented by a number of provisions applicable to all foods in particular circumstances or to certain categories of foods. In addition, there are a number of specific rules which are applicable to specific foods.

12. The court finds that the FIC Regulation has direct effect and contains specific requirements, which apply, inter alia, to food supplements, such as the claimant's products. In view of the above, the court rules that the Minister cannot be followed in his position that the presentation criterion and the Gnw are nevertheless applicable and that the claimant's products must be considered medicinal products on that basis.'

39.1 Introduction

In an earlier contribution to this book,[385] I described the NVWA's perennial policy line, which was followed by case-law, whereby products with expressions referring to disease, its prevention or cure (medical claims) are routinely considered medicinal products, even though they are foodstuffs by their nature. It has led to honey, cooking oil, chocolate and margarine, among others, being qualified as medicinal products in the past, based on expressions made about these products (foodstuffs). The present ruling abandons this previously consistent line.

39.2 The case

In 2020, an inspector from the Netherlands Food and Consumer Product Safety Authority (NVWA) carries out inspections of Super Nature Products (SNP)'s website as part of the Glucosamine 2020 and Corona claims project. The inspector finds that SNP's use of four statements on its website about the underlying products made them medicinal products within the meaning of the Dutch Medicines Act (*Geneesmiddelenwet* (Gnw)).[386] The NVWA takes the position that it follows from Article 1(6) of the Gnw that if a product, as far as its characteristics are concerned, meets both the definition of medicinal product and the definition of a product in another statutory regulation, the Gnw applies in full with regard to that product.

---

385 See Chapter 25 of this book.
386 Article 1(1), preamble and (b) and 1°.

The NVWA therefore imposes an administrative fine totalling €7,500 on SNP for keeping in stock, offering for sale, selling, delivering, handing over, importing or exporting or otherwise bringing medicines outside the Dutch territory without a marketing authorisation,[387] and for advertising these products.[388]

39.3 The review

Unlike in previous, similar, cases,[389] the District Court of East Brabant rules in this case that the products in question are clearly exclusively food supplements and therefore Article 1(6) Gnw does not apply. The court considers the question of whether the presentation criterion has been met in assessing whether these products fall under the definition of medicinal product within the meaning of the Gnw to be irrelevant; as there is no doubt that the products are food supplements, the presentation criterion is disregarded.

The NVWA argues that if a product were to fall under both the definition of medicinal product and the definition of food supplement, the Medicinal Products Directive should be followed; it follows, as interpreted in the Van Bennekom judgment, that a product should (also) be regarded as a medicinal product if it 'implicitly, but no less emphatically, creates the impression tothe averagely well-informed consumer, by the manner of presentation, that it possesses therapeutic or prophylactic properties' (in other words, has the effect of a medicinal product).

The court does not follow the NVWA in this respect, referring to point 7 of the recitals to the (amended and currently applicable) Medicinal Products Directive (2004/27/EC):

> With the same objective of clarifying situations, where a given product comes under the definition of a medicinal product but could also fall within the definition of other regulated products, it is necessary, in case of doubt and in order to ensure legal certainty, to state explicitly which provisions have to be complied with. Where a product comes clearly under the definition of other product categories, in particular food, food supplements, medical devices, biocides or cosmetics, this Directive should not apply. [...]

The court also referred to the FIC Regulation, which has direct effect and contains specific regulations, which apply to food supplements, among other things. In view of this, in the court's opinion, the view cannot be followed that, in addition to these

387 Violation of Article 40(2) Gnw.
388 Violation of Article 84(1) Gnw.
389 See contribution by Irene Verheijen in Chapter 25 of this book.

provisions, the presentation criterion and the Gnw apply, on the basis of which the products should be considered medicinal products.

39.4 New approach by the court

Unlike other courts before it, the district court did not start its assessment with the definition of 'medicinal product', but with the recitals of the amended Medicinal Products Directive, which by means of directive-compliant interpretation also apply to the Gnw. The court first examined whether a different product category was at issue. Now that the products 'clearly exclusively' fall under a different product category, the court no longer bothers assessing whether the products fall under the definition of medicinal product based on the manner of presentation. As a result, the hierarchy provision also does not apply. In fact, the court first asks the preliminary question: 'Does this product clearly and exclusively belong to a different product category?' before proceeding to examine the definition of medicinal product or the hierarchy provision. This is in line with the position taken by many food lawyers: 'no doubt, not a medicinal product'.[390] The NVWA's position, referring to the Van Bennekom judgment, is also swept aside by the court, citing the recitals of the amended (and currently applicable) Medicinal Products Directive. This also deviates from previous case-law, which regularly refers to this judgment for the elaboration of the presentation criterion. Finally, the court attaches value (as also frequently argued by food lawyers) to the fact that there is specific legislation (the FIC Regulation) that provides for a ban on medical claims on food products. Why does such legislation exist if a medical claim on a food product directly turns that product into a medicinal product?[391]

39.5 Is the court's approach correct?

Although the ruling was received with cheers in the food sector and, as indicated, is in line with what has been described in the literature, the approach taken is critically examined in an annotation to the ruling.[392] In it, the question is raised as to whether the court correctly arrived at this (otherwise entirely justified!) ruling. Referring to the Kwizda Pharma judgment,[393] the annotator suggests a roadmap in which the starting point does seem to be the definition of 'medicinal product':

390 See Chapters 5 and 25 of this book.

391 See for an interesting elaboration on the court's approach, annotated by Sarah Arayess in IER 2022/46.

392 Annotated by Koosje van Lessen Kloeke in JGR 2023/3.

393 See Chapter 19 of this book.

- Step 1: Presentation criterion: in the eyes of the average consumer, is the product a medicinal product by presentation?
- Step 2: Presentation criterion: does the manufacturer or seller have the intention (intent) to 'present' the product as a medicinal product?
- Step 3: Does the product also meet the definition of a product subject to the application of other European legislation or, in other words, what is the nature of the product?
- Step 4: Given the nature and all characteristics of the product, is there any doubt as to whether it is a medicinal product or a product from another regulated product category? Does European legislation on the other product category provide protection for consumers in relation to misleading information, advertising and the manner of presentation, such as a ban on medical claims?

Steps 1 to 4 apply cumulatively; this is followed by another step 5 and 6 to test whether the challenged conduct(s) violate one or more provisions of the Gnw.

In the annotator's opinion, the court skips steps 1 to 3 by starting with the question of whether there is doubt about the product category and ruling that if there is not, the test against the definition of medicinal product and the hierarchy clause can be omitted.

By following the 4-step plan, which is something the East Brabant District Court did not, and taking into account recital 1 of the preamble to the Medicinal Products Directive (2001/83/EC) that the amended directive seeks to create a codification, value can nevertheless be attached to case-law on the definition of medicinal product contained in the old Medicinal Products Directive (65/65/EEC). Obviously, this should be supplemented by case-law on Article 2(2) of the currently valid Medicinal Products Directive, which was only inserted in 2004, as well as case-law on subsequently introduced comprehensive European product legislation for food, medical devices and cosmetic products.

In step 4, based on all the characteristics of the product, it should be determined whether it clearly falls under the definition of a product from another regulated product category or whether there is any doubt about it. If it clearly falls under the definition of another regulated product, then there is no doubt and the product falls outside the scope of the Medicinal Products Directive and therefore outside the Gnw. If there is any such doubt, then the product does fall within the scope of the Medicinal Products Directive and therefore also of the Gnw.

39.6 The outcome

The roadmap advocated may constitute a more nuanced approach than the East Brabant District Court took to reach the same conclusion, also taking into account older (codified) case-law such as the Van Bennekom judgment. Roadmap or not, the conclusion remains: if there is no doubt, it's not a medicinal product. In other words: whichever path is chosen, it is certain that the outcome will not be the one the NVWA arrives at in accordance with application of established policy: if there is a medical claim, it is by definition a medicinal product and there is consequently a violation of Article 40(2) and Article 84(1) Gnw.[394] With correspondingly much higher fines than can be imposed under food legislation.

39.7 A new direction in case-law?

Is this a change in the assessment of such products? The Hague District Court has also issued two judgments, ruling that the foodstuffs in question – baby food – cannot be classified as medicinal products by presentation.[395] There, too, there was no doubt about whether the products were foodstuffs; nevertheless, the court examined the advertising and sales channels of the product, and ruled on that basis that there was no question of medicinal products by presentation. However, there have also been several rulings following the 'classic format', where food products have been classified as medicinal products by presentation.[396] Unfortunately, therefore, these still seem to be exceptions. Therefore, a change in case-law cannot (yet) be presumed at the time of writing.

For the time being, the Netherlands Food and Consumer Product Safety Authority (*NVWA*) maintains its position that a product bearing the indication of a medical claim must, by definition, be classified as a medicinal product. The NVWA has therefore lodged an appeal against the ruling with the Administrative Jurisdiction Division of the Council of State. We food law lawyers naturally eagerly await the Division's verdict. It will be interesting to see how they will apply the recent case-law of the European Court of Justice, in which the Court emphasised that if a product clearly meets the conditions stipulated in another definition (e.g. food supplement, cosmetics or medical device), the Medicinal Products Directive does not apply.[397]

394 https://www.nvwa.nl/onderwerpen/claims-levensmiddelen/verbod-opmedische-claims

395 The Hague District Court, 28 June 2022, ECLI:NL:RBDHA:2022:6294 and ECLI:NL:RBDHA:2022:6295.

396 Rb. Midden-Nederland 8 April 2022, ECLI:NL:RBMNE:2022:1729; Rb. Zeeland-West-Brabant 22 June 2022, ECLI:NL:RBZWB:2022:3370; Rb. Zeeland-West-Brabant, 21 October 2022, ECLI:NL:RBZWB:2022:6136; Rb. Amsterdam, 6 December 2022, ECLI:NL:RBAMS:2022:7416; Rb. Rotterdam, 10 January 2023, ECLI:NL:RBROT:2023:75. See also: Barbara Mutsaers and Roeland Carolus, *Kroniek jurisprudentie levensmiddelenrecht 2022 up to and inc. first quarter 2023*, Waar & Wet, no. 4, June 2023.

397 CJEU 19 January 2023, cases C-495/21 and C-496/21, ECLI:EU:C:2023:34 (L. GmbH & H. Ltd v Bundesrepublik Deutschland) ('Nasal spray'), see Chapter 18 of this book. Also relevant in this context are CJEU 2 March 2023, case C-760/21 (Kwizda Pharma), ECLI:EU:C:2023:143, see Chapter 19 of this book.

Hopefully, this will put an end once and for all to the discussion, in which the kitchen is confused with a pharmacy.

Irene Verheijen
Lawyer – Partner at Legaltree, Amsterdam

CHAPTER 40

Fipronil – Enforce MRL exceedance? If so, by what means?

Annotated by Bernd van der Meulen

Trade and Industry Appeals Tribunal 21 April 2020 ECLI:NL:CBB:2020:296
Rotterdam District Court 14 February 2019, ECLI:NL:RBROT:2019:1056
The Hague Court of Appeal, 2 March 2021, ECLI:NL:GHDHA:2021:281

Fipronil crisis. Fraud. Recall. Traceability. MRL. Conversion factors. Food safety. Liability for regulatory failures.

Main considerations Trade and Industry Appeals Tribunal (*CBb*):

Scope of traceability obligation

> NVWA (*the Netherlands Food and Consumer Product Safety Authority*) wanted [...] to know [...] whether the risk of fipronil contamination had increased or decreased due to the processing of the eggs. The charge of 4 September 2017 could therefore only be met after doing an MRL [maximum residue limit] risk calculation.
> In doing so, the charge goes beyond the limits of the information requirement under Article 18(2) and (3) of Regulation 178/2002.

Application condition recall obligation

> The respondent [...] took the view that the claimant had reason to believe that concentrated egg products that it manufactured and delivered did not comply with the food safety requirement. This food safety requirement concerns the MRL for eggs, taking into account the method of processing mentioned in Article 20(1) of Regulation (EC) 396/2005.
> The CBb further considers relevant in this context that on 11 August 2017, an opinion was published by the Bureau for Risk Assessment & Research (BuRO) on – among other things – whether there was a risk to public health from egg products that incorporated eggs containing fipronil. The answer to that question, after a worst-case calculation of three products in which many eggs or egg yolks are used, was that the public health risks of fipronil from the consumption of egg products were negligible. From this, the CBb concludes that it was not (or no longer) the minister's intention to force recall at Wulro.

 | HTTPS://DOI.ORG/10.1163/9789004751699-042

Scope of recall obligation

At the hearing, the Minister stressed that Article 19 of Regulation 178/2002 includes the obligation for an operator who has reason to believe that a foodstuff produced or processed by him is not in compliance with food safety requirements to inform his customers.
The Trade and Industry Appeals Tribunal (*CBb*) is of the opinion that the enforcement of compliance with an information obligation by the food business operator to customers based on Article 19(1) of the Regulation is only possible in the context of a recall obligation.
To the extent that the Minister considered himself competent to enforce mere customer information, the reasoning rests on a misinterpretation of Article 19 of the Regulation and is unsound.

Main considerations of The Hague Court of Appeal:

2.26 Regulation 396/2005 (on residue levels of plant protection products/biocides) sets maximum levels of pesticide residues (maximum residue limit, hereinafter MRL) in or on food and feed of plant and animal origin. Only when residue levels in food of animal or plant origin do not exceed the maximum standard may they be placed on the market.
2.27 Although an MRL itself is not a standard used to indicate public health risk (the toxicological limit values ADI and ARfD are for that purpose, and are discussed below) it does serve to protect the health of consumers. MRLs should, according to European regulations, be set at least at the lowest level necessary to protect the most vulnerable consumers. The MRL for fipronil in eggs is set at 0.0050 mg per kg egg.
3.11 The Court of Appeal is of the opinion that the allegations of LTO et al. do not hit the mark. First of all, it should be noted that the Court of Appeal does not follow LTO et al. in arguing that the NVWA's assessment that there was no acute danger to public health was based on an incorrect assumption. At issue here is the memorandum of Dr. Sijm, the deputy director of BuRO, [...]. In that memorandum, it was assumed that a company using fipronil to control red mite in chickens will ensure that the dose of fipronil per chicken is sufficient to control red mite, but not higher than the dose at which toxic effects to the chicken can be expected. Based on this assumption, the memorandum then calculates the maximum expected level of fipronil in an egg to be 0.00002 mg per kg adult per day.
3.16 The Court of Appeal further agrees with the District Court that NVWA's duty as regulator does not extend to protecting the economic interests of poultry farmers. Where NVWA had found that there was no acute threat to public health, it was entitled to follow the criminal route instead of enforcing the MRL standard in the (economic) interest of the poultry farmers. What is important here is that under Article 17 GFL, poultry farmers have the primary responsibility to ensure that the eggs they produce meet the requirements of food law (such as the MRL standard).

40.1 Introduction

This book has discussed several international food safety crises in which the Netherlands was more than incidentally involved, notably the horse-meat scandal[398] and the MPA crisis.[399] This contribution deals with the fipronil crisis, which gave rise to evaluation studies by the Sorgdrager Commission[400] and the Dutch Safety Board[401] and to quite some case-law. This note discusses an administrative law case (Rotterdam District Court and – especially – the Trade and Industry Appeals Tribunal (*CBb*)) and a civil law case (The Hague Court of Appeal). The administrative law case concerns the use of enforcement tools. It addresses the meaning of Articles 14, 18 and 19 of Regulation 178/2002 (food safety, traceability and recall). The civil case concerns liability in the event of failed supervision. An important cross link between the two cases concerns the relationship between MRL exceedance and food safety.

In the administrative case, the Minister (or on his behalf the Netherlands Food and Consumer Product Safety Authority (*NVWA*)) imposed obligations on a company processing eggs that received eggs contaminated with fipronil. These obligations included informing customers and performing calculations for the purpose of applying the processing factor. Regulation 396/2005 (residue levels of plant protection products/biocides), and also other regulations setting limits, essentially set limits for unprocessed products. Processing may affect the concentration of the undesirable substance. For example, if water is extracted during processing, the concentration may decrease if the undesirable substance disappears with the water or increase if the water does not take the undesirable substance with it. The conversion factor is used to determine whether or not processed product not mentioned as such in the regulation complies with the limit. Information on the impact of processing is needed to make the calculation. The Minister places an obligation on the company to generate this information.

40.2 The Fipronil crisis

The introduction to the Sorgdrager Commission report begins with the following two paragraphs:

> On 22 July 2017, a report appeared in the daily newspaper De Stentor that fipronil had been found in eggs in Belgium. The fipronil allegedly came from a Dutch company

398 See Chapter 29.

399 See Chapter 31.

400 See: https://www.tweedekamer.nl/kamerstukken/brieven_regering/detail?id=2018Z12302&did=2018D35848. See also: https://www.tweedekamer.nl/kamerstukken/detail?id=2019D12648&did=2019D12648

401 https://www.onderzoeksraad.nl/nl/page/4791/opkomende-voedselveiligheidsrisico-s

> tackling red mite. Red mites (also known as poultry mites) are a common problem on poultry farms. The red mite is a small insect that sucks blood from the chicken. The animal emerges in large numbers at night to attack the chickens. Soon after the report appeared in *De Stentor*, news spread in the national media and it became clear that fipronil was also present in Dutch eggs.
> The detection of the presence of fipronil in eggs in the summer of 2017 had a great impact both in the Netherlands and abroad. On consumers, who did not know whether eating eggs was harmful or not; on poultry farmers, who suffered the financial and emotional consequences; on employees of the Netherlands Food and Consumer Product Safety Authority (NVWA), who saw their organisation come under the full spotlight.

Fipronil is a drug used to control pests on pets. Its uses include flea collars for companion animals. It is not authorised for use on food-producing animals. The cause of the fipronil crisis is that the drug was used illegally by a cleaning company in poultry farming because of its great effectiveness and in the absence of alternatives.

40.3 Enforcement priority

The Sorgdrager Commission points to one of the reasons that the problem took on the scale of a crisis as being that the NVWA (in particular its criminal investigation branch IOD) deliberately allowed illegal activities to continue for quite some time in order to build a stronger criminal enforcement case.

The Sorgdrager Commission addresses a number of its recommendations to the NVWA, including the following: 'Guarantee that food safety is given the highest priority and not weighed against other values and interests. Ensure that the primary importance of food safety is shared throughout the organisation and all employees are aware of it'; and 'Fraud in food chains has received a lot of attention in recent years. Fraud needs to be tackled energetically. But the requirement for faster and more frequent use of criminal law resources should not result in violations with serious consequences being allowed to continue for a longer period of time for the purpose of gathering criminal evidence. Intensified exchange and coordination between surveillance and investigation is needed to ensure food safety and combat fraud effectively.'

These recommendations are in line with the enforcement priorities as required under Article 137(1) of the Official Controls Regulation 625/2017: 'When acting in accordance with this Chapter, the competent authorities shall give priority to action to be taken to eliminate or contain risks to human, animal and plant health, animal welfare or, as regards GMOs and plant protection products, also to the environment.'

In food law, the interest of consumer protection is paramount (see also Article

5 of Regulation 178/2002). The interest of criminal investigation and prosecution is secondary to this.

In the judgment of the Court of Appeal, liability of the State (NVWA) is denied on the grounds that there had been no acute danger to public health and that enforcement of the MRL would only serve to protect the economic interests of poultry farmers, which would not be the task of NVWA.[402]

40.4 Public health risk: condition for implementation of recall

A key concept in the condition for implementation of Article 19 is 'food safety requirements'. As discussed in this book in Chapters 28 and 33 there has been a difference of opinion in the literature as to whether this term generally refers to (all) food safety requirements, or to Article 14 of Regulation 178/2002 which is entitled: 'food safety requirements'. The prevailing thinking now is that it refers specifically to Article 14. In the administrative law case, the Minister appears to have taken the opposite view as evidenced by this consideration: 'The respondent [...] took the view that the claimant had reason to believe that concentrated egg products manufactured and supplied by it did not comply with the food safety requirement. This food safety requirement concerns the MRL for eggs'.

The Trade and Industry Appeals Tribunal (*CBb*) infers from the assessment by BuRO that the public health risks were negligible. For the CBb, this is sufficient reason to assume that the minister was not seeking a recall. This once again confirms that Article 14 (unsafe food) is the standard for applicability of the recall obligation under Article 19 GFL.

In the civil case, the State insists that the exceedance of the MRL did not result in harmfulness and hence no administrative enforcement obligation. The court gives its blessing to this.

40.5 MRL

Does exceeding a legal limit result in the unsafety of the food concerned? It is generally accepted that food is unsafe only when the conditions for applying Article 14

402 I hope that someone will write something about this remarkable position. Don't the economic interests concerned with violation of food law requirements belong to the interests that the NVWA is there to defend? Unfortunately, there is no time or space here. In the past I would have pointed out that interests of the business world pursuant to Article 5 of Regulation 178/2002 do not belong to the aims of food law. From the perspective of speciality principle that could cause problems. This ruling is an unexpected example of that. See Bernd van der Meulen, *The function of food law. On objectives of food law, legitimate factors and interests taken into account*, European Food and Feed Law review, 2010 p. 83-90.

of Regulation 178/2002 are met. That provision does provide for a legal presumption of safety in respect of compliance with legal standards (in paragraphs 7 and 9), but does not provide for a legal presumption of unsafety when a standard is exceeded (on this subject, see also Chapters 28 and 33 in this book). If an MRL is exceeded, the product in question may not be placed on the market. Before other legal consequences arise under Article 14 or 19 of Regulation 178/2002, the conditions of application of those provisions must be met. To establish that, risk analysis is generally required.[403]

In the Basic Food Safety Data Sheets,[404] a policy rule, the NVWA puts it as follows:

> The health-based limit for single or short-term exposure is called the "Acute Reference Dose" (ARfD). This health-based limit is used, for example, in determining whether there is a health risk when the maximum residue limit of a plant protection product is exceeded. The ARfD indicates the maximum amount of a particular plant protection product that consumers may ingest at one time without risk to health. (p. 106)
> Health-based limits are thus always set on the safe side. For example, to maximise consumer protection, the calculation method to determine the possible exceedance of the ARfD assumes the worst-case scenario, namely that someone eats a fairly large portion of a food with that high content. (p. 107)
> In the case of genotoxic substances that can be carcinogenic, in principle no safe limit can be derived, because in theory a single molecule can already cause damage to DNA and thus form the basis for a subsequent tumour. Instead of the health standard, one therefore usually works with the so-called 'margin of exposure' (MOE). This is the difference between the actual exposure to a substance and the exposure at which there is a risk of adverse effects. For genotoxic substances, this MOE is ideally greater than 10,000. (p. 107)
> In practice, this means that most MRLs are at a lower level than is necessary from a health perspective, i.e. are usually (much) lower than the health-based limits. Moreover, in good agricultural practice, residue levels are often well below the MRL. If an MRL is exceeded, the food does not comply with the law and is considered unfit under Article 14 of the General Food Law (EC No 178/2002). However, there is only a health risk if the health limit, known as the Acute Reference Dose (ARfD), is also exceeded. (p. 115)

---

403 So, for example, the Standing Committee. Standing Committee on the Food Chain and Animal Health (SCoFCAH) now Standing Committee on Plants, Animals, Food and Feed (SCoPAFF) Guidelines for the implementation of Articles 11, 12, 14, 17, 18, 19 and 20 of Regulation (EC) No 178/2002 on General Food Law, 26 January 2010, https://food.ec.europa.eu/system/files/2016-10/gfl_req_guidance_rev_8_en.pdf.

404 https://www.nvwa.nl/documenten/organisatie/hoe-de-nvwa-werkt/publicaties/staat-van-voedselveiligheid-2018.

> If the MRL is exceeded, the relevant product is not compliant with the law and further marketing must be stopped. The supplier and customer must be informed by the company where the violation was found. If, in addition to the MRL, the ARfD is also exceeded, there may be a health risk and the product should be instantly withdrawn from the market and a recall action initiated immediately. (p. 122)

40.6 Enforcement tools

In the administrative case, the Minister imposes some obligations on an egg-processing company that received contaminated eggs. These include the obligation to perform calculations to determine whether or not the processed product complies with the limits in the light of the conversion factors. Another obligation is to inform customers about the fipronil contamination. The Minister bases the first obligation on Article 18 of Regulation 178/2002 and the second on Article 19. Article 18 concerns the traceability obligation, Article 19 the recall obligation. In both cases, the Trade and Industry Appeals Tribunal rules that the obligation imposed cannot be based on the provision in question. Article 18 only imposes an obligation to provide traceability data upon request, not to conduct research to generate additional data. Article 19 deals with recall. The obligation in the first paragraph to inform customers is derived from the recall. Since there is no recall obligation and the Minister did not impose it because there was no danger to public health, no obligation to inform buyers followed from Article 19 either.

What is striking is that the Minister could easily have chosen a different legal basis. The Official Controls Regulation 625/2017 provides an open set of tools. Article 138 states in the opening words of the second paragraph, among other things:

> When acting in accordance with paragraph 1 of this Article, competent authorities shall take any measure they deem appropriate to ensure compliance with the rules referred to in Article 1(2), including, but not limited, to the following:

The obligations the Minister wanted to impose cannot be brought under any of the categories that then follow, but they apparently do fall under 'any measure they deem appropriate'. Unless it is to be assumed that these are only measures that have a legal basis elsewhere, the European legislature is straying very far here from the standpoint of the rule of law. A blank cheque is given to the executive in the Member States to devise their own enforcement tools. I can understand that the legislator realises it cannot foresee every practical situation, but the extent to which additional/open powers are provided for would, in my view, have been better left to national legislators. Now the European legislator is thwarting the trias politica in the Member States.

Be that as it may, in this case the Minister could have made things a lot easier for himself by relying on this power (or the corresponding provision in the preceding Control Regulation 882/2004).

One possible reason why the Minister/the Netherlands Food and Consumer Product Safety Authority (*NVWA*) generally favours Article 19 of Regulation 178/2002 is that this provision creates obligations by operation of law that reduce the regulator's liability risks.[405]

40.7 Relevance to practice

The fipronil crisis shows that there are weaknesses in the application of food law in stress situations. The legal system works like a well-oiled machine. However, it is important to take specifically those instruments from the toolbox whose conditions of implementation are fulfilled in practice and to extract from them the legal consequences they can actually deliver.

The Sorgdrager Commission and the Dutch Safety Board delivered very critical reports. According to the Minister and the civil court, there was nothing to worry about. Apparently, what we still have to learn is to explicitly state that certain risks are accepted. This requires investment in the fine art of risk communication.

Bernd van der Meulen
Chairman NVLR (Dutch Food Law Association www.nvlr.nl)

---

405 See also on this subject Chapter 33 of this book.

PART III

Advertising code

Introduction to the Dutch self-regulatory system

Ebba Hoogenraad and Lisanne Steenbergen

This introduction is to give background information to better understand the origin, relevance and the process of the Dutch self-regulatory system.

The Netherlands has a strongly developed and respected self-regulatory system. More than 60 years ago the Stichting Reclame Code ("SRC") was founded to deal with the self-regulation system of advertising in the Netherlands. The creation of the SRC was a result of the communal effort of the advertising industry (advertisers, advertising agencies and the media). The rationale of these players was that it would be better to self establish a self-regulatory foundation and draw up rules to protect the consumer against misleading practices or other undesirable or inappropriate forms of advertising, than to wait for the legislator to impose rules.

This idea has evolved into the creation of the Dutch Advertising Code ("NRC").

Dutch Advertising Code

The Dutch self-regulatory system consists of a general Dutch Advertising Code ("NRC") that is applicable to all advertising and numerous special codes aimed at specific types of advertising, such as advertising relating to alcohol ("RVA") , foodstuffs ("RVV"), children and young people ("KJC") and sustainability ("CDR" and before February 2023: "MRC"), social media and influencer marketing ("RSM"). The special advertising codes are created on the initiative of the relevant industry, for example the FNLI (Federation of the Dutch Food Industry) is the so-called 'code holder' and initiator of the RVV.

Important EU legislation (such as the Unfair Commercial Practices Directive, Directive 2005/29) is also incorporated in the Dutch Advertising Code (NRC). The Advertising Code Committee (ACC) also checks whether the ads are compliant with (EU) law (e.g. FIC Regulation 1169/2011, Claims Regulation 1924/2006 etc).

This makes the self-regulatory system in the Netherlands widely appreciated and respected. The accessibility with which you can file a complaint at the SRC and receive a considered and reviewed opinion on the permissibility of an ad, contributes to the good image of the SRC.

Complaint handling

The SRC prides itself of being very accessible. This translates to who can file a complaint: everyone who finds that an advertisement is not in order, because it is misleading or incompliant with other specific rules. Think of: rules about advertising for unhealthy foods aimed at children under 12 years old (in the RVV), rules on how

 | HTTPS://DOI.ORG/10.1163/9789004751699-043

to properly disclose the relevant relationship between the advertiser and the influencer on social media (e.g. in case of a paid relationship, gifting of a product etc). This results therein that consumers file complaints, but also NGOs and competitors find their way to the SRC. Consumers can file a complaint free of charge (in the first instance), while companies are required to pay a fee.

After a complaint is submitted, it will be assessed for admissibility. If is not admissible, the Chairman can directly dismiss the complaint. The complainant will be informed of this decision and may object to it with the Advertising Code Committee "ACC".

Complaints that are admissible are divided based on their complexity. The Chairman handles that simple cases by himself (without a hearing) and the more intricated cases are presented in front of the 'full' ACC, which consists of the Chairman and representatives of the column advertisers, column channel and column creation.

The complaint is send to the advertiser, who can react to the complaint by writing a statement of defence. Usually the written part the proceedings will end there, but in some cases at the discretion of the Chairman/ACC the parties will be given another opportunity to make further written comments.

When the case is presented in front of the ACC a hearing is planned. Both parties can attend the hearing (but this is not mandatory). The hearing is open to the public, unless one of the parties puts forward compelling objections. In practice it is quite rare that third parties attend an ACC-hearing. At the hearing parties are given the opportunity to make another verbal statement and react to the other parties. During the ACC-hearing the chairman and other members can also ask questions to the parties. After the hearing it is up to the ACC to determine whether the advertisement is in accordance with the NRC.

Recommendation

By means of a written decision the ACC decides to dismiss or uphold the complaint. When a complaint is upheld, the advertiser receives a recommendation to no longer advertise in this manner. The unsuccessful party can appeal a decision of the ACC by submitting a written statement of appeal at the Board of Appeal ("BoA") of the ACC.

All decisions of the ACC that have become final, and all Board of Appeal decisions are published in the on the website of the SRC. In severe cases, the ACC may issue an ALERT, which involves the secretariat of the SRC issuing a public press release in associated media. This is typically done when an advertiser repeatedly violates the same regulation or when there is a (potential) risk to human health.

The ACC (and for that matter also the Chairman of the ACC and the Board of Appeal) can only issue (non-binding) recommendations to refrain from certain advertising. The ACC cannot impose administrative fines, corrections or prohibit the use of an advertisement. Regardless, the recommendations of the ACC are generally followed by advertisers (high compliance rate of 96% in 2023). The ACC may not formally prevent the broadcast of an advertisement, the broadcasting (television and radio) of a commercial that received a recommendation from the ACC is effectively not possible. This is due to a collaboration of the SRC and the Dutch public broadcasting agencies, which has been incorporated in the Dutch Media Act. This means that advertisements that have received a recommendation from the ACC (and BoA) will not be accepted to be broadcasted on TV and radio by the STER (responsible for the advertising space on the channels of the Dutch Public Broadcasting Agency NPO) and regional and local public broadcasting agencies. Rulings of the ACC on misleading advertising are also highly regarded by Dutch civil courts.

Compliance

Within the SRC, the independent Compliance Department checks whether advertisers act in compliance with decisions of the ACC and the Board of Appeal. If not, they request the advertiser to do so. If the advertiser refuses to comply, this will be published on the website of the SRC. In this way, non-compliant advertisers will be brought to the attention of third parties, such as government regulators.

Advantages of self-regulation

The general advantage of the Dutch highly developed self-regulatory system is the large support it gets from the relevant industries. The special advertising codes are the initiative of big players of the industry behind it, such as trade organisations. As mentioned the Federation of the Dutch Food Industry is code holder for the RVV. The STIVA (Foundation for Responsible Alcohol Consumption) is code holder for the Advertising Code for Alcoholic Beverages (RvA), but also the Advertising Code for Alcohol-free and Low Alcohol Beverages (RvAVA).

Another advantage is that the (special) advertisings codes are (relatively) easy to amend in light of current affairs, changing social views and technological developments. Because the initiative is with the industry they are able to easily make changes, in comparison to EU an Dutch legislation that can take years to be adjusted.

Besides the social developments that are quickly incorporated in the advertising code, the ACC also includes developments in EU case law in their recommendations. If the complaint gives reason thereto the ACC assesses whether an advertisement is contrary to the law. The ACC takes into account the EU laws (FIC, Claims Regulation etc), but also regulary refers to EU case law such as the ECJ-decision in Teekanne. This emphasizes the relevance of the role of the SRC and the rulings of the ACC and BoA. Especially because there is no big threshold for parties to file a complaint at the ACC.

SRC in the EU context

The SRC is a member of the European Advertising Standards Alliance ("EASA"). The SRC works together with other self-regulatory authorities/systems in the Europe, amongst others with the UK Advertising Standards Agency ("ASA"). Within this collaboration the SRC is one of the most influential parties.

Ebba Hoogenraad
Partner, food law and advertising at Hoogenraad & Haak, Amsterdam

Lisanne Steenbergen
food law and advertising lawyer at Hoogenraad & Haak, Amsterdam

CHAPTER 41

Chupa Chups – Misleading health label

Annotated by Marjolein Driessen

Advertising Code Committee (RCC) 10 June 2003, file 030198, IER 2003, 78 cf. E.H. Hoogenraad (*Keesom/Chupa Chups*)

Advertising Code Committee 2 December 2004, file 04.0456/RCC Board of Appeal 12 January 2005, IER 2005, 32 annotated by E.H. Hoogenraad (*Consumers' Association/Chupa Chups*)

Misleading, Article 9 (currently Article 10) Dutch Advertising Code (NRC) (use of citations), Article 4 Code for Confectionery (link between consumption of confectionery and health, Article 10 (currently Article 11) NRC (appearance in advertising by authoritative persons), Article 13 NRC (misleading claim for children)

41.1 Keesom/Chupa Chups

The case involves the following TV commercial for Chupa Chups lollipops:

> At the hairdresser's, a boy sits next to 'Roberto Carlos World Cup Champion'.
> Carlos says: 'Would you like a Chupa Chups too?'
> The boy: 'My mum won't let me.'
> Carlos: 'Come, let's go see her.'
> At home at the front door: Boy: 'Hi mum.'
> Carlos: 'Hi ma'am.' When invited inside, Carlos explains: 'Look Chupa Chups is much more than a lollipop. It provides glucose, an energy source for your head.'
> At which point, Carlos and the boy point with a Chupa Chups lollipop to their heads and say '*koppie koppie*' [smart cookie].

Finally, the voice-over says: 'Chupa Chups, more than just a lollipop. Recognised by the study centre for nutrition and dietetics CESNID' and appears – along with a logo – on screen: 'Recognised by study centre for nutrition and dietetics CESNID'.

According to complainant Keesom, by referring to the organisation CESNID (unknown to Dutch consumers), the commercial wrongly suggests that Chupa Chups fits within 'the Good Food Guidelines' and appears as if the Dutch Association of Dieticians approved the lollipop, which is not the case.

Chupa Chups denies that suggestion. The Spanish organisation CESNID (Centre for Higher Studies in Nutrition and Dietetics), which is affiliated to the University of Barcelona and collaborates with the Spanish Association of Dieticians and Nu-

 | HTTPS://DOI.ORG/10.1163/9789004751699-044

tritionists (AEDN), has produced a scientifically backed paper titled 'Glucose and Brain' at Chupa Chups' request. It follows from this that glucose is the main source of energy for the brain. So, there is no question of deception, according to Chupa Chups.

The RCC rules that the advertisement is not misleading: the complainant has not refuted (with reasons) that the claim 'Chupa Chups is much more than a lollipop, it provides glucose, a source of energy for your head ('*koppie koppie*')' is correct. Nor does the advertisement contain the suggestion that the Chupa Chups lollipop fits within 'the Good Food Guidelines', nor that the Dutch Association of Dieticians would have approved the lollipop. However, the Committee does find that the reference to the foreign organisation CESNID violates Article 9 NRC (currently Article 10 NRC). It follows from that article that citations should be used with great caution when it comes to advertising intended for the general public. This is not the case with the citation 'Recognised by study centre for nutrition and dietetics CESNID', because it is not clear what this recognition refers to.

All in all, Chupa Chups is thus recommended by the RCC to stop advertising in this way.

Ebba Hoogenraad was right to point out in her note to this judgment in the magazine IER that – contrary to what the RCC's decision suggests – the complainant did not complain about the correctness of the claim that Chupa Chups contains glucose c.q. that glucose is an important source of energy for the head, and furthermore that it is not up to the complainant to substantiate/prove that a claim is incorrect: it is up to the advertiser to prove that claims made are correct and not misleading (via the reversal of the burden of proof of Article 6:195 of the Dutch Civil Code). In that respect, therefore, a somewhat strange decision by the RCC.

It is not the only Chupa Chups commercial that the RCC has examined.

41.2 Dutch Consumers' Association/Chupa Chups

In 2004, not long before Chupa Chups passed into Italian-Dutch hands (of candy manufacturer Perfetti-Van Melle), people complained about a Chupa Chups TV commercial starring Daphne Deckers. She is known as the wife of tennis player Richard Krajicek, but also as a photo model, and through her books, articles and appearances on TV programmes, she enjoys fame as a 'health mother'.

The complaint concerns the following commercial:[406]

> Daphne Deckers is playing the mother of three children. First she sits in the room behind her computer and the children are playing wildly around her, but a little later

406 As described in the RCC ruling: IER 2005, 32.

the children, with a lollipop in their mouths, sit with her on the sofa. Finally, she takes a lollipop herself.
When Daphne Deckers says, 'In our house, it's fruit', an apple, cherry and strawberry – seemingly drawn by children – appear on screen, after which the viewer sees whole strawberries disappearing into the lollipops, as it were.
The text spoken by Daphne Deckers reads as follows:
'As a mother, I know exactly what my children are doing. We don't snack whenever we feel like it! In our house, it's fruit! That's why they like Chupa Chups. This is because of the addition of real pulp and there's no fat in it. So, if they want to snack, I let them have Chupa Chups. And I'll have one too.'
Voice-over: 'New. Chupa Chups with real pulp and 0% fat.

Several complaints were received about this commercial. This note discusses the Consumers' Association's objection, which can be summarised as follows:
- The commercial is misleading (in contravention of Article 7 NRC): the suggestion is made that Chupa Chups are healthy (at least not unhealthy), while they consist of 87.1% sugar, which is converted into fat in the body. The advertisement is not 'as clear and complete as possible' as required, as the harmful ingredient sugar is not mentioned.
- The commercial violates Article 10 of the NRC (currently Article 11 NRC: because of appearances on radio/TV programmes authoritative/trustworthy persons are not allowed to appear in a commercial themselves): the average consumer sees Daphne Deckers as a 'role model mother', partly because she regularly appears in TV broadcasts and writes about health, family and children.
- The commercial targets children and therefore violates Article 13 NRC.
- The commercial violates Article 13.2 NRC: minors are being encouraged to buy the lollipops because the commercial will make them think that Chupa Chups are not harmful to health.
- The commercial violates Article 1 Code of Confectionery (CVZ): Daphne Deckers promotes the eating of the lollipops.
- The commercial violates Article 4 CVZ: it establishes a (prohibited) link between sweet consumption and health.

Chupa Chups challenges the complaint(s), stating that the commercial points to the changed and improved taste of the lollipops, which are widely known to contain sugar. The addition of pulp and the absence of fat sets Chupa Chups apart from other products. All further information is given in the ingredient declaration. A toothbrush appears on the screen (as per Article 6 CVZ). Daphne Deckers is known as a photo model and actress; her books and articles in magazines have no educational purpose. And finally, the commercial is aimed at mothers of children, not at children themselves nor is there any question of encouraging excessive consumption or establishing the connection prohibited by Article 4 CVZ.

The RCC rules that the emphasis in the commercial is on the presence of pulp and the absence of fat, while it does not mention that the lollies consist of more than 80% sugar and only 3% fruit pulp. This falsely suggests that the lollies are healthy, which is misleading. It also establishes a link between the consumption of sweets and health, in contravention of Article 4 CVZ. The fact that it is generally known that lollies contain sugar and that the amount of sugar is also stated on the packaging, as Chupa Chups argues, does not change this. In addition, the RCC finds that the commercial at least partially addresses children who are being misled about the properties of the lollies. That violates Article 13.1 NRC. And: Daphne Deckers enjoys authority/confidence with a certain audience group, among others through her appearance in various programmes. Now that she plays along in the commercial, there is also a violation of Article 10 (currently 11) of the NRC. The RCC rejects the complaint only where it concerns Article 13.2 NRC (inciting children to buy) and Article 1 CVZ.

Chupa Chups appeals.

41.3 Board of Appeal (CvB)

According to Chupa Chups, the RCC wrongly failed to assess the misleading question from the perspective of the 'average consumer' who reads the list of ingredients and corrects impressions created by advertising on that basis. By mentioning 'with fruit', this average consumer will not think that the lollies do not contain sugar now that this is on the ingredient list: it is common knowledge that lollies contain sugar; the advertisement repeatedly mentions 'snacking' and also the 'teeth-brushing' symbol shows that they are not sugar-free lollies. The reference to the pulp merely links the fruit to the flavour. Furthermore, emphasising the absence of fat is allowed now that the lollies are competing with chocolate products, etc. According to the proposed Claims Regulation, the ban on claims such as 'low fat' or '0% fat' for sugary products is violates the fundamental principle of nutritional science that there are no good or bad foods, only good and bad diets, Chupa Chups said. Moreover, according to Chupa Chups, fame is not synonymous with authority and confidence, and Daphne Deckers only shares her own experiences as a mother in her books and on TV. Finally, the commercial is aimed only at parents, not children.

The Board of Appeal ruled as follows:

- the advertisement as a whole suggests that eating Chupa Chups is more or less responsible from a health point of view. This violates Article 4 CVZ;
- the advertisement is not misleading because of the emphasis on the presence of fruit and the absence of fat. After all, it is common knowledge that lollies contain sugar;
- the advertisement is addressed to parents, not to children so there is also no violation of Article 13.1 NRC; and

– it cannot be said that Daphne Deckers enjoys authority or trust because of her appearance on TV programmes.

In other words, only the validation of the complaint under Article 4 CVZ remains.

It is and remains a remarkable judgment. It's worth noting that the Claims Regulation – like the Advertising Code for Foodstuffs ('RvV') – had not yet been introduced at the time of the commercial, nor had the EU judgment Teekanne[407] (see Chapter 11 of this book) been delivered yet. The question is whether the Board of Appeal would have reached a different verdict now. In my opinion, there's no question.

For example, the '0% fat' claim is not allowed if it is not a distinguishing feature within the relevant group of products (Article 4 RvV – in combination with the Claims Regulation). And that is the case here: lollies generally do not contain fat, so '0% fat' is not a distinguishing feature of Chupa Chups compared to other lollies. That Chupa Chups is or should be specifically compared to chocolate products is neither made clear nor obvious. Incidentally, Ebba Hoogenraad correctly points out in her note to this CvB ruling that the '0% fat' claim was already in violation of the *Warenwetbesluit Etikettering van Levensmiddelen* (Article 29) in 2005 as well, but that article did not play a role in the proceedings.

In addition, since the Teekanne judgment of the CJEU, impressions created by advertising can no longer be automatically corrected by the ingredient list. Thus, the suggestion that Chupa Chups are healthy, which is created by all the references in the commercial to fruit and 0% fat, cannot be automatically corrected by the fact that the ingredient list states that the lollipops contain 87% sugar (and therefore they are not that healthy after all and certainly not healthier than other lollipops). In my opinion, too much emphasis is placed on all the 'good' properties of the lollies ('lots of fruit, no fat') for it to be corrected via the ingredient list.

Apart from the new and sophisticated rules on the basis of which it is less easy to make an admissible (nutrition or health) claim, I am of the opinion that the commercial is inadmissible because of Daphne Deckers' appearance: precisely because of the various books and articles she has written on motherhood, children and parenting and her ensuing appearances on TV, it is very plausible that she does enjoy a certain authority or trust, at least among part of the Chupa Chups advertisement's target audience: mothers/parents.

Marjolein Driessen
Lawyer/partner at Legaltree, Rotterdam

---

407 CJEU 4 June 2015, ECLI:EU:C:2015:361, IER 2015/53 (Teekanne) annotated by E.H. Hoogenraad and B. Duivenvoorde.

CHAPTER 42

Koffieleutjes – Comparative Advertising

Annotated by Sarah Arayess

Advertising Code Committee (RCC) 27 July 2007, 2007/07.0332 (*Koninklijke Verkade/Koninklijke Peijnenburg*)

Articles 2 and 7 Dutch Advertising Code, Article 3 Advertising Code for Foods, Articles 9 and 13(1) Claims Regulation, nutrition claims, nutrition information, health claims, labelling, comparative claims misleading.

42.1 Summary

It should be clear to the average consumer which products are being compared when claims such as '40% less saturated fat' are made. Explanations of the comparison should be legible and displayed in a visible place.

42.2 Complaint and main considerations

In 2007, Peijnenburg launches a new biscuit: *Koffieleutjes*. The introduction of these new biscuits is not received enthusiastically by its competitor Verkade. Verkade files a complaint and leaves no stone unturned concerning the *Koffieleutjes* packaging. The accusations? The claim '40% less saturated fat' is misleading and placing this message within a heart shape would constitute an inadmissible health claim. Verkade believes that the packaging falsely suggests that this new product contains 40% less saturated fat than either a previous version of this product or, , since no reference is made to the comparison segment, all other biscuits in the similar market segment. The heart shape around the text reinforces this suggestion. There is no previous version of this product. The product does not contain 40% less saturated fat than similar biscuits in the 'biscuits' segment such as café noir, biscuit, wheat biscuit, filled biscuit and the caramel waffle.

The claim '40% less saturated fat' back on pack has an asterisk referring to the text: 'At least 40% less saturated fat than Speculaas and Sprits. Koffieleutjes 6.0g/100g* – Speculaas 10.2g/100g* – Sprits 12.5g/100g*'. Verkade finds this text and explanation is illegible, hidden (under an opaque sealing edge that must be lifted aside)and therefore misleading The claim 'source of fibre' on the front of the packaging is also unacceptable, according to Verkade. Peijnenburg defends itself extensively and in detail. The claim and explanation are correct, clearly visible and

 | HTTPS://DOI.ORG/10.1163/9789004751699-045

easy to read. The heart shape was used to support the nutritional claims and is not a health claim. The claim 'source of fibre' meets the requirements of the Claims Regulation.

The Advertising Code Committee (*Reclame Code Commissie: RCC*) ruled that the claim '40% less saturated fat' is misleading because it is not clear what the frame of reference is. The average consumer does not know how much fat different cookies and biscuits contain and does not know which other biscuit the *Koffieleutje* should be compared with. The fact that the claim '40% less saturated fat' has an asterisk on the back of the packaging does not lead to a different outcome. The text explaining that a comparison is only being made with *speculaas* (type of ginger biscuit) and *sprits* (type of shortbread) is so limited that it fails to rectify the misleading impression created by the claim. Moreover, the explanation is printed in small, poorly contrasting, difficult-to-read letters in a location that was not readily visible, namely under the sealing edge.

The claim '40% less saturated fat' and the accompanying explanation is also in violation of Article 23 of the *Warenwetbesluit Etikettering van levensmiddelen* (Commodities Act Decree on Labelling of Foodstuffs), says the RCC. This article stipulates that the indications and statements referred to in that decree must be clearly visible and easily legible. It is also in violation of Article 9 of the *Warenwetbesluit Voedingswaarde-informatie* (Commodities Act Decree on Nutrition Labelling), which stipulates that the statements regarding nutrition labelling and nutrition claims must be displayed together in one conspicuous place.

Placement of the saturated fat claim in a (vague) heart shape was not considered a health claim by the RCC.

In contrast, the claim 'source of fibre' in violation of the Claims Regulation. The packaging of the *Koffieleutjes* mistakenly stated that it contained 2.0 grams of dietary fibres per 100 grams. In reality this turned out to be more than 3 grams (enough to claim 'source of fibre' under the Claims Regulation). According to the RCC, however, the packaging and not the reality had to be taken as a basis and therefore this constituted a violation of the Claims Regulation. In short: the packaging violated Articles 2 and 7 of the NRC.

42.3 Discussion

This decision by the RCC concerns, among other topics, the coming into force of the Claims Regulation in 2007. Meanwhile, the Claims Regulation is rapidly approaching its 20th birthday. On the one hand, a lot has changed in the intervening period; on the other hand, some discussions around the Claims Regulation play an undiminished role anno 2026.

In its ruling, the Advertising Code Committee (*Reclame Code Commissie: RCC*) first discusses the alleged deception within the meaning of Article 7 NRC. The

RCC's opinion that the claim '40% less saturated fat' is misleading is explained in detail. The claim gives consumers no insight into the frame of reference of the claim. Consumers do not directly see what products are being compared. The explanatory text on the back is also inadequate because it does not provide enough information. In addition, the explanation has disappeared under the seal edge of the packaging and is printed unclearly. In itself, the explanatory text is quite clear: *Koffieleutjes* contain at least 40% less fat than *speculaas* and *sprits*. The source and the numbers are also mentioned. So the claim is explained quite clearly, which makes the RCC's conclusion remarkable. Did the RCC instead actually mean to say that the (broad) claim becomes too limited by the explanatory text?

In her note in the journal *Intellectuele Eigendom en Reclamerecht* (IER), Ebba Hoogenraad raised the point that the RCC in this case did not test whether there was a permissible comparative nutrition claim within the meaning of Article 9 of the Claims Regulation. Does this perhaps have to do with the wide transitional period of the Claims Regulation? Article 9 of the Claims Regulation states that in the case of comparative nutrition claims only foods of the same category may be compared, taking into account a range of foods in that category. But the Claims Regulation does not clearly define what is meant by 'the same category'. To this day, this point remains a challenge in the industry. After all, what exactly is 'the same category'? Are sweets and chewing gum the same category? Is liquorice the same category as wine gums? And how many products should you compare, to meet the requirement of 'a range'?

The claim '40% less saturated fat' and the accompanying explanation also violates Article 23 of the Dutch Food Labelling Commodities Act Decree and Article 9 of the Dutch Nutritional Value Information Act Decree. Due to the entry into force of the FIC Regulation (1169/2011/EC), both these national decrees have been repealed and therefore no longer exist.

The strict rule that claims about the nutritional value of a product must be placed next to or in the nutrition table has now been abandoned. Under certain conditions, it is possible to repeat nutrition claim(s) or nutritional information on the front of the packaging. Also the latter remains a challenge in 2024 / 2025, as in some cases the Netherlands Food and Consumer Product Authority (NVWA) reportedly considers a nutrition claim front of pack (such as a protein content claim) as an unlawful repetition of nutritional information. This clearly illustrates that, despite the FIC-Regulation and the Claims Regulation having been in force for several years, interpretative questions continue to arise.

At the time, the Claims Regulation was just fresh out of the oven. Peijnenburg uses the claim 'source of fibre', one of the authorised nutrition claims in the new Claims Regulation. However, the amount of fibre stated on the packaging was incorrect (and lower than in the actual product) and not enough to lawfully make the claim. Since, the RCC is reviewing the packaging – and not the actual product – the only logical conclusion was that the claim violates the Claims Regulation!

The 'clumsy' oversight in this case illustrates how challenging it can sometimes be to fully comprehend new rules upon their entry into force.

Then finally, the 'little heart outline' as background to the claim '40% less saturated fat'. The complainant argued that this heart could be interpreted as a health claim. But the RCC was not buying it: the heart is too vague and cannot be construed as a health claim. Meanwhile, the RCC's judgment has proved to be incorrect. The *Deutsches Weintor* judgment (Chapter 9 of this book) already confirmed that the definition of a claim should be interpreted broadly. A heart can also be an (implied) health claim.

Sarah Arayess
Lawyer food and advertising law / founder Say legal, Amsterdam

CHAPTER 43

Vitaminwater – It's all in the name

Annotated by Sarah Arayess

Advertising Code Committee 16 April 2009, file 2009/00196 (*Lemmens/Vrumona-Vitaminwater*)

Articles 7 and 8 Dutch Advertising Code, foods, deception, name, Consumers' Association, health

43.1 Summary

In the food sector, complaints are often made about allegedly misleading product names or labelling. In this case, the complainant found that the name Vitaminwater gives the impression that it is a spring or mineral water with additives. However, sugar appears to have been added to the product, making it a soft drink. This is misleading, in the complainant's view. Or is it?

43.2 Complaint and main considerations

The Advertising Code Committee (*Reclame Code Commissie: RCC*) summarises the complaint clearly: the complainant objects to the product name *Vitaminwater* and announcements about this product in the Distrifood brochure. The brochure contains more information about the product, which is presented as 'more than water!'. According to the Advertising Code Committee (*RCC*), it is clear that this is a composite product (*samengesteld levensmiddel*). In the case of a composite product, the product name alone cannot be relied upon, says the RCC. This is because the product name is often only an indication of the main ingredients and/or flavours. Consumers know this and will often read the list of ingredients before making a purchase decision.

The list of ingredients is clear, says the Advertising Code Committee (*Reclame Code Commissie: RCC*). Sugar has been added to the product, making it effectively a soft drink. The advertiser makes the additional argument that *Vitaminwater* is a so-called 'functional water' or 'water+'. The Advertising Code Committee (*RCC*) considers that the average consumer is familiar with this specific product group. It is therefore not necessary to mention the word 'soft drink' on the packaging or in the leaflet. The name is neither legally required nor is the absence of it misleading. After all, the list of ingredients is sufficiently clear and the brightly coloured packag-

 | HTTPS://DOI.ORG/10.1163/9789004751699-046

ing also makes it clear that this is a soft drink. The advert in Distrifood magazine is aimed at professionals. This audience knows very well what kind of product is being advertised. The Advertising Code Committee (*RCC*) concludes that the advertising is not misleading. Neither the packaging, nor the advertising in the leaflet.

43.3 Discussion

Complaints about misleading product names recur regularly over the years. Until about the year 2015, these complaints were often dismissed under the so-called 'labelling doctrine'. Basically, the labelling doctrine means that interested consumers always read the list of ingredients. So: if the list of ingredients is correct, then it may be acceptable to exaggerate on the rest of the label. This is the explanation for the long-standing practice of images of excessive fruit and product names like 'mango juice' when in reality the product contains mainly orange juice. Until 2015, that is, when the Court of Justice ruled in the Teekanne case.[408] In very short: a statement on the packaging (name, image or even colour) can be so misleading that it cannot be rectified by a correct list of ingredients. When deciding on misleading claims, the entire packaging must be examined.

At the time, the Netherlands Food and Consumer Product Safety Authority (NVWA) seemed to have taken this ruling as a need to enforce the rules more strictly: suddenly, companies that had been marketing the same packaging for years were facing enforcement actions by NVWA. Mainly food labels referring to fruit (in word or depiction) were addressed. NVWA is of the opinion that in such situations, the product should include:

- Actual fruit(s);
- Extracts of the fruit(s); or
- Natural flavours of the fruits.

If this is not the case, this should at least be made known with a disclaimer ("Fruit flavour").

Slowly, a shift emerged in the market. If a product does not contain strawberry, but only the flavour of strawberry, the term 'strawberry flavour' will appear on the label. Nevertheless, care must be taken to ensure that Teekanne is not interpreted incorrectly and too strictly. Teekanne, after all, still says that consumers look at the list of ingredients. The list of ingredients is still relevant in this assessment. The entire packaging must be considered when assessing whether a package is misleading.

Then back to a product name again. So even a name can be misleading under

408 This classic case is discussed in depth in Chapter 11 of this book.

certain circumstances. Consumer organisations like Foodwatch also used the Teekanne ruling as justification for their claim that some brands, should change their packaging. In 2016,[409] for instance, Foodwatch argued that the name *Vitaminwater* should no longer be used. The addition of vitamins would give the product a healthier image when in fact the product contains a lot of sugar. Also the *Consumentenbond* (a consumer organisation) paid considerable attention to *Vitaminwater* in 2017; the name would suggest that the product was healthier than it actually was. This even led to a petition to manufacturer Vrumona to get a name change.[410] To this day, *Vitaminwater* is still called *Vitaminwater*. It remains to be seen whether this will still be the case 10 years from now.

Quite a lot has changed between the decision made in 2009 and the writing of this book in the original Dutch edition. Labelling is looked at a lot more strictly and is also going through a change. Being clear(er) to consumers about the contents seems to be a new trend. If a complaint about '*Vitaminwater*' were made today, it might very well not be dismissed so quickly. On the other hand, relatively little has changed. Consumers and consumer organisations still complain about misleading product names and packaging. And the list of ingredients remains important. The rest of the packaging definitely plays a role in the overall assessment as well.

Sarah Arayess
Lawyer food and advertising law / founder Say legal, Amsterdam

409 https://www.foodwatch.org/fileadmin/foodwatch.nl/Onze_campagnes/Misleidende_Marketing/Vitamine_Claims/Files/Voedselindustrie_misleidt_consument_met_vitamine_verrijkt_junkfood.pdf.

410 https://www.consumentenbond.nl/nieuws/2017/sourcy-vitaminwater-50-miljoen-suikerklontjes-per-jaar

CHAPTER 44

Unox smoked sausage – Enjoy more often

Annotated by Ebba Hoogenraad

Advertising Code Committee 4 December 2012, file 2012/00998 (*Unox smoked sausage*)

Article 7 Advertising Code for Food Products, Article 16 Dutch Advertising Code, excessive consumption, foodstuff, comparative nutrition claim, energy value.

44.1 Summary

The phrase 'Enjoy more often' (in Dutch 'Extra vaak genieten') used for an extra lean Unox smoked sausage is in violation of Article 7 of the Dutch Advertising Code for Food Products, as it encourages higher consumption.

There are many 30% less fat, sugar and salt products around. Good for your health. However, advertising must not suggest that one can therefore eat extra amounts of these products. That would be an exercise in futility. Where should we draw the line? In any case, 'Enjoy more often' is not allowed.

44.2 Complaint and main considerations

The magazine '*Allerhande*' contains an advert for the Unox extra lean smoked sausage, with 30% less fat. 'So you can enjoy more of it!'

The complainant's objection is that the advert suggests that one can eat more of this low-fat smoked sausage, which is actually bad for your health.

The fact that the smoked sausage contains less fat is, of course, fine. The problem lies with the phrase 'so enjoy more'. The Advertising Code for Food Products stipulates in Article 7 that an advert for a food product with a lower energy value than the original food product is not allowed to encourage people to consume more of it than of the original food product with a higher energy value.

The primary defence of Unilever (the producer of the sausage) is that it is a permissible comparative nutrition claim: the new smoked sausage contains 30% less fat. And, furthermore, Unilever states that there is no suggestion of better health, or more consumption, but simply of being able to enjoy the sausage more often.

The Chairman of the Advertising Code Committee (Reclame Code Commissie: RCC) makes short shrift of this: 'enjoy more often' will be perceived by consumers as an incentive to eat smoked sausage more often than before, meaning more of the

 | HTTPS://DOI.ORG/10.1163/9789004751699-047

extra lean Unox smoked sausage. Why? Because it is lower in fat and has an extra lean composition. Therefore, overall, consumers will eat more smoked sausage than before. This is a violation of Article 7 of the Advertising Code for Food Products. According to the Chairman of the Advertising Code Committee (RCC), it is not relevant that Article 7 refers to a lower energy value, and the advert mentions a reduced fat content. After all, consumers will associate the latter with a lower energy value to some extent. The complaint is upheld.

44.3 Discussion

Choosing a healthier product, but subsequently eating more of it, which in fact results in consuming the same amount of calories: that's not the way to go. Which is why the advertising industry has agreed that adverts for products with fewer calories should not encourage you to eat more of them. The legislator is not very strict,[411] but the Advertising Code for Food Products is.

Enjoy more

This ruling clarifies two things. Although 'enjoy more' is not necessarily an explicit call to consume more, the Chairman does interpret it that way. If an advertiser wants to emphasise that the new version of the product tastes extra delicious, it is best to literally phrase it that way. Because 'enjoy more' of the extra lean composition of the smoked sausage with less fat can be understood as a call to consumers to buy the leanUnox sausage more often.

Lower energy value: broader application in the spirit of the code

The other notable outcome of this decision is that, although the wording of Article 7 is quite tightly formulated in this sense, the ruling is not just about fewer calories. The Chairman finds a way around this: consumers will associate the statement 30% less fat and extra lean 'to some extent' with lower energy value. And thus it falls within the scope of Article 7 of the Advertising Code for Food Products. This case is an example of interpreting a provision in the spirit of the Advertising Code. This corresponds to Article 16 of the general Dutch Advertising Code (NRC): the NRC should not only be applied literally, but also in accordance with the spirit of the code. Which is not to be confused with extensive interpretation: an interpretation outside what the industry parties have agreed on as part of self-regulation. Extensive interpretation has no place in self-regulation.

---

411 The Claims Regulation 1924/2006 only bans claims that encourage or condone excessive consumption of a food (Art. 3(c)).

No extensive interpretation in self-regulation

The industry stakeholders determine the content of the obligations contained in an advertising code. If the content should be different/stricter, that is for the 'code holders' to decide. The Advertising Code Committee and the Board of Appeal do not play a role in this. A dogmatic argument applies here: in self-regulation, the decision-makers must stay within the limits of the text of the regulation. An extensive interpretation 'contra legem' – as is quite often applied by judges in legislation – is out of the question in self-regulation. This would be a contravention of the principle of self-regulation. It also violates legal certainty. The industry stakeholders have deliberately agreed what is and what is not allowed in food advertising. The decision-makers should apply the regulations, explain them and clarify texts where necessary. It is not up to decision-makers such as the Advertising Code Committee and the Board of Appeal to step outside the frameworks agreed by the industry stakeholders. That is up to the stakeholders themselves. Unlike the lengthy process of legislation, self-regulation can be adjusted within a very short period of time.[412] This eliminates an essential argument for applying extensive interpretation. Back in 2007, in a case on alcohol advertising, the Board of Appeal rejected a request from complainant STAP (Dutch Institute for Alcohol Policy) for extensive interpretation ('contra legem') in the Amstel Gold Race case. In the annotation of this ruling in the Dutch journal *Intellectuele Eigendom en Reclamerecht* (Intellectual property and advertising law), Prof Madeleine De Cock Buning argued that extensive interpretation in self-regulation is contrary to the principle of self-regulation. She states that extensive interpretation implies modification of self-regulation outside the framework of what was considered acceptable by the parties involved in its formation.[413,414]

Who knew an extra lean smoked sausage could cause such a stir?

Ebba Hoogenraad
Partner, food law and advertising law at Hoogenraad & Haak, Amsterdam

412 And if the industry agrees among themselves, it can even be applied retrospectively. A good example is the entry into force of the unfair commercial practices provisions implementing the Unfair Commercial Practices Directive through the far-reaching amendments laid down in the new Articles 7 and 8 of the Dutch Advertising Code. These were declared applicable *with retrospective effect* as of 1 February 2008 to statements made as of 12 December 2007.

413 Board of Appeal, 19 June 2007, file 1459/07.0011 (*Amstel Gold Race*); annotated by Prof Madeleine De Cock Buning in the Dutch journal 'Tijdschrift Intellectuele Eigendom en Reclamerecht (IER)', no. IER 2007/91 (only available in Dutch).

414 For more on extensive interpretation, see in the Dutch journal 'Tijdschrift Intellectuele Eigendom en Reclamerecht (IER) no. 2009/57, Ebba Hoogenraad (only available in Dutch).

CHAPTER 45

Amstel Radler 0.0 – Alcohol-free advertising

Annotated by Ebba Hoogenraad

Advertising Code Committee 9 July 2014, file 2014/00435 (*Amstel Radler* 0.0, *Laurens ten Dam*)

Article 30 paragraph 3 Alcohol Advertising Code (old version), Article 1 Dutch Advertising Code, alcohol, alcohol-free beer, food.

45.1 Summary

Does Amstel Radler 0.0 fall under the definition of 'alcohol advertising'? No. Therefore, the TVC (television commercial) featuring professional cyclist Laurens ten Dam drinking a beer (Amstel Radler 0.0%) while descending a mountain is permitted. In the commercial, the alcohol-free beer is shown only with the addition '0.0%'. The possible association with the alcoholic brand Amstel does not transform the advertising into alcohol advertising.[415]

The European branch of the Dutch Foundation for Alcohol Prevention (*Stichting Alcohol Preventie* (STAP)), called EUCAM (European Centre for Monitoring Alcohol Marketing), strongly objects to a top athlete promoting an alcohol-free beer when that alcohol-free beer carries the name Amstel. In their view, the name alone would still amount to advertising for the well-known alcoholic version of Amstel. Under alcohol advertising rules, top athletes are not allowed to appear in advertising. The ACC provides clear guidelines: since '0.0%' is shown throughout the commercial, it does not qualify as alcohol advertising. The fact that viewers may associate the alcohol-free beer with the alcoholic Amstel beer does not change this conclusion.

45.2 Complaint and main considerations

The TVC shows the peloton in a mountain stage of the Tour de France. Laurens ten Dam, who leads this peloton, 'drifts back to the team car, where he appears to be handed a bottle of beer. Ten Dam toasts with the team manager and takes a sip from the bottle. Moments later, he also hands out some of these bottles, pulled from

415 This ruling is annotated by A. Alkema in IER 2014/70.

 | HTTPS://DOI.ORG/10.1163/9789004751699-048

the back of his jersey, to other cyclists in the peloton. In the commercial, two commentators can be heard, initially assuming it is a water bottle, then realizing with surprise that it is beer. They joke that Ten Dam is treating the others to drinks. In the closing shot of the TVC, two packs of Amstel beer 0.0% Radler appear on screen. Viewers then see an opened bottle of Amstel beer 0.0% Radler alongside a filled glass and a sliced lemon. On screen, the text reads: 'Amstel 0.0% Radler'.

EUCAM argues that the TVC – along with a similar radio advertisement – should also be classified as alcohol advertising. The Amstel logo is visible, and the word 'beer' is used. According to EUCAM, this violates Article 30 paragraph of the Advertising Code for Alcoholic Beverages (RVA) (old version), which prohibits advertisers from hiring athletes actively playing sports at the highest senior level (European Championships, World Championships and the Olympic Games) for radio, cinema and television commercials and print communications that use scripted, staged situations. In this case, cyclist Laurens ten Dam was hired.

The ACC questions whether this constitutes (also) alcohol advertising. In the TVC, viewers and listeners are is briefly misled to think that top athletes are drinking beer. However, the setting in which it takes place, a (mountain) stage of the Tour de France, is so unrealistic that it is clear to the average consumer that beer containing alcohol is not actually being drunk.

Since in both the TVC and the radio advertisement the logo or the name Amstel mentioned only in combination with '0.0%', it is sufficiently clear that the advertisement refers to the non-alcoholic drink Amstel 0.0% Radler. The ACC rejects an appeal to two other alcohol-related cases, as those situations are not relevant here.

The mere use of the word 'beer' does not in itself constitute advertising for alcoholic beverages. The advertiser has sufficiently demonstrated that 'beer' is a term regulated in the Beer regulation (in Dutch: '*Bierverordening Productschap Dranken* 2003'), which also includes non-alcoholic variants. Under the Beer Regulation, Amstel 0.0% Radler may and must be described as 'alcohol-free beer'.

The voice-over text – 'The natural mix of Amstel beer and lemon water is now also available alcohol-free. Amstel 0.0 Radler' – may evoke associations with alcoholic Amstel beer. However, these associations arise from the fact that the advertiser markets both alcoholic and non-alcoholic Amstel beer. The advertiser is free to produce both types. Any possible association with other beers produced by the advertiser therefore does not mean that this is, in this case, advertising for alcoholic beverages. Moreover, the voice-over text unmistakably refers to the alcohol-free variant of Amstel Radler.

It is concluded that there was no advertising of alcoholic beverages and the complaint is therefore dismissed.[416]

416 This case was handled for Heineken Nederland by Ebba Hoogenraad.

45.3 Discussion

A thorny issue was settled in 2014: does advertising for the 0.0 variant of an alcoholic brand of beer fall under the Alcohol Advertising Code? The answer is no. Provided that, as in this case, the advertising consistently and clearly concerns the 0.0 variant. Here, the number 0.0 (%) appears repeatedly. EUCAM argues that since the word Amstel is mentioned/the Amstel logo appears on screen., the advertisement also constitutes alcohol advertising. This position is understandable in itself. However, it could also be counterproductive: the 0.0 variants are specifically marketed to give consumers the opportunity to choose the non-alcohol variant more often. EUCAM's reasoning would mean that manufacturers who already market an alcoholic beer brand would still fall under all the restrictions of the Alcohol Advertising Code, even when promoting a non-alcoholic variant, which can be consumed by pregnant women, at work, and by designated drivers. That cannot be the intention. The ACC is clear in its considerations: the mere association with the alcoholic beer brand does not, in itself, make it alcohol advertising.

EUCAM invokes the *Bavaria malt* decision. In that case, actor Mickey Rourke – known for his role as an alcoholic writer in the film *Barfly* – enters a hotel and orders a beer at the bar: 'Hey man, I need a drink, give me a fucking beer.' The bartender hesitates between Bavaria and Bavaria 0% and decides to serve Rourke a alcohol-free Bavaria, without telling him. In the company of several others, Rourke is then served multiple (non-alcoholic) beers. Shortly after he goes to a hotel room with a woman, a minibar is thrown out of the window, and an angry Rourke, wearing a bathrobe, storms back into the bar, shouting: 'Man, what kind of shit hotel is this: alcohol-free beer in the mini bar?! I'll have one of yours!' The bartender then secretly pours him another Bavaria 0%. Here too, the question arose of whether this is 'alcohol advertising'. In this TVC, the bartender opens a fridge containing both the non-alcoholic and the alcoholic versions of Bavaria side by side. The ACC ruled that, although the emphasis in the TVC is unmistakably on the non-alcoholic variant, attention is also explicitly drawn to the alcoholic Bavaria beer. Both types are shown several times, and the bartender explicitly chooses between them. Therefore, the Advertising Code for Alcoholic Beverages does apply in this case. As a result of this decision, Bavaria also violated the Dutch Media Act, because the TVC was broadcast before 9 pm.

The Laurens ten Dam TVC focuses exclusively on the 0.0 variant and therefore the ACC considers a different outcome justified.

The use of the term 'beer' does not make the commercial alcohol advertising. After all, 'beer' is the mandatory term under the Beer Regulation.[417]

---

417 https://wetten.overheid.nl/BWBR0014243/2006-01-14. The Beer Regulation Product Board Drinks 2003, which includes the definition of 'beer' and 'alcohol-free beer' and which is referred to in the ruling, expired on 1 January 2025.

Partly in response to the Laurens ten Dam decision, beer brewers decided to consistently use the 0.0 statement in words and images in advertising for advertising alcohol-free beer, and avoid any reference to the alcoholic variant. In addition, advertising of alcohol-free beer is not directed at individuals under the age of 18.

In the National Prevention Agreement of December 2018, it was agreed that the industry association STIVA, together with the Dutch Brewers, would draft a separate advertising code for responsible marketing of alcohol-free beer. This means that advertising for alcohol-free beer would also be subject to regulation.

Postscript: since 15 October 2020, the Advertising Code for Alcohol-Free and Low Alcohol Beer (RvAAB) has applied to advertising for alcohol-free beer and advertising for low-alcohol beer. Such advertising may not specifically target minors, even if it concerns 0.0 or low-alcohol beer. And since 1 April 2024, the Advertising Code for Alcohol-Free Variants of Alcoholic Beverages (RvAVA) has come into effect, replacing the RvAAB. This code applies not only to beer but also other alcohol-free variants.

Ebba Hoogenraad
Partner, food law and advertising law at Hoogenraad & Haak, Amsterdam

CHAPTER 46

Shelf tags – Organic food labelling

Annotated by Silvia Gawronski

Annotated by Silvia Gawronski

Advertising Code Committee 4 November 2014 (2014/00653)

Advertising Code. Advertising Code Committee (*Reclame Code Commissie: RCC*). *Organic food labelling. Art 23 Regulation 834/2007 (old) and Regulation 2018/848 (new) on organic production and labelling of organic products. Shelf tags with the text 'organic' misleading for consumers.*

Main consideration:

> Article 23 of Regulation (EC) No 834/2007 on organic products and labelling of organic products and repealing Regulation No 2092/91, provides that the terms listed in the Annex to that Regulation, including 'organic', may be used only if the requirements laid down in or pursuant to that Regulation have been met. The advertiser has acknowledged that the requirements of Regulation (EC) No 834/2007 have not been met in respect of the products mentioned by the complainant. On this basis, said products cannot be considered 'organic' and the advertiser has acted in breach of Article 23 of Regulation (EC) No 834/2007 by calling the products 'organic' on the shelf tags.

46.1 Introduction

On 1 January 2022, the new Organic Regulation – (EU) 2018/848 of the European Parliament and of the Council of 30 May 2018 on organic production and labelling of organic products and repealing Council Regulation (EC) No 834/2007[418] – came into force. This regulation replaces the previously existing European Regulations 834/2007 and 889/2008 and from that moment on[419] contains the main rules for producing, processing, marketing as well as labelling organic products in the European Union (EU).

However, the rules on the mandatory and optional use of the Organic logo and

418 14 June 2018, OJ L 150.

419 The new Organic Regulation was accepted in 2018 and was set to come into force on 1 January 2021. Due to the outbreak of the COVID-19 pandemic, however, it was decided for Regulation 2020/2042 to postpone the entry into force of the new Organic Regulation by one year until 1 January 2022. All other terms outlined in the new Organic Regulation were also postponed by one year. It is also useful to know that products made before 1 January 2022 were also brought onto the market after that date.

the designation 'organic' are not new. It was and still is prohibited to advertise a product as organic if the product does not comply with the applicable European rules. But the labelling rules are clarified with the entry into force of the new 2018 Organic Regulation.

The proposal for the new Organic Regulation dates back to 2014 – the year of this decision by the Advertising Code Committee (Reclame Code Commissie: RCC). In particular, the explanatory memorandum to the proposal[420] stated that under the old regulation 'Production rules do not sufficiently take into account evolving consumer and citizen concerns and expectations', but also that the 'labelling rules are complicated'. For this reason, the standards from the various regulations have been merged and clarified on a number of points.

It is important for everyone involved in organic production, processes or products to be familiar with these rules in order to avoid regulatory violations – and sometimes even decertification of the organic product by the regulator Skal with all the ensuing damage. But thus also to avoid an instruction from the RCC for misleading advertising. This case before the Advertising Code Committee (RCC) involved, as I understand it, shelf tags in an Albert Heijn supermarket,[421] but in practice, consumers frequently encounter tags on shelves and counters that refer to 'bio', 'eco' or 'organic'. If the product (and the company) do not comply with European rules, this is not only a violation of the Agricultural Quality Act, but it also violates Article 2 of the Advertising Code and is therefore misleading.

46.2 The contested advertisement

The case that came before the RCC in 2014 involved so-called shelf tags with the text 'organic' followed by, among other things, the description of the product and the corresponding price. In particular, it concerned products such as fair trade peanuts, nut mix, cashew nuts, pineapple slices, green tea with lemon and rooibos with spices, but also non-fair trade coffee, jars of sprouts, recycled toilet paper and jars of baby food.

The Advertising Code applies to all advertising irrespective of the medium used, unless the Code explicitly stipulates otherwise. Advertising is defined as: any form of public and/or systematic direct or indirect commendation of goods, services and/or ideas by an advertiser or, either wholly or partly, on their behalf of him, with or without the help of a third party (Article 1, Advertising Code). So shelf tags, or other signs, cards on the counter, or (chalk) signs at the entrance of a café are also included.

420 Proposal for a REGULATION OF THE EUROPEAN PARLIAMENT AND OF THE COUNCIL on organic production and labelling of organic products, COM/2014/0180 final – 2014/0100 (COD).

421 'AH in overtreding met biologische schapskaartjes', reclameboek.nl

46.3 The complaint

The complainant in this case claimed that the products she mentioned where the shelf tag said 'organic,' were not organic products. The complainant felt that the advertiser had thereby misled consumers and undermined confidence in organic products. The complainant also felt that this constituted unfair competition. In addition, she considered the expression to be in violation of the Dutch Advertising Code and the Environmental Advertising Code.

46.4 Albert Heijn's defence

Albert Heijn stated in its defence that shelf tags marked 'organic' were used for the products mentioned by the complainant. This deviated from shop instructions and the list of items where the relevant shelf tag was allowed to be hung. According to Albert Heijn, this was a one-off implementation error in a specific shop. Albert Heijn tightened its shop instructions in response to the complaint. To substantiate this, Albert Heijn also submitted copies of said shop instructions and the tightened instructions for organic products.

46.5 The judgment of the Advertising Code Committee (RCC)

The judgment of the RCC was crystal clear, and therefore highly relevant for all those other entrepreneurs claiming to sell 'bio', 'eco' or 'organic' products. The RCC referred to Article 23 of the old Organic Regulation 834/2007, which stipulates that the terms listed in the annex to this regulation, including 'organic', may only be used if the requirements laid down in or pursuant to that regulation are met. Albert Heijn had acknowledged that the requirements of Regulation (EC) No 834/2007 had not been met in respect of the products mentioned by the complainant. The RCC continued: 'On this basis, said products cannot be considered "organic" and the advertiser has acted in breach of Article 23 of Regulation (EC) No 834/2007 by calling the products "organic" on the shelf tags. This regulation is binding in all its parts and is directly applicable, as a result of which it is to be regarded as law for the purposes of the Dutch Advertising Code. Therefore, the chairman is of the opinion that the advertisement was made in violation of the law as referred to in Article 2 of the Dutch Advertising Code (NRC). As the advert is not admissible for this reason, the other grounds of complaint raised by the complainant do not require further consideration.'

46.6 The decision of the Advertising Code Committee (RCC)

Based on the above, the chairman considers the advertisement to violate the provisions of Article 2 of the Dutch Advertising Code. However, the RCC is of the opinion that Albert Heijn has made it sufficiently plausible that it has taken appropriate measures to prevent repetition, so that it can be assumed that this was a one-off error. For this reason, in line with Article 12(5) of the Advertising Code Committee's and the Board of Appeal's regulations, a concrete recommendation was omitted.

46.7 Rules on labelling in the new Organic Regulation

This ruling makes it clear without prejudice that consumer deception lurks when foods are advertised as 'bio', 'eco' or 'organic' while not complying with those regulations. What follows is a brief explanation of what these rules look like under the new Organic Regulation.

Labelling

The rules on 'Labelling' in the new Organic Regulation apply to 'any words, particulars, trade marks, brand name, pictorial matter or symbol relating to a product that are placed on any packaging, document, notice, label, ring or collar that accompanies or refers to that product'. This follows from the definition of the term in Article 3(52) of the new Organic Regulation. The rules therefore cover more than just the label of the product.

Use of bio(logical), eco(logical) and organic

If products, their ingredients or feed materials used in their production are described in the labelling, advertising or commercial documents in terms that give the buyer the impression that the products, ingredients or feed materials have been produced in accordance with the new Organic Regulation, then they must also comply with the new Organic Regulation. This follows from Article 30 of the new Organic Regulation. In particular, words like organic, ecological or derivatives or diminutives thereof, such as 'bio' and 'eco', are only allowed if the product is actually organic. Therefore, this also applies – given the broad definition of 'labelling' – to company and trade names.

Article 30(5)(b) of the new Organic Regulation does allow these terms to be used in the ingredients list, so for example: [water, sugar (organic), apple, sultanas (organic)], or in the sales description if the processed ingredients are at least 95% organic and meet the requirements of Annex II Part IV (these are production rules for processed food of the new Organic Regulation). Then again, the Organic logo may not be used (see below). Similar requirements apply to animal feed under Article 30(6) of the new Organic Regulation. If certain substances such as plant protection products, fertilisers or soil conditioners are allowed to be used in organic farming, the labels of those products may also refer to them (Article 31 of the new Organic Regulation).

The Organic logo

The Organic logo is often called a quality label, but strictly speaking it is an official attestation within the meaning of Articles 86 and 91 of EU Regulation 2017/625, which in that Regulation is called a 'certificate'. This is confirmed by Article 33(2) of the new Organic Regulation.

The use of the Organic logo is separate from the use of the designation 'organic'. For both, use is allowed if the requirements are met, but it is not mandatory. If the Organic logo is allowed to be used, the product may also be called 'organic', but there are products that are called 'organic' but do not carry the Organic logo. The use of the Organic production logo of the European Union is regulated by Article 32 of the new Organic Regulation.

For pre-packaged products produced in the European Union, the Organic logo is mandatory under Article 32(1)(b) of the new Organic Regulation. The EU organic production logo may be used – optionally – in the labelling and presentation of products complying with this regulation and in the advertising of these products under Article 33(1) of the new Organic Regulation. The use of the Organic logo for EU organic production is optional for products imported from third countries (Article 33(3) new Organic Regulation). However, the organic logo must in all cases comply with the model and requirements in Annex V of the new Organic Regulation.

Compulsory indications on the label

Besides the use of the organic designation and the Organic logo, the new Organic Regulation also prescribes a number of compulsory indications that must be put on the label. These rules are also found in Article 32 of the new Organic Regulation. The most important of these are:

(a) the code number of the control authority or control body to which the operator that carried out the last production or preparation operation is subject shall also appear in the labelling; and
(b) in the case of prepacked food, the organic production logo of the European Union referred to in Article 33 shall also appear on the packaging, except in cases referred to in Article 30(3) and points (b) and (c) of Article 30(5).

2. Where the organic production logo of the European Union is used, an indication of the place where the agricultural raw materials of which the product is

composed have been farmed shall appear in the same visual field as the logo and shall take one of the following forms, as appropriate:

(a) 'EU Agriculture', where the agricultural raw material has been farmed in the Union;
(b) 'non-EU Agriculture', where the agricultural raw material has been farmed in third countries;
(c) 'EU/non-EU Agriculture', where a part of the agricultural raw materials has been farmed in the Union and a part of it has been farmed in a third country.

Moreover, the words 'EU' or 'non-EU' may not be displayed in a more prominent colour, font size and typeface than the name of the product.

The indications must be marked in a conspicuous and highly visible place. Moreover, practical implementation rules on the use, presentation and size of the indications are expected to follow.

46.8 Supervision and enforcement Skal

The new Organic Regulation takes direct effect in the Dutch legal system; there is therefore no transposition into national laws or regulations. However, supervision of compliance and enforcement in the event of violations is regulated nationally, in the Agricultural Quality Act and the Economic Offences Act. It goes beyond the scope of this annotation to set out the complicated system of disciplinary law, administrative law supervision and enforcement and criminal investigation and prosecution. Most importantly, supervision and enforcement are mainly in the hands of the Skal Foundation, a so-called independent administrative body. Skal has a clear and accessible website,[422] which also includes the label checker[423] as a handy tool for the entrepreneur and lawyer.

Silvia Gawronski
partner in International Trade & Commodities at Van Traa law, Rotterdam

422 skal.nl
423 'Bio-checker', skal.nl.

CHAPTER 47

Fruit on bread – With no added sugar

Annotated by Ebba Hoogenraad

Advertising Code Committee 3 June 2015, file 2015/00350

Article 7 Regulation (EU) No 1169/2011 (FIC Regulation), Article 8 Regulation (EU) No 1924/2006 (Claims Regulation), Article 2 Dutch Advertising Code, Articles 7 and 8 Dutch Advertising Code, foodstuff, nutrition claim, no added sugars, fruit concentrate, misleading advertising

47.1 Summary

'Leev Fruit on Bread contains 98% organic fruit and no added sugar'. This sounds delicious and healthy. But it is misleading. In addition to the 66% fresh fruit, the product contains about 31% fruit concentrates. These contain sugar, which consumers will not realise. Moreover, the concentrates are added for their sweetening properties. And for that reason, the claim 'no added sugar' is not allowed.

For a long time, consumers were unaware that fruit concentrate also contains sugars. The Leev Fruit on Bread (in Dutch: 'Fruit op Brood') was the first to raise awareness. While fruit concentrate is made from fruit, it also contains a high amount of sugar. When fruit concentrate is used, as in this spread, specifically to make the product taste sweeter, the term 'no added sugar' is prohibited. It was not until 2019 that Foodwatch (a non-profit campaigning organisation in the food sector) began widely challenging the incorrect use of 'no added sugar' in the food industry. This Leev Fruit on Bread case was the first. Nearly four years later, the Chairman of the Advertising Code Committee (*Reclame Code Commissie; RCC*) provided a standard assessment in the case regarding Albert Heijn's Squeeze Fruit Apple and Pear.[424]

47.2 Complaint and main considerations

Leev Fruit on Bread positions itself on the Albert Heijn supermarket website, www.ah.nl, as a fruit-packed sandwich spread, containing as much as 98% fruit with sweeteners coming only from fruit. Tastier and healthier.

---

424 Case file 2018/00621.

 | HTTPS://DOI.ORG/10.1163/9789004751699-050

The complainant considers this misleading, as the spread does not in fact consist of 98% fruit. Though the main ingredient is fruit (e.g. 66% raspberry), the spread also contains 32% fruit concentrates. It is misleading to state that the product is made entirely out of fruit. Moreover, the 31% apple concentrate is largely composed of sugar, therefore the claims of 'no sugar added', and 'only sugars from fruit' are not allowed.

Leev's argues that fruit concentrate is both a sweetener and fruit. In addition, the percentages of ingredients are clearly stated on the label. Therefore, the advertisement is not misleading according to Leev.

The Advertising Code Committee (*RCC*) divided the case into two issues. The claim 'contains 98% fruit' is an ingredient claim. The 98% also includes 31% fruit concentrate. Leev states this has a sweetening function and the Advertising Code Committee (*RCC*) takes the view that the fruit concentrate cannot be considered 'fruit'. Consumers will not expect that one-third of the '98% fruit', is actually carbohydrates or sweetener. The average consumer will also not know that fruit concentrate consists almost exclusively of carbohydrates. Therefore, the consumer will form an overly positive impression of the product. In principle, an advertisement may refer to fruit concentrate. However, if the concentrate consists mainly of carbohydrates and is intended for its sweetening effect, the false impression must not be given that the concentrate can be equated with the original fruit. The misleading impression is not sufficiently corrected by the other correct information on the product (ingredient declaration and nutrition table). The statement '98% fruit' is in breach of Article 7 of Regulation 1169/2011, the FIC regulation,[425] and thus also violates the law as referred to in Article 2 of the Dutch Advertising Code (NRC).

Can the statement 'no added sugar' be used? No. This claim is a nutrition claim under the Claims Regulation. This claim is subject to the condition that the product does not contain any added mono- or disaccharides or any other food used for its sweetening properties. Since Leev stated that the fruit concentrates were added because of their sweetening effect, 'no added sugar' is not allowed (contrary to Article 8 paragraph 1 of the Claims Regulation[426] and therefore in violation of the law as referred to in Article 2 NRC).

---

425 Regulation (EU) No 1169/2011 of the European Parliament and of the Council of 25 October 2011 on the provision of food information to consumers, amending Regulations (EC) No 1924/2006 and (EC) No 1925/2006 of the European Parliament and of the Council, and repealing Commission Directive 87/250/EEC, Council Directive 90/496/EEC, Commission Directive 1999/10/EC, Directive 2000/13/EC of the European Parliament and of the Council, Commission Directives 2002/67/EC and 2008/5/EC and Commission Regulation (EC) No 608/2004.

426 Regulation (EC) No 1924/2006 of the European Parliament and of the Council of 20 December 2006 on nutrition and health claims made on foods.

47.3 Discussion

The Leev case started a discussion that has resulted – so far – in a series of complaints from Foodwatch to the Advertising Code Committee about products with added sugar or sweetening ingredients. Many products containing 'healthy' syrups are nevertheless marketed as 'sugar-free' or 'no added sugars'. This is officially in violation of the Claims Regulation, but the NVWA (the Dutch Food and Consumer Product Safety Authority) has other enforcement priorities.

According to the Claims Regulation, the claim 'with no added sugars' – or any other claim having the same meaning for consumers – is only allowed: 'where the product does not contain any added mono- or disaccharides or any other food used for its sweetening properties'.[427] Concentrated fruit juice is often added. This means that the claim 'no added sugar' cannot be used unless the manufacturer can prove that the concentrated fruit juice is added for a reason other than its sweetening properties.

What could those reasons be? The main reason for adding sugar is for its sweet taste. But there are other reasons: as a flavour enhancer (for example, cinnamon in yogurt tastes stronger after adding a spoonful of sugar), as a filler (sugar provides 'bulk' to a product), or to create a shiny brown crust. Sugar also contributes to structure and texture (in meat products) or serves as a stabilising binding agent. But in the end, sugar is sugar.

With regard to the Claims Regulation, only sugar added because of its sweetening properties is relevant when assessing whether is the claim 'no added sugar' may be used.

In the Leev case, the advertiser already explicitly stated on the jars and on the website that the fruit concentrate was added for its sweet taste. Therefore, it is crystal clear that the phrase 'no added sugar' is not allowed.

The Advertising Code Committee acts as a guardian for consumers by ruling that a claim such as 'with 98% fruit', while one third of it consists of added sugar in the form of fruit concentrate, is misleading. Consumers will not realise that fruit concentrate is essentially added sugar. Those who choose Leev, and think they are being healthy and responsible, will unwittingly spread extra sugar on that crunchy Sunday morning sandwich.

More than three and a half years later, Foodwatch has once again drawn public attention to this hidden sugar issue. In a series of complaints, Foodwatch has challenged various claims about added sugar.[428] Many manufacturers have stated in their defence that they will adjust their packaging. From now on, products like

427 Regulation (EG) No 1924/2006 on nutrition and health claims and the Annex.

428 For more on this series of ACC/Foodwatch cases, see also an article in Waar & Wet no 2, 2019, Christine Grit, 'Suikers onder de RCC-loep'.

Rinse apple syrup[429] will no longer state that no sugar has been added. In the cases of Hipp children's biscuits[430] and Kellogg's crunchy muesli bars,[431] the advertisers did not succeed in demonstrating (sufficiently convincingly) that the sugar was added for a reason other than sweetness. The dates in the muesli bars were after all mainly added for sweetness, and not for texture.

This series of complaints did leave the food industry with a fundamental guideline. In the Albert Heijn squeeze fruit Apple and Pear case,[432] with added pear concentrate, the Chairman of the ACC provided a clear framework. The mere intention of the manufacturer to add fruit concentrate for a purpose other than sweetening is not sufficient, although there are indications that the Claims Regulation was intended in this way.[433] According to the Chairman, a case-by-case assessment must be made: is the addition of fruit concentrate used in a form or to an extent compatible with the use of the nutrition claim 'no added sugar'? The nature, composition and properties of the product all play a role. The ultimate question is whether it is misleading to the consumer who sees the words: no added sugar.

Pear fruit concentrate has been added to Albert Heijn Knijpfruit for permanent viscosity. It adds only a small amount of extra sugar (1 gram extra sugar per 100 grams of product). So the drink does not become significantly sweeter. This limited effect supports the conclusion that the pear fruit concentrate was indeed for consistency rather than sweetness. Therefore, the claim 'no added sugar' may remain on this packaging.

With this guideline in hand, the food industry can move forward. It all began with Leev- Bio-Fruit, which marked the start of a consumer awareness process regarding added sugar (via fruit concentrate, among other things).

Ebba Hoogenraad
Partner, food law and advertising law at Hoogenraad & Haak, Amsterdam

---

429 Case file 2018/00622.

430 Case file 2018/00620 (1).

431 Case file 2018/00624 (1).

432 Dossier 2018/00621.

433 The Chairman quotes the European Handbook on Claims Regulation, Commentary on the EU Health Claims Regulation, Meisterernst/Haber, p. 144–145 as well as a ruling in the German Landes Gericht Hamburg, LG Hamburg Urt v 9.5.2008-408 O 55/08 BeckRS 2009, 13988, ECLI:DE:LGH-H:2009:0509.408O55.08.0A.

CHAPTER 48

Holy Soda – 'Natural' claims

Annotated by Sarah Arayess

Advertising Code Committee (RCC) 15 September 2015, 2015/00862 (*Holy Soda*)

Article 7 Regulation 1169/2011, Article 2 Dutch Advertising Code (NRC), Article 8.2 opening words and (b) of NRC, claims, foodstuffs, natural, additives.

48.1 Summary

Vague claims like 'natural' are causing headaches for both manufacturers and bodies like the RCC. The RCC ruled that the claim '100% natural' may be used for a product containing steviol glycosides (stevia). Explanatory texts ensure that consumers can put the claim in context.

One of the most commonly used claims in recent years is undoubtedly 'natural'. There are no legal rules for this claim. The only rule to fall back on is the general rule that advertising must not mislead. But when is an ingredient or product actually natural? Does the food have to be picked right off the tree, or is some processing allowed?

48.2 Complaint and main considerations

The complaint objects to the statement '100% natural' on the packaging of soft drink Holy Soda. The product contains steviol glycosides, an 'E number' derived from the plant stevia. The complainant believes that the claim 'natural' should not be used in these circumstances.

The Advertising Code Committee (*RCC*) starts by outlining the (standard) assessment framework for food products: consumers whose decision to buy a food product is (partly) guided by the contents of the packaging of that food product will look at the entire packaging including the label. The label shows the words 'sweetener' and an arrow with the text: 'naturally sweetened with stevia extract'. The list of ingredients says 'sweetener (steviol glycosides)', with an arrow after 'steviol glycosides' saying 'this comes from a plant'.

The RCC expects the average consumer to interpret the statement '100% natural' to mean that Holy Soda contains only natural ingredients. The complaint concerns only the ingredient steviol glycosides. Is the claim 'natural' also correct for this particular ingredient? The RCC rules that the statements 'naturally sweetened

 | HTTPS://DOI.ORG/10.1163/9789004751699-051

with stevia extract' and 'this comes from a plant' make it sufficiently clear in which context 'natural' should be seen, in so far as this indication refers to the sweetener steviol glycosides.

In addition, the advertiser submitted a page from the Netherlands Food and Consumer Product Safety Authority (*NVWA*), which shows that the NVWA considers the claims 'With natural sweetener' and 'Sweetened with natural ingredients' to be permissible. This is sufficient reason for the RCC to reject the complaint. After all, the statements are considered permissible by the NVWA and the list is not exhaustive. In this context, the Advertising Code Committee (*RCC*) does not consider the statement '100% natural', viewed in the context of the entire label, to be incorrect.

48.3 Discussion

Fairly vague terms like 'natural' and 'artisanal' are not defined in law. In marketing, legal and compliance departments, this causes a great many headaches. When can you call a product natural? Is this about the origin of the ingredient, or rather the way it is processed? Can an ingredient that comes from a plant but that has later gone through an 'artificial' process still qualify as 'natural'?

This decision by the Advertising Code Committee (*RCC*) shows the importance of the context in which the claim is used. Additions on the label that further explain or clarify the claim help to form this context. In this case, this specifically involves the mentions of 'naturally sweetened with stevia extract' and the arrow with the text 'this comes from a plant', which further explains the ingredient sweetener (steviol glycosides). These additions help consumers understand in what way the ingredient is natural: namely, the ingredient comes from a plant. In this particular case, of course, the NVWA's guideline also helped. The list of permitted entries for the ingredient steviol glycosides is straightforward and convinces the RCC.

There are often collective sighs from within the industry: 'If only the NVWA would give us some more guidance on what it thinks is allowed, it would save us a lot of ambiguity'. This case is a good example of that. Netherlands Food and Consumer Product Safety Authority (*NVWA*) speaks out clearly about the permitted and non-permitted labelling with regard to the ingredient steviol glycosides. Both the advertiser *and* the RCC adhere to this policy.

Because no set rules exist for the use of the claim 'natural', additional pressure is placed on the advertiser. For each ingredient, evidence must be available as to why this ingredient is natural. The Flavouring Regulation does contain a rule for the use of the term 'natural'. It includes detailed requirements regarding origin and method of preparation.

Two years later there is another ruling on the claim 'natural'[434]. That case con-

434 See chapter 51 of this book.

cerned the use of the term 'natural' for chewing gum. The Advertising Code Committee (*RCC*) noted that the correct use of the claim 'natural' depends on the origin and method of treatment of the ingredient. The moment the ingredient meets the strict rules for flavourings stipulated in the Flavouring Regulation, then the claim 'natural' can confidently be used. The Flavouring Regulation is thus analogously applied to ingredients other than flavourings in that case. If the ingredient does not meet these regulations, then the advertiser must still show, in its own way, that the ingredient has undergone a different method of preparation that does not impair the consumer's understanding of the claim. If the advertiser succeeds in this, the claim may of course be used.

In summary, the term 'natural' seems to refer to both the origin and the method of preparation of a product or ingredient. Of course, there are many variations on this. For instance, the claim 'made with natural ingredients' is already much less broad. Another complaint, also in 2017, about the claim 'made with ingredients of 100% natural origin' got the RCC thinking. The complainant felt that sugar is not a natural ingredient. However, the advertiser shows that each product does indeed have a natural origin. Sugar comes from sugar beet, after all. The complaint is therefore dismissed.

We haven't heard the last word on the claim 'natural' (and its variants). Back in 2018, the European Consumer Organisation BEUC called for strict rules on vague claims such as 'natural'. As of yet, this call has not been heeded. So until then, it remains the responsibility of the producer/advertiser to be able to substantiate the claim properly and in detail. As with other advertising, in fact.

Postscript: the discussion on 'natural' remains topical, as witnessed in the legal battle between Perfetti (Mentos) and BenBits, which advertises its 'natural and plastic-free' BenBits chewing gum. With the preliminary result that first the Board of Appeal (*College van Beroep: CvB*) upheld the Code Committee's verdict with the justification that BenBits hadn't sufficiently substantiated that the relevant gum bases are free of synthetic substances (CvB 6 April 2022, 2021/00281 – CvB). In court, things turn out differently for Perfetti, as BenBits has since submitted new studies. In summary proceedings, first the Court in preliminary relief proceedings, and then the Court of Appeal,[435] determined that further expert evidence is required, and there is no room for this in summary proceedings). In the context of a balancing of interests, Perfetti was awarded zero; the injunction against BenBits was dismissed.

It seems that this may not be the end of this dispute in proceedings on the merits. So it seems we will have to wait even longer for the building blocks for the legal admissibility of the term 'natural'.

Sarah Arayess
Lawyer/founder Say Legal

435 Amsterdam Court of Appeal 23 March 2023, ECLI:NL:GHAMS:2023:725 (Perfetti v BenBits).

CHAPTER 49

Green Coffee – Health claims on hold

Annotated by Barbara Mutsaers

Board of Appeal, CvB recommendation, file number 2015/00916, date 2 February 2016.

Article 13 Regulation (EC) 1926/2004, Article 7 Dutch Advertising Code (NRC), Claims, health claims, botanical claims, botanicals, weight loss and weight management.

Main considerations:

> The Board of Appeal agrees with the Advertising Code Committee that as long as the European Commission has not decided with regard to certain (components of) foodstuffs whether a health claim submitted for this purpose will be placed on the Community list of permitted claims and this claim is 'on hold' (hereinafter: the authorisation procedure), an advertiser may mention that claim in advertising during the authorisation procedure. However, if, in advertising, the advertiser in fact anticipates the outcome of the authorisation procedure by claiming a concrete and actual effect already during the 'on hold' status without reservation in advertising, the advertiser should, in the event of a reasoned challenge, plausibly demonstrate this effect ahead of the decision on the inclusion of the claim in the Community list of authorised claims. If an advertisement for a food in this way claims an effect on health that is not made plausible, the advertisement should be considered misleading for that reason. [...]
>
> The above does not preclude the possibility of using a health claim that is 'on hold': if an advertiser clearly makes the reservation in advertising that the claim will be used while an authorisation procedure is still pending, and the claim is therefore conditional in that respect, it is unlikely that they already need to make the claim plausible during the authorisation procedure. After all, in that case, on the basis of that reservation, it will be clear to consumers that the claimed effect has not (yet) been officially established.

49.1 The case

The case involves an advertisement for 'Green Coffee Plus Extra Strong' on Vitaminesperpost.co.uk. It includes the claim 'stimulates fat burning' and states 'Green Coffee Plus is a powerful formula for people who want to stimulate fat burning'.

 | HTTPS://DOI.ORG/10.1163/9789004751699-052

The complainant filed a complaint with the Advertising Code Committee (RCC). Green Coffee Plus Extra Strong consists of green coffee extract, green tea extract and artichoke extract. The complainant considers the advertisement misleading and therefore unfair. The RCC handled the case and ruled on 17 November 2015 that it considers the advertisement in violation of Article 7 of the Dutch Advertising Code (NRC). It recommends the advertiser to stop advertising in such a way. Vitaminesperpost.nl subsequently appeals.

49.2 Opinion of the Board of Appeal

The aforementioned claim is a health claim within the meaning of Regulation 1924/2006[436] (hereinafter the 'Claims Regulation'). Indeed, claims related to weight loss and weight control, which may include the stimulation of fat burning, should be considered as health claims under this Regulation. Pursuant to the Claims Regulation, only claims that have been approved by the European Commission and placed on the Community list of permitted claims as referred to in Article 13(3) of the Claims Regulation may be used. An approved claim does not exist in this case. Besides the list of permitted claims, the Claims Regulation also has a category of products for which the use of health claims is provisionally allowed under certain conditions. This mainly concerns claims linked to certain botanical substances for which a timely application for the use of a health claim has been submitted. These claims fall under a transitional regime ('on hold'). In this regard, Article 27(5) of the Claims Regulation states: 'Health claims as referred to in Article 13(1)(a) may be made from the date of entry into force of this Regulation until the adoption of the list referred to in Article 13(3), under the responsibility of food business operators provided that they comply with this Regulation and with existing national provisions applicable to them [...]'.

The ingredients of the present product are botanicals. The Board of Appeal (CvB) assesses whether an application for a health claim has been submitted for these ingredients. The CvB finds that among the stated ID numbers submitted by Vitaminesperpost.nl in appeal, which correspond to ID numbers of the European Food Safety Authority (EFSA):

- for green tea – on which Vitaminesperpost.nl has specifically based its claim on appeal – (or its extract with EGCG) a claim for 'weight management' and 'fat metabolism' is 'on hold'.[437]
- for Cynara Scolymus (artichoke), a claim for 'weight control' is 'on hold'.[438]
- for Green coffee extract for 'weight management' a claim is 'on hold'.[439]

436 Regulation (EC) No 1924/2006 of the European Parliament and of the Council of 20 December 2006 on nutrition and health claims made on foods.

437 Under ID numbers 1112, 1124, 1544, 1806 and 3698

438 Under ID number 4395.

439 Under ID number 2031.

These claims are subject to the transitional regime of Article 27(5) of the Claims Regulation. If a health claim falls under this transitional regime, it must comply with certain requirements during the period it is 'on hold'. The important requirements are as follows:

- health claims may not be misleading (Article 3 Claims Regulation),
- health claims should be based on generally accepted scientific evidence where the advertiser must justify the use of the claim (Article 6 Claims Regulation),
- health claims must comply with national regulations during the transition period (Article 27(5) Claims Regulation).

The CvB agrees with the Advertising Code Committee (*RCC*) that as long as the European Commission has not decided whether a health claim made for certain (components of) foodstuffs will be included in the Community list of permitted claims and this claim is 'on hold' (hereinafter: the authorisation procedure), an advertiser is allowed to mention this claim in advertising during the authorisation procedure. However, if, in advertising, the advertiser in fact anticipates the outcome of the authorisation procedure by claiming a concrete and actual effect already during the 'on hold' status without reservation in advertising, it must, in the event of a reasoned challenge, make this effect plausible. If an advertisement for a food product in this way claims an effect on health that is not made plausible, the advertisement must be considered misleading for that reason. The Board of Appeal (*CvB*) takes into account that the Claims Regulation pursues a high level of consumer protection (cf. Article 1 of the Claims Regulation) and takes as a starting point that advertisers have to substantiate claims (cf. recital 17 of the Claims Regulation). The average consumer will expect that a product for which a certain effect on health is firmly and unreservedly claimed does indeed have that effect.

The above does not preclude the possibility of using a health claim that is 'on hold': if an advertiser clearly makes the reservation in advertising that the claim will be used while an authorisation procedure is still pending, and the claim is therefore conditional in that respect, it is unlikely that they already need to make the claim plausible during the authorisation procedure, according to the Tribunal. After all, in that case, on the basis of that reservation, it will be clear to consumers that the claimed effect has not (yet) been officially established.

Vitaminsperpost.nl refers in its appeal to (page 22 of) the so-called 'Guidance document' of the European Commission, from which it would follow that claims may be used if they are 'not implausible'. However, the mere fact that a claim falls under the transitional regime does not imply that already on that basis certain conclusions can be drawn regarding the claimed effect, according to the Board of Appeal (*CvB*). Vitaminesperpost.nl further states in the notice of appeal that it can make the claim 'plausible' by means of examples. The CvB states that this shows that Vitaminesperpost.nl also assumes that a disputed 'on hold' claim still has to be made plausible if, as in this case, a concrete and actual effect on health is claimed ('stimulates fat burning') and this effect is disputed and substantiated. The fact that

Vitaminesperpost.nl does nuance its claim to some extent, by stating that the advertised product is not a miracle cure but a supporting product, does not detract from this requirement. After all, Vitaminesperpost.nl also claims in so many words that its product as such stimulates fat burning.

On the basis of the above, the assessment on appeal is limited to the question whether Vitaminesperpost.nl has made the claimed effect of 'stimulates fat burning' sufficiently plausible in relation to one or more of the three ingredients to which the claim relates. This is not the case according to the CvB. On the contrary: with regard to green tea and green coffee, it has emerged that EFSA has now published a 'Scientific Opinion' to the effect that no 'cause and effect relationship' can be established between the consumption of these (components of) foods and fat burning. This basically negative opinion by EFSA implies a substantive rejection of its substantiation for this claim. While this does not detract from the 'on hold' status of the claim, it does imply that the claimed effect cannot apparently be demonstrated. With regard to the separate 'abstracts' and the 'Master file' of the Committee for the Inspection of Phytoproducts submitted by Vitaminesperpost.nl in appeal, the Board of Appeal (*CvB*) therefore suffices to rule that these documents, in view of the aforementioned opinion of EFSA, do not make the claim sufficiently plausible. Regarding Cynara Scolymus (artichoke), EFSA has not yet published an opinion. However, Vitaminesperpost.nl itself, according to what was stated on its behalf during the oral hearing, does not consider its effect on stimulating fat burning convincing. The CvB rules that Vitaminesperpost.nl has therefore also in that respect not made the claimed effect plausible.

As Vitaminesperpost.nl has not been able to make the claimed effect of the ingredients plausible, its claim 'stimulates fat burning' goes further than can be considered justified on the basis of the available data. The CvB therefore upholds the Commission's contested decision.

49.3 Status of health claims for botanicals

In September 2010, the European Commission decided that it was not possible to continue assessing health claims for plant and herbal substances, also known as botanicals. EFSA was therefore asked to discontinue its assessment of applications for claims for botanicals, and put them, along with those already assessed, 'on hold'. Indeed, it was noted that certain herbal substances can be present in the composition of a traditional herbal medicine as well as in food. However, different legal requirements apply between traditional herbal medicines and food health claims. It could therefore be the case that a claim in respect of the same substance is treated differently, which could – among other things – lead to consumer confusion. The European Commission and Member States needed more time to decide how they were going to address this issue, which led to the decision to suspend claims for botanicals.[440] However, Member States do not seem to be able to agree among themselves on how to resolve the issue, to the dismay of a number of parties. At one point, for example, Bionorica SE (hereinafter 'Bionorica') and Diapharm GmbH & Co. KG (hereinafter 'Diapharm') even turned to the General Court of the European Union because they felt it was taking too long. In so doing, Bionorica and Diapharm asked the General Court to find that the European Commission had unlawfully failed to order EFSA to assess the health claims for botanical substances with a view to adopting the final list of permitted health claims in accordance with the procedure laid down in Article 13(3) of the Claims Regulation. By the order of 16 September 2015, Bionorica v Commission[441] and the order of 16 September 2015, Diapharm v Commission,[442] the General Court dismissed the actions of Bionorica and Diapharm respectively and ordered them to pay the costs. Both parties appealed against those orders, which cases were subsequently joined.[443] In its judgment of 23 November 2017, the Court of Justice held that (i) the General Court's order of 16 September 2015, Bionorica v Commission, should be set aside, but the action for failure to act brought by Bionorica SE in Case T-619/14 should be dismissed as inadmissible, and (ii) the appeal in Diapharm's other case – Case C-597/15 P – should be dismissed.[444]

The obstacle seems to be that Member States cannot agree on a solution. The European Commission seems to have picked up the issue again, perhaps even by amending the Claims Regulation. As of now, though, there is a lack of clarity and a risk of differences arising within the European Union, whereas it was the aim of the European Claims Regulation to try and prevent those differences.

49.4 Is the CvB ruling justified?

In the present judgment, the CvB basically ruled that there is deception when a health claim for a botanical that is 'on hold' is used, and:

a. there is no caveat in the advertisement; or
b. in the absence of a caveat, the claim cannot be made plausible by the advertiser.

The Inspection Council KOAG/KAG[445] provides the following suggestions for disclaimers that can be used in this context:

- Health claim pending European Commission approval.
- Authorisation procedure for this health claim is ongoing.

440 https://ec.europa.eu/commission/presscorner/detail/en/MEMO_11_868
441 T-619/14, not published, EU:T:2015:723.
442 T-620/14, not published, EU:T:2015:714.
443 C-596/15 P and C-597/15 P.
444 EU:C:2017:886.
445 https://www.koagkag.nl/uitgelicht/disclaimer-bij-gezondheidsproducten.

- Health claim evaluation is ongoing.
- Health claim pending European authorisation.
- Evidence for this health effect is still being assessed in European authorisation process.
- Health effect not yet officially established.

The disclaimer may be stated immediately after the full health claim. It can also be indicated with an asterisk (*) that the disclaimer only applies to part of the claim. Incidentally, the disclaimer does not have to be stated on the front of the packaging with the abbreviated claim. This would not benefit readability and could lead to confusion, according to the Inspection Board KOAG/KAG.[446]

But for the record, the CvB thus does not rule that a disclaimer is mandatory. Without a reservation/disclaimer, an 'on hold' health claim can also be used, but then the advertiser has a duty to make the claim plausible. I would like to explicitly point out here that the CvB is not a legislative body. The Advertising Code is and remains a matter of self-regulation. Of course, a disclaimer can prevent discussions in principle, so it is advisable to use it. Precisely because the bar for making a certain effect plausible can be high. In my opinion, not only can a party be required to demonstrate or make it plausible that the ingredient in question has a certain effect, but also the conditions of use must be taken into account. Is there enough of a particular ingredient in the finished product to have the claimed effect? Et cetera. Indeed, Article 5(1)(b-d) of the Claims Regulation states that health claims may only be used, if:

> (b) the nutrient or other substance for which the claim is made:
>
> (i) is contained in the final product in a significant quantity as defined in Community legislation or, where such rules do not exist, in a quantity that will produce the nutritional or physiological effect claimed as established by generally accepted scientific data, or
>
> (ii) is not present or is present in a reduced quantity that will produce the nutritional or physiological effect claimed as established by generally accepted scientific data;
>
> (c) where applicable, the nutrient or other substance for which the claim is made is in a form that is available to be used by the body;
>
> (d) the quantity of the product that can reasonably be expected to be consumed provides a significant quantity of the nutrient or other substance to which the claim relates, as defined in Community legislation or, where such rules do not exist, a significant quantity that will produce the nutritional or physiological effect claimed as established by generally accepted scientific data [...]

446 https://www.koagkag.nl/uitgelicht/disclaimer-bij-gezondheidsproducten.

Similarly, Article 5(3) stipulates that health claims relate to the food ready for consumption in accordance with the manufacturer's instructions.

What is striking about this ruling is that the CvB is already including EFSA's assessment in that assessment of plausibility. In my opinion, this is wrong (at the moment) and unfair. The CvB thus fails to recognise that, within the European Union, there is a debate about the standards against which health claims for botanicals should be tested. Surely it is no coincidence that the European Commission has asked EFSA to cease its work. And it's also worth noting, from the press release posted by the European Commission on its website on 5 December 2011, that the European Commission has also asked EFSA to put the already assessed claims for botanicals 'on hold'. That press release reads, 'EFSA was asked to discontinue its assessment on claims for botanicals and these, together with a number on botanicals already assessed, have been put on hold.'[447]

From that point of view too, I think the CvB's judgment is unfair.

This discussion also hits the sore spot: When have you, as an advertiser, provided sufficient evidence to arrive at the plausibility of a claim? On appeal Vitaminesperpost.nl had apparently submitted other documents. I do not know the content of the documents. However, the CvB ruled that those documents were insufficient, 'in view of the aforementioned EFSA opinion'. The impression is given that the CvB did not assess the content of those documents because the EFSA report had already ruled otherwise. If that were so, in my opinion, that is a judgment contrary to the Claims Regulation. After all, Article 5 of the Claims Regulation talks about generally accepted scientific evidence. EFSA obviously gives scientific opinions/opinions, but that does not mean that there cannot be other generally accepted scientific evidence.

For producers and advertisers, this ruling has certainly not made things easier, at least for those parties who (for marketing considerations) do not wish to use a disclaimer.

Even after this ruling, the CvB has continued that line. See also, for example, file number 2018/A00320/A.

Barbara Mutsaers
Lawyer and Industry Leader Food & Life Sciences at AKD law firm, Eindhoven and member of the editorial board of the magazine Waar & Wet (Sdu Publishers)

447 https://ec.europa.eu/commission/presscorner/detail/en/MEMO_11_868. For more on claims on hold see also Chapter 16 of this book and also the thesis by Karin Lenssen, *The role of non-scientific data in risk analysis. The case of botanical health claims*, Maastricht 2023, https://cris.maastrichtuniversity.nl/en/publications/the-role-of-non-scientific-data-inrisk-analysis-the-case-of-bota.

CHAPTER 50

Unsweetened almond drink – Maltodextrin added for texture

Annotated by Sanne Heestermans

Advertising Code Committee (RCC) 24 January 2017, file number 2016/00924.

Nutrition claims, unsweetened, Claims Regulation.

Main considerations:

> 1. In the opinion of the Committee, the indication 'unsweetened' in the name of the UHT almond drink has the same meaning for the average consumer as 'with no added sugar'. Thus, the term 'unsweetened' qualifies as a nutrition claim within the meaning of Regulation (EU) No 1924/2006 on nutrition and health claims made on foods (hereinafter: Claims Regulation). Pursuant to Article 8 of the Claims Regulation and its Annex, the claim that sugars have not been added to a food – and any claim likely to have the same meaning for the consumer – may only be made where the product does not contain any added mono- or disaccharides or any other food used for its sweetening properties.
> 2. Maltodextrin is a carbohydrate and not a mono- or disaccharide. The Committee understands that maltodextrin, taken in isolation, is capable of serving as a sweetening ingredient, on the understanding that – also taking into account what the advertiser states about the limited sweetening power of maltodextrin – its sweetening effect depends on the quantity processed in the product. It follows from what has been submitted in defence that the quantity of maltodextrin in this case is insufficient for a sweetening effect and serves only as a texture enhancer. On this basis, the indication 'unsweetened' must be considered correct.

50.1 The case

The packaging of the product Melkan UHT almond drink included the expression 'unsweetened with added calcium and vitamins B12, B2 and D2'. The complaint in this case focused on the use of the term 'unsweetened'. According to the complainant, the term 'unsweetened' was misleading in this case because the almond drink contains maltodextrin. The complainant believed that the lesser-known designation 'maltodextrin' was mentioned on the packaging to avoid the term 'sugar'.

In its defence, the advertiser argued that the nutrition statement explicitly in-

 | HTTPS://DOI.ORG/10.1163/9789004751699-053

cluded the proportion of sugar and carbohydrates in the almond drink, so that there was no misleading avoidance of this term. In that context, the advertiser points out that no mono- or disaccharides have been added to the almond drink: maltodextrin is, after all, a carbohydrate. The maltodextrin would have been added to improve the texture of the drink and not to make it sweeter.

50.2 Background

The European Regulation (EC) No 1924/2006 on nutrition and health claims made on foods (the 'Claims Regulation') applies to nutrition and health claims made in commercial communications.[448] The Claims Regulation states that the term 'nutrition claim' includes, inter alia, the claim which states or suggests that a food has certain beneficial properties with respect to a lower energy value or contains substances in reduced amounts or not at all.[449]

Furthermore, the Claims Regulation stipulates that only the nutrition claims listed in the Annex are allowed,[450] and this in so far as these nutrition claims comply with the conditions imposed on them. For example, the Annex stipulates that the nutrition claim 'with no added sugar' is only allowed if no mono- or disaccharides or other foods used for their sweetening properties have been added to a product. This also applies, incidentally, to any claim likely to have the same meaning for consumers.

50.3 Advertising Code Committee: use of term 'unsweetened' in this case not misleading

The Dutch Advertising Code Committee (RCC) considers that, for the average consumer, the term 'unsweetened' will have the same meaning as the nutrition claim 'with no added sugar'. For this reason, the RCC is of the opinion that if the producer of the almond drink wants to use the term 'unsweetened' on the packaging, this is only permitted if the conditions are met as named in the Annex to the Claims Regulation. In this case, this meant that no mono- or disaccharides or other foodstuffs included for their sweetening properties were allowed to be added to the almond drink.

Maltodextrin is not a mono- or disaccharide, but a carbohydrate composed of organic compounds of glucose. This substance has a slightly sweet taste. The RCC therefore had to determine in this case whether this substance had been added to

448 Article 1(2), Claims Regulation.
449 Article 2, para 2(4), Claims Regulation.
450 Article 8(1), Claims Regulation.

the almond drink because of its sweetness or for other purposes. After all, if the maltodextrin had been added because of its sweetness, the claim 'unsweetened' should not be made.

The RCC considered that the substance maltodextrin can be used to give a product a sweeter taste, but that this is not its only function. Maltodextrin – like xylitol – is also added to products as a texture enhancer. The RCC considered that the sweetening effect of maltodextrin depends on the amount of this substance added to a food product. In the RCC's view, the amount of maltodextrin added to the almond drink in question was insufficient to sweeten the drink and was therefore added as a texture enhancer. For that reason, the mention of the term 'unsweetened' on the packaging was therefore not misleading, in the RCC's opinion.

50.4 Sweeteners as foodstuffs

In this case, the RCC states its opinion that the substance maltodextrin was not added to the almond drink for its sweetening properties. If this had been the case, the use of the nutrition claim 'unsweetened' would not have been allowed, according to the RCC. This is notable because within Europe there is currently a debate about the addition of such sweeteners and making the claim 'with no added sugar'.

As mentioned, the Annex to the Claims Regulation states that the nutrition claim 'with no added sugar' is only allowed if no mono- or disaccharides or other foods included for their sweetening properties have been added to a product. Certain sugar substitutes such as xylitol do not qualify as mono- or disaccharides. So the question with these sugar substitutes is whether they qualify as foods used for their sweetening power. After all, if they do, then the claim 'with no added sugar' should not be made.

The Claims Regulation[451] defines 'food' as any substance or product that is intended or reasonably expected to be consumed by humans. This definition is so broad that it also includes sugar substitutes such as xylitol. This is notable because (certain) sugar substitutes also qualify as sweeteners within the meaning of Regulation (EC) No 1333/2008 on food additives, which sets out the rules for adding sweeteners to food.

The above leads to uncertainty in the EU about when the nutrition claim 'with no added sugar' is allowed if sweeteners are included in food. For example, the Belgian consumer group *Test Achats* has already filed a complaint with the Belgian

451 Article 2(1) of the Claims Regulation refers for this definition to Regulation (EC) No 178/2002 laying down the general principles and requirements of food law, which includes a definition of 'food' in Article 2.

food authority,[452] because a product containing stevia had the claim with no added sugar' on it. According to *Test Achats*, this was not allowed because stevia would qualify as a foodstuff added to the product for its sweetening power.

As a result of this complaint, several European food authorities expressed their views on this. The Danish food authority carried out an investigation and concluded that it had not been the intention of the European legislator to include sweeteners among 'foods added for their sweetening power'. The Swedish authorities followed suit by stating their view that 'with added sugar' refers only to mono- and disaccharides, including sugars from honey, fruit juice and fruit concentrates. The Italian version of the RCC took the position that the nutrition claim 'with no added sugar' may be made if the product contains sweeteners. This was later confirmed by several courts in Italy. The Irish authorities indicated that they agreed with this Italian opinion.

This discussion was also held within the 'European Commission's Working Group meeting on health claims'. This working group published a bulletin in December 2016[453] stating – in brief – that it is up to individual Member States to take a position in this discussion. The position of the Dutch Food and Consumer Product Safety Authority (NVWA) is unclear to date, although the RCC seems already to have taken its (strict) position in this case.

50.5 Context of the claim

In addition, the RCC ruled in the present case that the term 'unsweetened' will be understood by the average consumer to mean that the product does not contain added sugars. This is striking for the following reason.

In an earlier case,[454] a file was submitted to the RCC about the claim 'less sweet' made on the packaging of the yoghurt drink Optimel Pure. Although this yoghurt drink tastes less sweet than the regular variant, this less sweet variant does contain more sugar. The complainant in this case pointed out that the use of the claim 'less sweet' wrongly implies that the product also contains less sugar. However, the RCC ruled in this case that the claim 'less sweet yoghurt drink' does not refer to the sugar content, but to the taste.

Whereas the RCC ruled in the case at the centre of this annotation that the claim 'unsweetened' would be understood by consumers as the claim 'without added sugar', the RCC previously held that the claim 'less sweet' clearly referred only to the flavour of the drink. This difference can be explained by the RCC looking not only

---

452 For more information on this case, see https://www.foodnavigator.com/Article/2018/02/06/No-added-sugar-plus-asweetener-Pure-and-simply-illegal-says-Test-Achats.

453 For more info, see http://www.reading.ac.uk/foodlaw/pdf/2016-Claims-Update-December.pdf.

454 Advertising Code Committee (RCC) 6 November 2013, file number 2013/00971.

at the ‘bare’ claim, but also at the context in which this claim was made. The claim ‘less sweet’ was part of a preceding main commercial that clearly referred to the taste of the yoghurt drink. Thus, when using claims, which are likely to have the same meaning to a consumer as the nutrition claims listed in the Annex of the Claims Regulation, it is very important to pay attention to the context in which these claims are made.

Sanne Heestermans
Head of Legal & Government relations YoungOnes

CHAPTER 51

BenBits – Claim 'natural'

Annotated by Jasmin Buijs

Advertising Code Committee (RCC), 20 March 2017, file no. 2016/00917
Advertising Code Committee (RCC), 13 January 2022, and
Board of Appeal (CvB), 6 April 2022, file no. 2021/00281

Law articles: Article 7 Food Information for Consumers (FIC) Regulation, Article 2 Dutch Advertising Code (NRC), Article NRC. Natural, misleading food information, traditional food preparation processes.

Main considerations Advertising Code Committee (*RCC*) of 20 March 2017:

> Regulation 1169/2011 on food information for consumers contains no specific rules on the use of the term 'natural' in combination with the nature of the product or the ingredients used. Nor does it prescribe the circumstances under which a product or the ingredients used may be called 'natural. Such rules can, however, be found in Regulation 1334/ 2008 on flavourings and certain food ingredients with flavouring properties for use in and on foods. This regulation defines a natural flavouring substance as: 'a flavouring substance obtained by appropriate physical, enzymatic or microbiological processes from material of vegetable, animal or microbiological origin either in the raw state or after processing for human consumption by one or more of the traditional food preparation processes listed in Annex II.' This list of traditional food preparation processes includes heating, drying, evaporation, fermentation, grinding, infusion, microbiological processes, distillation, emulsification and extraction. This benchmark developed by the European Commission is apparently intended to clarify that certain preparation processes must not impair the natural character of a flavouring substance, as well as to clarify which preparation processes these are. It must be assumed that the European Commission followed the expectations of the average consumer when creating this list. The Dutch Advertising Code Committee notes that the complainant, Cloetta, did not allege that the flavourings used in BenBits' chewing gum do not meet the definition of 'natural flavourings' according to Regulation 1334/2008. To this extent, the Dutch Advertising Code Committee assumes that Cloetta has no objection to the use of this term, taking into account that the flavourings used have not lost their natural character due to the preparation process applied.

 | HTTPS://DOI.ORG/10.1163/9789004751699-054

Main considerations Board of Appeal (CvB) of 6 April 2022:

> In the context of the principal appeal, the following claims need to be assessed: 'natural chewing gum', 'natural gum' and 'natural gum base'. Perfetti objects to the use of the term 'natural'. Regarding the demarcation between 'natural' and 'non-natural', the Appeals Tribunal considers it sufficient to assess whether the gum base of BenBits' chewing gum contains synthetic substances. The average consumer will interpret the claim 'natural' to mean that there are in any case no synthetic, industrially manufactured petrochemical components present in the product. The absence of these ingredients is the core message of the claim 'natural'. The message is namely that the chewing gum is a natural product, which is reinforced by other claims in the same advertisement, including those referring to the environment such as 'Bits of nature' and 'plastic-free', combined with the symbol of a sea turtle. The mere presence of synthetic ingredients in the gum base means that (in the perception of the average consumer) the product as a whole can no longer be called 'natural' in an absolute sense. This also applies to the claims on the packaging – taking into account the assessment framework that follows from CJEC 16 July 1998, C-210/96, ECLI:EU:C:1998:369 (*Gut Springenheide*) and CJEC 4 June 2015, C-195/14, ECLI:EU:C:2015:361 (*Teekanne*) – as well as for claims made via Instagram and Facebook.

51.1 The cases

For some time now, many consumers have been favouring natural food products in their search for healthy food. But what exactly is 'natural'? Various interpretations can be assigned to this term, such as unprocessed food, food without synthetic substances, and organic food. In European food law, there is no harmonised definition of the term 'natural'. This annotation describes the development of the interpretation given to the term 'natural' within Dutch (self-)regulation. It focuses on the BenBits cases handled by the Advertising Code Committee RCC) and its appeals tribunal, the CvB. These cases concern chewing gum advertised as 'natural' by chewing gum manufacturer BenBits. The first case dates back to 2017. In this case, the RCC provides for the first time clear guidance on the use of the term 'natural'. This annotation starts with a brief outline of the use of the term 'natural' before the BenBits ruling. It then discusses the first BenBits case, followed by an overview of developments after 2017, in particular with regard to private standards. In 2021, a complaint against BenBits about its 'natural' chewing gum was again submitted to the RCC, which was followed by proceedings at the CvB, a preliminary injunction before the Amsterdam District Court and an appeal case before the Amsterdam Court of Appeal.

51.2 '100% 'natural' before BenBits

As mentioned, there is no harmonised definition of 'natural' in European food law. Instead, food companies are supposed to follow the general rules on fair information practices. Article 7 of Regulation 1169/2011 on the provision of food information to consumers FIC Regulation) prescribes that food information shall not be misleading. To determine whether food information is misleading, the likely interpretation of the information by the average consumer has to be taken into account. Back in 1998 the Court of Justice of the European Communities (CJEC) taught us in the *Gut Springenheide* case that this average consumer is 'reasonably well informed and reasonably observant and circumspect'.[455]

Taking into account the likely view of the average consumer, the Advertising Code Committee (*RCC*) ruled in various cases that it must be clear, and therefore often explained, to consumers what is meant by the claim made (see e.g. RCC) 2 November 2016, file no. 2016/00808 regarding 100% sustainable fish). In the absence of such information, the average consumer could be enticed to make a decision on a transaction which he/she would otherwise not have made. That an explanation of generic terms used is essential to not mislead is even more relevant in case of non-legally defined terms (such as 'natural'). That food companies are however not entirely free to communicate their own interpretation of such term has become clear from several cases handled by the RCC prior to the BenBits case. These cases with regard to use of the term 'natural' are discussed in the next section. When it comes to using absolute (100%) claims, such as '100% natural', the RCC has shown in several of its rulings that such claims are often considered misleading. 100% must be demonstrably 100%, which is usually not the case (see e.g. RCC 9 March 2020, file no. 2020/00059 regarding 100% compostable coffee cups).

In a case dated 7 January 2008, the Advertising Code Committee (*RCC*) rules that an ingredient that, although derived from natural raw material, cannot be labelled 'natural' if it is subject to several 'conversions' before being added to the product.[456] The RCC repeats the tenor of this ruling in a case dated 18 October 2013, in which it rules that the claim 'natural' is understood to refer not only to the food itself or its ingredients, but also to the production process used.[457] In this case, the food supplement ProstEase (containing among other things the mineral zinc) is advertised as 'natural'. The complainant alleges that 'minerals containing zinc must undergo considerable processing, to isolate zinc' and that it is therefore misleading to advertise the food supplement as 'natural'. The producer did not refute the complaint, as a consequence of which the RCC rules that the use of the term 'natural' is mis-

455 CJEC 16 July 1998, C-210/96, ECLI:EU:C:1998:369 (*Gut Springenheide*).
456 RCC , 7 January 2008, file no. 2008/07.0458 (*Truly fruitsap*).
457 RCC , 18 October 2013, file no. 2013/00311 (*ProstEase*).

leading. The RCC does not further elaborate on the scope and significance of the alleged 'considerable processing'.

The RCC also deals with a claim regarding the processing of 'natural' ingredients in a case dated 15 September 2015.[458] This case concerns a 'Carbonated soft drink with apple and raspberry, with sweetener' that is advertised as '100% natural'. 'Naturally sweetened with stevia extract' is additionally stated on the label. According to the complainant, steviol glycosides are not 100% natural as they are extracted from stevia and have an E number. However, the RCC rejects the complaint. As indicated by the advertiser, the Netherlands Food and Consumer Product Safety Authority (NVWA) has published a list of permitted and non-permitted claims relating to steviol glycosides. 'With natural sweetener' and 'Sweetened with natural ingredients' are indicated on this list as permitted claims.[459] The Advertising Code Committee (*RCC*) therefore does not provide new criteria for '100% natural' claims in this case, but aligns with claims approved by the NVWA.

In a case dated 22 February 2017, the claim 'made from ingredients of 100% natural origin' is at issue.[460] The case concerns a can of Simply Cola from producer RedBull. In response to the complaint, the advertiser refers to the website where the origin of each individual ingredient is explained. The RCC rules that this explanation does indeed make it sufficiently plausible that the ingredients are of natural origin. In this case, the RCC does not address the complainant's argument that the processed ingredients are not 100% natural because they are 'developed by the food industry'. The complainant refers here specifically to the added sugar, glucose syrup and caramel sugar syrup contained in the product. Possibly, the RCC considers it irrelevant which production processes were used, since the claimed 'naturalness' in the advertisement in question is specifically linked to the origin of the ingredients. The ruling therefore leaves unanswered the question of whether ingredients can still be referred to as 'natural' after processing.

51.3 BenBits 1.0: the use of traditional food preparation processes

In the BenBits case of 20 March 2017, the RCC provides clear guidance on the use of '100% natural' claims. Cloetta, a competitor of BenBits, states in its complaint that not all processed ingredients in BenBits' chewing gum can be referred to as natural because they are allegedly manufactured using chemical processes. Moreover, Cloetta states that the origin of salmiak salt is not natural.

According to the Advertising Code Committee (*RCC*), the average consumer

458 RCC, 15 September 2015, file no. 2015/00862 (*Holy Soda*).

459 The list of permitted and non-permitted claims relating to steviol glycosides can be found in the NVWA Handboek Etikettering van levensmiddelen (NVWA Manual for the labelling of foods).

460 RCC , 22 February 2017, file no. 2017/00859 (*Simply Cola*).

'understands by "natural ingredient" first and foremost the naturally occurring manifestation of ingredients [...], and secondly ingredients that may have lost their original manifestation due to a preparation process, but the natural character of which is not impaired by such a preparation process.' In other words, according to the RCC, consumers would assume that a '100% natural' claim covers both the ingredients used in the food and the preparation process used. For the follow-up question of what constitutes a natural preparation process, the RCC refers to Regulation 1334/2008 (the Flavouring Regulation). Annex II to the Flavouring Regulation contains a list of traditional food preparation processes, including chopping, slicing, peeling, and microbiological processes.

By interlocutory decision, the RCC gives BenBits the opportunity to demonstrate that the processed ingredients can be considered natural. It can do so either by adhering to the preparation processes listed in the Flavouring Regulation, or by demonstrating that the preparation processes in some other way do not detract from the designation 'natural ingredient' in the likely view of the average consumer. BenBits was, however, not able to sufficiently demonstrate the naturalness of, among other things, the chewing gum's main ingredient xylitol. Neither was the chewing gum manufacturer able to refute the complaint regarding the alleged natural origin of salmiak salt. The RCC therefore considers BenBits' claim '100% natural' to be misleading.

51.4 Developments in private standards

Although the term 'natural' is not legally defined, the BenBits ruling of 2017 provides clear guidance. Within a year of the BenBits case, a definition of the term 'natural' is laid down in NVN-ISO/TS19657:2018, a technical specification for provisional application. This technical specification draws on the Codex Alimentarius Guidelines on the Use of Flavourings (CAC-GL 66-2008). Technical specification NVN-ISO/TS19657:2018 sets out a number of cumulative conditions to be fulfilled for food ingredients to be considered 'natural'. Firstly, the source material shall consist of one or more of the following materials: plants, algae, fungi, animals, microorganisms, mineral deposits or seawater. Secondly, the food ingredient shall be obtained from source materials by physical and/or enzymatic and/or microbiological processing. Where such techniques are not available for food safety or regulatory reasons, other techniques may be used provided that the constituting component(s) of the food ingredient is/are not altered. The technical specification emphasises that enzymatic and/or microbiological processing shall not be used to deliberately produce substances that do not occur in nature.

If the BenBits case had taken place after the publication of NVN-ISO/TS19657:2018, it would not have been surprising if the RCC had adhered to it. It should be noted, however, that this technical specification is designed to create a level playing field and to provide for fair commercial practices at B2B level. The introduction of the technical specification stresses that it is not applicable in B2C communications because consumer expectations vary from country to country and from product to product. It is however likely that in practice, the technical specification will also prove useful in B2C communications. As the RCC assumed in the 2017 BenBits case, the European Commission sought to align with the expectations of the average EU consumer in its development of the list of natural food preparation processes in the Flavouring Regulation. Technical specification NVN-ISO/TS19657:2018 is not that different from the outcome of the BenBits case with reference to the aforementioned list. In other countries, such as the UK, interpretations of the term 'natural' similar to that provided by the Dutch Advertisement Code Committee can be found in self-regulation. For example, in a case dated 8 March 2017, the UK Food Standards Agency does not consider a product of Appy Food & Drinks to be 100% natural because a non-traditional enzymatic treatment had been applied to the glucose-fructose syrup in the product. Moreover, in the absence of a substantial description of the preparation method of the added calcium lactate, the UK Food Standards Agency does not consider the aforementioned ingredient natural either. It would be interesting to collect information on the interpretation of the term 'natural' from several EU Member States (of which the UK is now no longer one) to gain more insight into the 'average' EU consumer's understanding thereof.

51.5 BenBits 2.0: free of synthetic substances

In 2021, a complaint against BenBits' natural claim was again submitted to the RCC, and later appealed against before the self-regulatory Appeals Tribunal.[461] This time it is competitor Perfetti who turns against BenBits' 'natural' chewing gum. The discussion in this cases focuses on the question of whether the gum base used in the product contains synthetic substances. According to the CvB, the average consumer will understand the claim 'natural' to mean that the product does at least not contain any synthetic, industrially manufactured petrochemical ingredients. To give a ruling, it would therefore suffice to assess whether the chewing gum is free of such substances. Perfetti had BenBits' chewing gum analysed by the Technical University of Milan (Politecnico di Milano). The results showed that chemical substances were indeed present in the product, rendering the 'natural' claim misleading.

The fact that this latter BenBits case focusses solely on the question of whether the product is free of synthetic ingredients does not mean that the RCC has abandoned its earlier formula 'natural = natural origin + traditional food preparation processes'. The difference between the former and latter BenBits case lies in the

461 CvB (Appeals Tribunal), 6 April 2022, file no. 2021/00281 (*BenBits*).

message conveyed by the advertiser, given the overall expression of the advertisement and context of the claims. Given among other things, the claims 'Bits of nature' and 'plastic-free' on the packaging in the second BenBits case, the mere presence of synthetic ingredients in the gum base was sufficient to consider the term 'natural' misleading.

Since the composition of BenBits' chewing gum has changed and new studies have been conducted, BenBits takes the position that the CVB ruling is no longer applicable. Perfetti does not leave it at that and starts a preliminary injunction before the Amsterdam Court,[462] which procedure is followed by an appeal before the Amsterdam Court of Appeal.[463] In these proceedings, the relevant court and the court of appeal are more nuanced and rule that further factual research is required. There was obviously no room for this in the preliminary relief proceedings. Such in-depth investigation will have to take place in a possible proceeding on the merits, in which, the meaning that can be attributed to the qualification 'natural' in the given context can also be discussed. We haven't heard the last word on this matter.

51.6 Conclusion

This annotation aims to provide an overview of (self-regulatory) cases concerning the term 'natural' in the Netherlands, with the BenBits case of 2017 as an important milestone. For about ten years prior to the BenBits case, it was clear that the term 'natural' referred not only to the origin of the ingredients used (unless explicitly stated), but also to the preparation processes used. In the 2017 BenBits case, the Advertising Code Committee (*RCC*) provides a useful tool to establish which preparation processes are suitable as referred to in the list of traditional food preparation processes in the Flavouring Regulation. Shortly after the BenBits ruling in 2017, a similar interpretation of the term 'natural' is laid down in the technical specification NVN-ISO/TS 19657:2018. In 2022, the RCC and the Board of Appeal (*CvB*) rule in a second case against BenBits' 'natural' claim that a product containing synthetic ingredients cannot be referred to as natural. The absence of such ingredients is also an important criterion for consumer organisation Safe Food Advocacy Europe for the use of the claim 'natural'.[464] However, the other way around, a food that is free from synthetic substances is not per definition 'natural'.

Although not all questions regarding the concept of 'natural' have yet been answered, it has become increasingly clear when foods or their ingredients can be advertised as 'natural' in the Netherlands thanks to Dutch (self-regulatory) cases. A

462 Rechtbank Amsterdam (Amsterdam Court), 15 July 2022, ECLI:NL:RBAMS:2022:4065.

463 Gerechtshof Amsterdam (Amsterdam Court of Appeal), 21 March 2023, ECLI:NL:GHAMS:2023:725.

464 https://www.safefoodadvocacy.eu/natural-campaign/.

comparative study involving different EU Member States can teach us more about the interpretation of the term 'natural' EU-wide. This is very relevant for food companies, since their market is usually not limited by national borders.

Jasmin Buijs
Specialisation life sciences law, including food & feed law
Attorney-at-law at Axon Lawyers

CHAPTER 52

HappyHealthy – Advertising or information?

Annotated by Sarah Arayess

Advertising Code Committee (RCC) 22 December 2017 file 2017/00767 (*HappyHealthy*)

Article 1 Dutch Advertising Code (NRC), nutrition claims, foodstuffs, social media, Relevant Relationship

52.1 Summary and discussion

An important decision on the interface between information and advertising. A vlog in which a coconut oil from The White Pearl is heavily spotlighted while no relationship exists between The White Pearl and the vlogger is not advertising within the meaning of Article 1 of the Dutch Advertising Code.

52.2 Complaint and main considerations

An important phenomenon has emerged in the food world in the last five years: advertising via social media channels. In a vlog on the website www.happyhealthy.nl, slimming coach, nutritionist, dietary psychologist and orthomolecular therapist Oscar Helm (roles as described on the website) compares different types of coconut oil. The conclusion of the vlog according to the complainant: almost all coconut oils are inferior to The White Pearl coconut oil. According to the complainant, this is inaccurate and misleading for several reasons. Moreover, the complainant notes that the videos indicate that the vlogger did not receive any money for the vlog. The complainant questions this. After all, social influencers are never free, are they? The defence is clear: the vlogger never received any remuneration or other reward from The White Pearl for the advertisement. There was no kind of collaboration or plan to do so. The White Pearl confirms this when asked.

The Advertising Code Committee (*RCC*) concludes that the vlog is not advertising. It has not been established that the recommendation of the coconut oil in the vlog was made by or on behalf of the advertiser. No advertising and therefore no substantive review of the vlog by the RCC. So we will not know whether the recommendation is in line with advertising rules according to the RCC.

 | HTTPS://DOI.ORG/10.1163/9789004751699-055

52.3 Discussion

The line between advertising (strict advertising rules) and information (freedom of expression) remains a hot issue. With the advent of newer media forms, the grey area between advertising and information has become increasingly complicated. A decision has to be made for each specific advert. The term advertising is interpreted broadly in self-regulation. With the advent of the Social Media Advertising Code in 2014, an attempt was made to provide clear guidance at least for advertising on social media.

Once a Relevant Relationship exists between advertiser and influencer ('disseminator'), then it is advertising. A Relevant Relationship exists when there is a relationship between the advertiser and disseminator aimed at (arranging for) the dissemination of advertising via social media that may influence the credibility of the advertisement, in return for payment or (the chance of) some benefit. In that case, the relationship between advertiser and influencer should be disclosed. For example, by explicitly stating, 'I got this product from' or by using a hashtag such as #spon. But if a blogger truly posts content on his or her own initiative, then this is information and therefore freedom of expression, right?

Another complaint case before the Advertising Code Committee (*RCC*), earlier in 2017, revolves around a vlog advertising Nutrilon infant formula (VT without recommendation, 19 June 2017, 2017/00387/I). It is not permitted to advertise infant formula. Both the vlogger and the manufacturer of the advertised product denied that there was any advertising or a relationship between the two. The vlogger indicated that she was simply extremely happy with the product and wanted to tell her followers about it. When asked, Nutricia said that it closely follows the strict rules for infant formula and never advertises or commissions advertising for this product category.

Nevertheless, the chairperson took the strict (and, in our view, incorrect) position that this was advertising. Another factor was that the chairperson ex officio looked at the vlogger's website and drew the conclusion that the website is commercial in nature. For instance, the vlogger actively appeals to advertisers to contact her for a collaboration. The advertising violates the applicable rules for infant formula and the vlogger is responsible, the conclusion said. Nutricia, manufacturer of the infant formula, is obviously not responsible. It is not involved in the advertising in any way.

Although the concept of advertising is broad, it is not so broad that such vlogs must also be considered advertising. After all, advertising is: *any public and/or systematic direct or indirect commendation of goods, services and/or ideas by an advertiser or wholly or partly on its behalf, with or without the help of third parties.* When vloggers express their own views in their own blog, this is not advertising on behalf of an advertiser. Fortunately, the incorrect ruling in the Nutrilon case was corrected with the HappyHealthy ruling.

Postscript: meanwhile, the rules for influencer marketing have again been further tightened in the Media Act and the (amended) Advertising Code for Social Media & Influencer Marketing. The DDMA (Data-Driven Marketing Association) has a very clear website where the influencer, advertiser, enforcer and consumer can check exactly what is allowed, and in what way the Relevant Relationship must be made clear in or with the vlog or blog (see www.influencerregels.com). And it is also clearly stated the following if there is no financial relationship between advertiser and influencer, but the influencer buys and promotes the product himself, then no disclaimer (like #ad) is needed.

Sarah Arayess
Lawyer/founder Say Legal

CHAPTER 53

Muesli – Serving suggestions

Annotated by Sarah Arayess

Advertising Code Committee (RCC)/Board of Appeal (CvB) 12 October 2018, file 2017/00460 (*Van Mook*)

Article 7 Regulation (EU) 1169/2011 (FIC Regulation), Article 2 Dutch Advertising Code (NRC), foods, deception, images, serving suggestion.

53.1 Summary

Food packaging often contains beautiful, enticing images of the product or serving suggestions. The product then looks just that little bit more colourful, tasty and attractive than in reality. Everyone knows that, right? Van Mook Muesli with sultanas and hazelnuts does contain a lot of hazelnuts, according to the image. So the disappointment was great when the number of hazelnuts in the pack could be counted on the fingers of one hand. Is this a typical case of misleading information or still really advertising exaggeration?

This book mentions on several occasions the quiet revolution in product packaging and names. Since the Teekanne ruling[465] of the European Court of Justice, industry, consumers and other stakeholders alike are looking ever more strictly at naming, graphics and use of colour on food packaging. Gone is the 'labelling doctrine' of yesteryear, according to which the interested consumer studies the list of ingredients and (provided the list is correct) will not easily be misled by exaggeration elsewhere on the label. This decision by the CvB is a good illustration of the implications for practice.

53.2 Complaint and main considerations

Van Mook Muesli with sultanas and hazelnuts is sold in partially transparent plastic packaging with the image of a logo containing a mill on the front. Also on the front is a picture of a bowl of muesli with a fresh green leaf on top, seven hazelnuts and some almonds. The list of ingredients on the back of the packaging states that

465 See Chapter 11 of this book.

 | HTTPS://DOI.ORG/10.1163/9789004751699-056

the product contains, among other things, '1% hazelnut pieces' and '11% sultanas (sultanas, sunflower oil, wheat flour)'. The complainants' complaint is fourfold but can be summarised as follows: the packaging portrays the product more beautifully than it is. It gives the impression that the product contains whole hazelnuts when it is only 1% hazelnut pieces. Moreover, the image shows almonds while the product does not contain any almonds at all.

The Advertising Code Committee (*RCC*) did not find the packaging misleading on this point. The complainant himself stated that he found five whole hazelnuts in the packaging. The fact that the picture might show slightly more hazelnuts than are usually contained in the packaging was considered by the Advertising Code Committee (*RCC*) to be admissible advertising exaggeration. The complainant did not leave it at that and appealed.

The Board of Appeal (*CvB*) applies the correct framework of review that follows from the Teekanne judgment. The average consumer, whose decision to purchase a product is determined by its composition, reads the list of ingredients first. However, a correct list of ingredients does not preclude that certain other indications on the labelling may mislead the average consumer, for example if they are ambiguous, contradictory or incomprehensible in practice. The entire packaging should be taken into account when deciding whether or not a package is misleading.

The name of the product ('Muesli with Sultanas and Hazelnuts') and the image on the packaging of a bowl containing hazelnuts and almonds gives the impression that the product contains almonds and a considerable number of hazelnuts, whereas in reality (according to the list of ingredients) it contains 1% hazelnut pieces and, according to the appellant, only a few whole hazelnuts. The illustration therefore does not offer a representative picture of the composition. The composition is furthermore not made clear on the packaging. Moreover, the word 'serving suggestion' appears in very small writing on the packaging. This will probably escape the average consumer. The fact that the packaging is transparent does not help either. The CvB does not think that consumers can be expected to inspect the packaging so carefully that they can conclude that the image is only a serving suggestion. In short, according to the CvB, it is blatantly obvious. The image is misleading as to the composition of the product. The list of ingredients can no longer correct this impression.

53.3 Discussion

This ruling was selected as one of the 20 most important food law rulings by the RCC/CvB, because it illustrates well the new-found direction. No longer is 'the sky the limit' in terms of images and names on packaging. Compared to 'pre-Teekanne', the packaging has to pass a slightly stricter deception test. The list of ingredients no longer takes precedence. But be aware that, even in the new era, the entire packaging is still relevant. The assessment should therefore not only look at the allegedly misleading image. The rest of the packaging is just as important, including the ingredient declaration.

In practice, this is sometimes forgotten. Consumer organisations, but also the regulator NVWA (Netherlands Food and Consumer Product Safety Authority), do tend to be too quick to label a name or image as misleading. The conclusion: companies should adjust their product packaging. Yet, even after Teekanne, the test has not changed that much at all. It is not the case that every statement or element individually has to pass the deception test. The Teekanne judgment is clear on this point: the entire packaging must be included in the assessment. This includes the list of ingredients. While the list of ingredients is no longer the determining factor in whether a packaging is misleading or not, it is still important.

The danger of applying too strict a test is that it creates a 'race to the bottom' effect and leaves no room for the exaggeration inherent in advertising. For producers, it is a search for the balanced middle ground: marketing may still play a role, but deception may not. And in between is the grey area. Cases like that of Van Mook muesli help point the way in this grey area. Whereas the RCC rejected the complaint, the CvB did find that there was deception. In this case, the outcome could well have gone the other way. The reasoning of the RCC – that the difference between five hazelnuts in the packet and seven hazelnuts on the packaging can be explained by usual advertising exaggeration – is not very far-fetched. Even in food law, it's okay to exaggerate a little in advertising.

Sarah Arayess
Lawyer/founder Say Legal

CHAPTER 54

Beter Leven – Sustainability claims

Annotated by Ebba Hoogenraad

Board of Appeal 6 February 2019, 2018/00826, Albert Heijn (*DoeMaarLekkerDuurzaam*)

Articles 7 and 8 Dutch Advertising Code, food, sustainable, environment, misleading with regard to animal welfare.

54.1 Summary

Can meat carrying the Beter Leven (Better Life) label be called 'sustainable'? The complainant believes that meat products, by definition, cannot be sustainable, because of the environmental impact caused by animals. In fact, more animal welfare, and thus often a longer life, results in additional environmental impact. 'Sustainable' is an umbrella term that can encompass environmental impact, animal welfare, as well as working conditions. In its branded content in Metro[466], the supermarket Albert Heijn explicitly linked the word 'sustainable' to animal welfare. Given the current state of affairs, this must be considered permissible. Despite the environmental implications, this cannot lead to the concept of 'sustainable/sustainable' being interpreted in a way that is narrower than the meaning the average consumer would attribute to it.

The 'DoeMaarLekkerDuurzaam' (freely translated: go ahead and be sustainable) campaign by Albert Heijn, Dutch National Postcode Lottery and Unilever encourages consumers to choose from a range of more sustainable products carrying sustainability labels. In a branded content interview in Metro, Albert Heijn describes the various products and the aspect in which each is considered more sustainable. Meat products with a Better Life label are described as a more sustainable alternative. The government, the Nutrition Centre and consumers interpret the umbrella term 'sustainable' as including – for the time being – animal welfare. Therefore, by current social standards, there is no case of misleading advertising.

466 A free Dutch newspaper.

 | HTTPS://DOI.ORG/10.1163/9789004751699-057

54.2 Complaint and main considerations

The complainant raises a fundamental question: can animal products be 'sustainable/more sustainable' given they are very damaging to the environment? The complainant argues that the current diet, rich in animal products, poses major problems for the climate and biodiversity, as well as air, soil and water pollution. This situation will only become more problematic when animals are given more space and live longer for animal welfare reasons. In fact, sustainability and animal welfare are diametrically opposed, unless the consumption of animal products is significantly reduced. Qualifying a product as sustainable because it is more animal-friendly, while being less beneficial for the environment, leads to a substantial dilution of the meaning of the term 'sustainable'.

Albert Heijn argues that 'sustainable/sustainability' is a catch-all term associated with socially responsible living, the environment, ecology and future-oriented thinking. The Dutch Government also uses the term in this way in the context of Sustainable Livestock Farming and the Nutrition Centre also use the term in this way. Albert Heijn refers to the guiding decision on sustainability of the BoA of 17 March 2015 (file 2014/00812, *Eneco duurzame warmte* (*sustainable heat*)) and the subsequent decisions of the Advertising Code Committee and the Board of Appeal. These decisions determined that the interpretation of the term 'sustainable' should be considered within the context of the advert and may vary from one advert to another. The *Metro* branded content states that 'sustainable' in meat products with the Better Life label is about animal and their welfare. For the average consumer, it is clear that in this context 'sustainable' refers to animal welfare. Consumers therefore understand that it is not about the environment, but about better living conditions for animals. Albert Heijn does, however, agree with the complainant that eating less meat is relevant to achieving a more sustainable diet. On the website doemaarlekkerduurzaam.nl, referred to in the advertisement, Albert Heijn encourages consumers to eat less meat and to choose vegetarian or vegan alternatives.

The ACC refers to the Eneco decision of the BoA: the concept 'sustainable' does not have a fixed meaning and must be assessed on a case-by-case basis, with its meaning potentially varying from one advert to another. In the advert, attention should be paid to the way this term is used and explained. Another factor is the interpretation that the average consumer is likely to give in the context of the overall advert. The ACC sees no reason to deviate from this established approach.

In the Committee's view, the average consumer regards 'sustainable' as a generally positive descriptive word that can refer to various aspects. The qualification 'sustainable' is not exclusively used in relation to the environment, but rather as an 'umbrella term' that can also include, in addition to environmental aspects, aspects of animal welfare or working conditions. Take, for instance, the 'sustainability policy' of banks, and the interpretation of the term 'sustainable' by the government (in the *Uitvoeringsagenda Duurzame Veehouderij* (Implementation Agenda for Sustainable Livestock Farming)) and the *Voedingscentrum* (Nutrition Centre) (on its website in relation to 'sustainable food'). Based on the circumstances mentioned by the complainant, it cannot be ruled out that the concept of 'sustainable/sustainability' will evolve in the future and be increasingly associated by the average consumer exclusively with the environment. At present, however, such a strict link between sustainability and the environment does not (yet) exist.

Next, the ACC examines whether it is clear to consumers what interpretation Albert Heijn gives to the term 'more sustainable' in this case. It is clear, in their opinion. 'More sustainable' clearly refers to animal welfare. The average consumer is not misled.

The BoA confirms this decision. The fact that certain sources (e.g. dictionaries) cited by the complainant associate the term 'sustainable' with a relatively low environmental impact, is insufficient to rule that the average consumer will now always associate this term with (protection of) the environment. Other sources, such as the Nutrition Centre's website, show that animal welfare can still be considered an aspect of sustainability. Since there is no reason to assume that the average consumer will now by definition perceive 'more sustainable' as a (form of) an environmental claim, the BoA considers, in line with the Eneco case, that this remains a broad concept, the interpretation of which may vary from one advert to the next, and must therefore be assessed on a case-by-case basis.

At the current stage, the BoA considers it is permissible to describe the term 'sustainable' in combination with the explicit reference to the Better Life label as a 'more sustainable' choice. This is regardless of the objections raised by the complainant on environmental grounds. Such objections cannot result in the term 'more sustainable' being interpreted in a way that is narrower than the average consumer's likely interpretation in the context of the overall impression of the advertisement. Albert Heijn leaves no doubt that the term 'more sustainable' as used in the advertisement refers exclusively to animal welfare. The complaint is also dismissed on appeal.[467]

54.3 Discussion

The umbrella term 'sustainable' is subject to change. The term 'sustainable' or 'more sustainable' can (still) refer to a variety of aspects: environment, animal welfare, working conditions and future-oriented thinking. The debate in this case is whether 'animal welfare' by way of Better Life labels on meat products can be called 'sustainable'. After all, according to the complainant, animals are very harmful to the

467 This case was handled for Albert Heijn Netherlands by Ebba Hoogenraad.

environment, and if better living conditions make them graze in pastures even longer, the animals are even an extra burden on the environment.

For several years, the Dutch National Postcode Lottery, Albert Heijn and Unilever have organised the *DoeMaarLekkerDuurzaam* campaign. Participants in the Postcode Lottery receive vouchers to purchase 'more sustainable products' at Albert Heijn. The campaign and television advertisement refer to 'More sustainable products, better for people, animals and the environment'. In the *Metro* interview (branded content), Albert Heijn says very explicitly that the Better Life labels on meat products focus on animal welfare.

The complainant raised a fundamental question: can animal welfare be considered sustainable? In other words, should the starting point be what the average consumer still associates with the word sustainable, or should a more limited explanation be given and is any campaign that refers to animal welfare as sustainable/more sustainable misleading?

The guideline is the Eneco decision,[468] which concerned sustainable heat. The term 'sustainable' has no fixed meaning, so its interpretation must be assessed on a case-by-case basis and may vary from one advert to another. When interpreting the term, consideration must be given to how it is used and explained in the advertisement, and also to the likely interpretation of the average consumer given the context of the advertisement as a whole. With that interpretation in mind, there is no doubt: the *Metro* text explicitly mentions that 'more sustainable' products refers to animal welfare. So there is no misleading advertising.

Or are things different given that there are now indications that, in fact, animal welfare could still have a negative effect on the environment? Both the ACC and the BoA conclude that the term 'sustainable' does not refer exclusively to the environment, but is an umbrella term that also includes aspects of animal welfare or working conditions. The ACC itself cited the 'sustainability policy' of banks as an example. Albert Heijn has pointed out that, in its Implementation Agenda for Sustainable Livestock Farming, the Dutch government refers to 'animal welfare' as part of achieving more sustainable livestock farming. The complainant asked the Nutrition Centre for its position. For now, the Nutrition Centre also regards animal welfare as sustainable.

The ACC is clearly keeping its options open: it cannot be ruled out that the concept of sustainability will evolve and be increasingly exclusively associated to the environment by the average consumer. At present, however, such a strict link between sustainability and the environment does not yet exist. The BoA also refers to 'the current state of affairs'. For now, the term 'more sustainable' cannot be interpreted more narrowly than the average consumer would interpret it at this moment. In other words, the current interpretation among consumers is the benchmark. It

468 BoA of 17 March 2015 (file 2014/00812, *Eneco duurzame warmte*).

is possible that, in time, 'sustainable' will refer exclusively to the environment. But this would require a shift in how consumers perceive the term. The outcome of this decision is that the law is responsive rather than directive. Information campaigns by politicians and NGOs are possibly going to trigger a more limited interpretation of the term sustainable in the future: sustainable = environment. Until then, sustainable remains an umbrella term. Advertiser bear a strict duty to make it very clear what aspect of sustainability a statement refers to. The advertiser cannot take this responsibility lightly.

Either way, it will always be helpful to eat a little less meat from time to time.

Postscript: on 1 February 2023, in the Netherlands, the Environmental Advertising Code was repealed and replaced by the Sustainability Advertising Code came into force. As of June 2023, the ACM (Netherlands Authority for Consumers and Markets) published an updated Sustainability Claims Guideline. One of the key requirements in both regulations is that absolute claims such as 'sustainable' must be specified in the advert itself. The advertisement must clarify in what respect the product or service is sustainable: does it concern a social (ethical) claim, such as animal welfare or working conditions, or does it relate to the environment or a product lifespan claim?

Ebba Hoogenraad
Partner food law and advertising law at Hoogenraad & Haak, Amsterdam

CHAPTER 55

CoolBest Skinny Orange – Deception by adding even more water

Annotated by Lisanne Steenbergen

Advertising Code Committee 11 June 2018, 2018/00287/I (*CoolBest Skinny Orange*)[469]

Article 2 Dutch Advertising Code (NRC), Articles 7 and 14 Regulation (EC) 1169/2011 (FIC Regulation), deception, labelling, juice from concentrate, CJEU Teekanne

55.1 Summary

The oranges adorning the packaging of CoolBest Skinny Orange and the word 'Skinny' raise questions for the complainant about i) the qualification of the product: orange juice or lemonade? And ii) the amount of water that has been added to the product.

The chairman of the Advertising Code Committee (ACC) handles both parts of the complaint together and agrees with the complainant that the packaging is misleading. Because of the images of oranges, texts such as 'Orange' and '100% natural ingredients', consumers will expect the product to consist (almost) entirely of orange juice. The word 'Skinny' and other statements on the packaging such as the ingredient list do not make it sufficiently clear for the consumer that a lot of (i.e. an extra portion of) water has been added to the drink.

Advertiser and manufacturer Riedel announces it will soon launch the product in new packaging. The front of the new packaging will feature the text, in large lettering: 'CoolBest 70% juice/30% water'. The chairman uses his ability (Article 12(5) of the Regulation of the Advertising Code Committee and the Board of Appeal (BoA) 'Regulation') to also consider the new packaging, although strictly speaking this new packaging falls outside the scope of the complaint (because it has not yet been put on the market and has therefore not been made public). The chairman expects that the average consumer, when being faced with the new packaging, will have no doubts about the composition of the product. The percentages on the front of the pack are clear, the other statements will also not confuse the average consumer. Conclusion: a recommendation in so far as necessary.

469 This ruling is annotated by Sarah Arayess in IER, no. 2018/59.

 | HTTPS://DOI.ORG/10.1163/9789004751699-058

55.2 Complaint and main considerations

The complaint focuses on the packaging of CoolBest Skinny Orange and the webshop of retailer Albert Heijn in which the product is offered. The complainant states that it is not clear from the packaging that the product contains 33% water. The packaging also gives the impression that the product is orange juice instead of lemonade. According to the complainant, the cause for the deception is the text on the packaging '100% natural ingredients' in combination with images of oranges and the fact that the product can be found on the orange juice shelf in stores and must be kept chilled in the fridge. The word 'Skinny' would also not communicate clearly that water has been added to the product. Furthermore, the statement on the addition of water as an ingredient is only provided on the side of the packaging and near the list of ingredients, but not on the front of the pack.

Advertiser Riedel did not react to the substance of the part of the complaint directed against the old packaging, but instead mentioned that the packaging will soon be changed to avoid ambiguity about the meaning of the word 'Skinny'. The new name of the product will be: 'CoolBest 70% juice/30% water'.

The complainant believes that the new packaging is also misleading. Because of the images of oranges on the packaging, consumers will expect the product to consist of 100% orange juice. In reality, the product consists of only 70% orange juice and 30% water. Additionally, the combination of the percentages would be confusing and the water content is not visually represented on the pack, unlike the oranges.

Riedel primarily expresses their opinion that the complainant expands the initial complaint with these comments, but nevertheless puts forward a substantive defence. According to Riedel, the amended packaging would comply with the FIC Regulation, the General Food Law Regulation and be in line with the Court of Justice of the European Union's decision in the Teekanne case. The new packaging informs consumers properly and comprehensively. The statement '70% juice/30% water' has a prominent place on the packaging and accurately reflects the composition of the product.

The chairman addresses the complaint using the standard legal framework as follows from the decisions of the Court of Justice of the European Union (CJEU). Whether food information (on the label) is misleading, must be assessed through the eyes of the CJEU Gut Springenheide[470] average consumer. This average consumer is expected to first read the list of ingredients.[471] And the golden rule from CJEU Teekanne: a correct declaration of ingredients on the ingredient list does not

470 CJEC 16 July 1998, C-210/96, ECLI:EU:C:1998:369 (*Gut Springenheide*)
471 CJEC 26 October 1995, C-51/94, ECLI:EU:C:1995:352 (*Commission v Germany*)

preclude the packaging from giving a misleading impression that the product contains an ingredient that it does not actually contain.[472]

The chairman concludes that the average consumer will expect the product to consist almost entirely of orange juice. This is due to the images of oranges and the statement 'Orange' combined with the statement '100% natural ingredients'. The word 'Skinny' will not make consumers aware of the fact that the product contains (a lot of) water. Nor does it clearly follow from the text on the back of the packaging that the product contains a significant amount of water. The consumer would only be able to tell indirectly from the list of ingredients what the water content is: 'fruit juice from concentrate (orange juice 63%, lemon)'. The chairman considers it likely that the average consumer will believe that water has been added to the concentrate to obtain the final product. The consumer will probably not realise that an extra portion of water has been added to further dilute the product.

Then, the chairman also assesses the new packaging. This is unusual because strictly speaking the new packaging is outside the scope of this complaint procedure. After all, the packaging has not yet been publicised or made available in the shops. Nevertheless, for practical reasons, the chairman considers the new labelling. This works out well for CoolBest because the new packaging is not misleading. Why? On the new packaging, 'Skinny Orange' has been replaced by the product's new name: '70% juice/30% water'. According to the chairman, the average consumer will have no doubts (anymore) about the composition of the product, as the percentages are explicit, large and prominently visible on the front of the packaging. Nor do the other statements and images confuse the average consumer. In short, from the new packaging it is clear that the product does not consist entirely of orange juice but also contains (extra) water. Therefore the chairman refrains from giving Riedel a recommendation to not advertise in this manner. This also applies to retailer Albert Heijn, as it is to be expected that Albert Heijn will adjust the information in its webshop in line with the new packaging.

55.3 Discussion

This chairman's decision contains at least two interesting points. On the substance of the decision, the amount of water that has been added to the product is one subject of discussion. The old 'Skinny' packaging does not sufficiently inform the average consumer about the (extra) amount of water, the chairman concludes. On the ingredient list of CoolBest Skinny Orange, water is declared as the second ingredient: 'concentrate (orange 63%, lemon), water, sweetener of natural origin (steviol glycosides)'. According to the chairman, it is not clear from the ingredient list that

472 CJEU 4 June 2015, C-195/14, ECLI:EU:C:2015:361 (*Teekanne*)

the final product contains two parts water. The question is whether that is actually correct?

CoolBest – like many other fruit juices on the Dutch fruit juice shelf in stores – is made from fruit juice from concentrate. To obtain fruit juice from concentrate, the fruit has to go through a concentration process. What does this process consist of? Let's suppose that the oranges used are grown in Brazil. They are harvested in Brazil and the juice is squeezed from the oranges. Then (part of) the water is extracted from the product, leaving a concentrate. The concentrate is transported by ship to the Netherlands in large tankers. The concentrate takes up less volume, allowing a higher volume of product (i.e. more concentrate) to be transported more effectively. In the Netherlands, the manufacturer then adds the extracted amount of water again, resulting in juice from concentrate.

The chairman could therefore have ruled differently in this case with regard to the list of ingredients. It can be assumed that the average Dutch consumer is familiar with fruit juice from concentrate (and the process preceding its production). From the list of ingredients, the average consumer can tell that another (extra) part of water and sweeteners are added to the fruit juice from concentrate to obtain the final product. Therefore, there would be no deception based on the ingredient list. What about the images and statements on the packaging? CJEU Teekanne teaches us that although consumers read the list of ingredients first, this does not mean that consumers cannot be misled by the other elements on the packaging. So it could well be that despite the correct ingredient declaration, the consumer is misled (insufficiently informed) by, for example, the word 'Skinny' and images of oranges. In other words, another route, with the same result.

The second noteworthy point about this decision of the chairman is of a 'procedural' nature. Strictly speaking, procedural law is the law that governs the process before the (state) court and thus facilitates the correct application of material (substantive) law. The Stichting Reclame Code (SRC) is a self-regulatory body and not a state court or governmental body. However, proceedings before the ACC of the SRC are similar to proceedings before the state courts in a number of aspects. All 'procedural' rules of the SRC can be found in the Regulation.

In this case, the chairman applied Article 12(5) of the Regulation. It was in that context that the new packaging 'CoolBest 70%/30%' was considered. Under Article 12(5), it is possible for the chairman to refrain from a recommendation or to make a recommendation in so far as necessary. The Article describes when such a decision is appropriate: '[...] if the person against whom the complaint is directed makes it plausible that, in response to the complaint, appropriate measures have been taken to make the contested advertisement(s) comply with the Dutch Advertising Code.'

Riedel has announced that it will replace the 'old packaging' of CoolBest Skinny Orange with the new packaging named 'CoolBest 70%/30%'. The complainant believes that the percentages on the front on this packaging are also confusing and (implicitly) submits a request to the chairman to review this packaging as well. Riedel protests, saying this would be an extension of the initial complaint, which relates exclusively to the old packaging. The chairman states that there is no – and cannot be – an extension of the complaint, because at the time of the complaint procedure the modified packaging has not yet been publicised in Albert Heijn's webshop nor is it available in the shops. If it is not an extension of the complaint, what can it be? The chairman considers that, in principle, he cannot rule on adverts that have not (yet) been used (ACC file 2017/00164), but in this case he does so anyway for practical reasons. In the end, things end somewhat well for Riedel, as the new – yet to be publicised – packaging is found to be not misleading. Thanks to Article 12(5), Riedel (only) receives a chairman's decision without a recommendation

Lisanne Steenbergen
food law and advertising lawyer at Hoogenraad & Haak, Amsterdam

CHAPTER 56

Optimel vla Vanille – The taste of vanilla, without the vanilla pod

Annotated by Ebba Hoogenraad

Board of Appeal 16 May 2019, 2018/00701 (*Optimel custard Vanilla/Foodwatch*)

Article 2 Dutch Advertising Code, Article 7 FIC Regulation 1169/2011, flavourings, vanilla, misleading food information.

56.1 Summary

Foodwatch argues that the average consumer cannot tell from the packaging of Optimel vanilla custard that the custard does not contain real vanilla (derived from the vanilla plant). Both the Advertising Code Committee (Reclame Code Commissie: RCC) and the Board of Appeal (BoA) conclude that statements such as 'mild vanilla flavour' on the side and 'aroma' on the ingredient list do not dispel the incorrect misleading impression created by the product name. Therefore the packaging is considered misleading. FrieslandCampina argues that the mention of 'flavour' on the list of ingredients and 'vanilla flavour' are in accordance with the requirements of the European flavouring legislation for indicating a product that does not contain vanilla from the vanilla plant. The BoA rejects this argument because the average consumer cannot be expected to have knowledge of the complex flavouring legislation. FrieslandCampina tries to substantiate with a consumer survey, stating that the average consumer understands that the product tastes like vanilla but does not contain vanilla from the vanilla plant. However, the ACC interprets the results to mean that a relevant proportion (26%) in fact does not understand that there is no vanilla plant involved. Without saying so explicitly, the BoA corrects the ACC's reasoning that a percentage mentioned in a consumer survey automatically proves consumer deception.

The BoA considers: leaving aside the question of whether the (abstract) assessment of whether the average consumer is being misled can and should be done by consumer surveys, the two surveys contradict each other.'

 | HTTPS://DOI.ORG/10.1163/9789004751699-059

56.2 Complaint and main considerations

Foodwatch complains about the packaging of FrieslandCampina's *Optimel custard Vanilla*. In addition to the text '*Optimel custard Vanilla*', the front contains an image of yellow custard. One side of the packaging includes the text mild vanilla flavour. Above the ingredient list on the back is the text '*Optimel Vanilla*', followed by 'pasteurised low-fat vanilla custard'. The ingredient list shows that the product contains a flavouring.

According to Foodwatch, the names '*custard Vanilla*' and '*low-fat vanilla custard*' give the average consumer the impression that the product contains vanilla. Since this is not the case, Foodwatch considers the packaging is misleading. Foodwatch invokes Teekanne, arguing that consumers are being misled despite the fact that the ingredients list shows that the product contains a flavouring (and not vanilla from the vanilla plant).

The core of FrieslandCampina's defence boils down to the following three (related) points:

i. *Vanillevla* (vanilla custard) is a typically Dutch product with a designation that has been used in the Netherlands for many years. For this reason, consumers will perceive the name '*vanillevla*' as a proper, clear and recognisable name for custard that tastes like vanilla.

ii. The packaging does not in any way suggest that vanilla pods have been used to obtain the vanilla flavour. Given the fact that the ingredients list states 'aroma' (flavouring) and the product is called '*custard Vanilla*', consumers will understand that a flavouring has been used to obtain the typical vanilla flavour of custard.

iii. It follows from the Teekanne ruling that in addition to the list of ingredients, the other elements of the packaging must also be considered. The packaging does not give the impression that the product contains vanilla: the colour yellow is dominantly present, no vanilla (pods) are depicted and the ingredient list says flavouring and not vanilla. The side refers to vanilla-flavoured custard. Taken together, these elements make it unlikely that consumers expect vanilla from the vanilla plant, what they do expect is that it tastes like vanilla, and that expectation is met.

In response to a question from the Chairman of the Advertising Code Committee (*RCC*), FrieslandCampina submitted a consumer survey to support its claim that it is obvious to the average consumer that no vanilla (fruit of the vanilla plant) has been added to a product called '*vanilla custard* or '*custard vanilla*'. This survey showed that 74% did not expect the vanilla flavour of *Optimel custard Vanilla* to come from the fruit of the vanilla plant, and 88% expected the flavour to come from a flavouring. Foodwatch disputes the survey and submits a counter survey.

The ruling of the Advertising Code Committee (RCC)

According to the ACC, the average consumer will to some extent expect the product to contain vanilla. The absence of vanilla pods on the packaging does not change this, nor does the statement 'mild vanilla flavour' on the side. A product containing vanilla from the vanilla plant could also be described as having a mild vanilla flavour, according to the ACC. Furthermore, the fact that the product does not contain vanilla from the vanilla plant would not be apparent from the label. Moreover, the ACC concludes that it has not been sufficiently established that the average consumer would know that the custard does not contain vanilla from the vanilla plant, but only has a vanilla flavour. For these reasons, the ACC considers the packaging to be misleading with regard to the composition of the product within the meaning of Article 7(1)(a)of the FIC Regulation and therefore in contravention of the law as referred to in Article 2 NRC.

Board of Appeal

On appeal, FrieslandCampina raises three grounds of objection:

a. The ACC wrongly failed to classify the statement 'mild vanilla flavour' as a flavour disclaimer. In doing so, the ACC disregards the entire flavour legislation and acts in contravention of case law and the enforcement policy of the Netherlands Food and Consumer Product Safety Authority (NVWA). The consequence would be that the flavour disclaimers that have been widely used in the market since CJEU the Teekanne ruling could no longer be used in the future. Furthermore, it is sufficiently clear to the average consumer that the products flavour is vanilla, due to the combination of flavour disclaimer on the side of the packaging, the mention of 'aroma' in the ingredient list, the absence of an image with vanilla pods and the additional text on the packaging about the flavour.
b. The ACC wrongly assumes that deception can be established by exact data from a consumer survey.[473] The results of a consumer survey are only an indication of how the average consumer thinks; they are not hard evidence. The ACC misunderstands that the deception test is an abstract assessment in which any deception must be determined using the interpretation of the average consumer as defined in Gut Springenheide.
c. The ACC decision reflects a misunderstanding of European food law and flavouring legislation. The ACC wrongly assumes that consumers, when reading

473 According to the FrieslandCampina consumer survey, 74% of those surveyed did not expect the vanilla flavour of Optimel vla Vanille to come from the vanilla plant fruit. The RCC incorrectly deduces from this that 26% of those surveyed (according to the RCC a significant percentage) have been misled, because they do expect the product to contain vanilla from the vanilla plant.

> 'flavouring', think that the product contains (flavouring made from) vanilla from the vanilla plant. If the product contains flavouring made from vanilla from the vanilla plant, the ingredient list will say 'natural vanilla flavouring' (natural y-aroma; flavouring with the specified flavour). Consumers are used to the term 'flavouring' in the ingredient list implying that a flavouring not made from the vanilla plant has been used. Moreover, the consumer survey found that 88% of respondents do not expect there to be vanilla in vanilla custard but a flavouring.

The BoA finds that the grounds of appeal do not stand and upholds the ACC's ruling that the packaging has violated Article 2 NRC.

As customary in such cases, the BoA first refers to European case-law including Gut Springenheide, Darbo and Teekanne.[474]

In response to the ground concerning the taste disclaimer, the BoA notes that the explicit reference to an ingredient in the product name on the front (*Optimel Vanilla custard*), unmistakably suggests that the product actually contains this ingredient. The front and side do not correct this suggestion. There is no notice on the front side showing that vanilla refers only to the flavour. When seeing the indication 'flavour' on the ingredient list on the back, consumers will not realise that the product does not contain vanilla from the vanilla plant. To understand this, the average consumer will need to have knowledge of the complex flavouring legislation; this level of knowledge cannot be expected from the average consumer.

The BoA considers it unlikely that the average consumer would understand from the other elements on packaging that the product does not contain vanilla from the vanilla plant. It is not clear to the average consumer that 'mild vanilla flavour' on the side should be an indicator of a product that does not contain vanilla as an ingredient. Moreover, if 'mild vanilla flavour' is meant as a nuance to the name *Optimel custard Vanilla* then it would have been logical to put this text on the front. In addition, the image of the yellow custard does not guide the consumer towards the correct understanding. Even when considering all the elements mentioned by FrieslandCampina together, the average consumer might think that *Optimel Vanilla custard* contains the ingredient vanilla.

Regarding the consumer surveys submitted by the parties, the BoA rephrases the ACC's conclusion: 'Leaving aside the question of whether the (abstract) test of whether the average consumer is being misled can and should be done on the basis of consumer surveys, both surveys contradict each other.' From this, the BoA concludes that vanilla custard is, in any event, not a customary name or established

474 CJEC 16 July 1998, C-210/96, ECLI:EU:C:1998:369 (*Gut Springenheide*); CJEC 4 April 2000, ECLI:EU:C:2000:184 (*Darbo*), CJEU 4 June 2015, C-195/14, ECLI:EU:C:2015:361 (*Teekanne*). The Teekanne ruling is discussed in Chapter 11 of this book.

term (requiring no explanation as to whether or not vanilla has been added as an ingredient).

The complaint is upheld.

56.3 Discussion[475]

This case highlights the tension between the legal reality of the requirements of the flavouring legislation, case-law and the enforcement policy of the Netherlands Food and Consumer Product Safety Authority (*NVWA*) on the one hand, and the interpretation of the average consumer on the other. The mere mentioning of 'flavouring' on the ingredient list together with the indication 'vanilla flavour' tells any lawyer with knowledge of food law and any QA manager that a flavouring has been added that makes the product taste like vanilla (not being natural vanilla flavouring). According to the ACC and the BoA, to the average consumer, it is less clear that no vanilla pods have been used. But is that really the case? Did the average consumer in 2019 not already understand that the indication 'flavour' means that the product tastes like the ingredient referred to, but does not contain the actual ingredient? FrieslandCampina tried to substantiate with examples that flavour claims are widely used on food packaging since the Teekanne ruling. Custard, confectionery, crisps and soft drinks: all contain a flavour claim. For the average consumer, this is therefore nothing unusual; knowledge of the 'complex' flavour legislation is not required. A simple trip through the supermarket is enough to get the idea.

For a while, it seemed that given the wording in the ACC decision, a red line might be drawn through any use of flavour disclaimers for products that do not contain the 'real' ingredient. Fortunately, the BoA has nuanced this, emphasising that it is specifically the placement of the 'mild vanilla flavour' flavour disclaimer on the *Optimel custard Vanilla* packaging that is not sufficient to prevent deception. With vanilla in the name, the placement of the flavour disclaimer on the side is insufficient, despite the word flavour in the ingredients list. Meanwhile, on the new packaging, the word 'vanilla flavour' appears on the front.

Another key issue in the ACC ruling was the legally and dogmatically incorrect approach to assessing the average Gut Springenheide consumer. This is an abstract assessment. Consumer surveys and percentages of respondents who are allegedly misled are not the decisive measure. The Gut Springenheide judgment itself was a response to the German practice of determining misleadingness for years based on the number of consumers found to be misled in surveys.

Flavour disclaimers have since become firmly established, and rightly so. Consumers need to know clearly what they are getting. The so-called pictorial rules in

475 This case was handled in appeal for FrieslandCampina by Ebba Hoogenraad.

the NVWA's labelling guidance provide very detailed guidelines on when fruit or flavour-giving ingredients may be shown on packaging, and in which cases a flavour disclaimer is required.

The fact that our grandparents already ate yellow vanilla custard, flavoured, without real vanilla, no longer carries any weight. And for good reason.

Ebba Hoogenraad
Partner food law and advertising law at Hoogenraad & Haak, Amsterdam

CHAPTER 57

Kinder Milk Slice – Health claim: certain properties and link

Annotated by Ebba Hoogenraad

Advertising Code Committee 24 February 2020, 2019/00675 (*Kinder Milk Slice*)

Article 2 of the Dutch Advertising Code (NRC), Article 2(5) and 10(1) of the EU Nutrition and Health Claims Regulation, (general) health claim, link between health effect and food.

57.1 Summary

Does the enthusiastic jumping of a girl in a TV commercial qualify as a health claim within the meaning of the Claims Regulation? The Chairman over the Advertising Code Committee (ACC) considered this question. The outcome: yes, the images show a clear link between a girl who is initially tired and, after eating Ferrero's snack, enthusiastically resumes playing. On appeal, the ACC again considers the images carefully and concludes otherwise. The girl shows happy and enthusiastic behaviour while the packaging of the snack is still closed. After taking two bites, the girl jumps up to continue playing. According to the ACC, these images do not give the impression that eating the snack has a positive effect on health (such as replenishing the used energy).

57.2 Complaint and main considerations

The complainant objects to a television commercial (TVC) by Ferrero promoting the snack *Kinder Melkschijfje* (Kinder Milk Slice in English-speaking markets). The TVC could wrongly give the impression that Milk Slice is a healthy snack.

In the TVC, children are shown playing and jumping across the screen. A tired girl plops down in a beanbag in the kitchen. Her mother hands her a (packaged) *Kinder Milk Slice*. Meanwhile, the voice-over says: '*And to quickly get back to playing, there is Kinder Milk Slice.*' The girl enthusiastically accepts the snack and takes a bite. In the next shot, she takes another bite and gets up to continue playing. '*Nice and light ... and back we go*', the voice-over adds.

The discussion initially focuses on whether or not the ingredients are (un) healthy. The case takes a different turn when the Chairman of the ACC asks Ferrero

 | HTTPS://DOI.ORG/10.1163/9789004751699-060

whether the commercial contains a (permissible) health claim within the meaning of the Claims Regulation, given the suggestion that *Kinder Milk Slice* would ensure the rapid replenishment of (spent) energy. Ferrero initially argues that this question has nothing to do with the original complaint. In addition, the commercial shows recognisable, normal behaviour of children who quickly get up again after a short break and are eager to resume their play activity. The enthusiastic behaviour is unrelated to the moment of eating and bears no connection to an alleged energy intake by the snack.

The Chairman sees the broad scope of the complaint as a reason to assess whether there is a health claim. In his view, the fatigue of the girl sitting down, eating the *Kinder Milk Slice* and then jumping up again to continue playing, shows that there is a clear link between the consumption of the Milk Slice and health. This connection is emphasised by the voice-over that can be heard when the girl drops into her seat tired: '*And to quickly get back to playing, there is Kinder Milk Slice.*' The pay-off that follows also seems to refer to this connection: '*Nice and light ... and back we go.*'

Ferrero appeals, successfully, to the Chairman's decision that this constitutes an unauthorised general health claim within the meaning of Article 10(1) of the Claims Regulation.

The ACC overturns the Chairman's decision and dismisses the complaint. Contrary to the Chairman's conclusion, the ACC finds that the necessary link between eating the Kinder Milk Slice and the effect on health is lacking. The ACC notes that the girl takes the Milk Slice from the beanbag with a smile and enthusiasm and then revives. At that moment, the packaging is still closed. Then she clearly enjoys taking two bites and jumps up to continue playing. The message seems to be: the Milk Slice is tasty and convenient, and playtime does not have to be interrupted. These images do not give the impression that eating the Milk Slice has given the girl energy to continue playing. It is not suggested that the children (have to) eat a Milk Slice to replenish the energy expended.

57.3 Discussion

Young children are energetic. They run this way and that across a lawn, race down a hill on a skateboard and kick a ball recklessly around the garden. But children also get tired of all that playing. And sometimes they feel like a tasty snack, maybe a *Kinder Milk Slice*? Ferrero outlines this scenario in its television commercial. After playing, a girl plops down contentedly in a beanbag in the kitchen and is handed this Milk Slice by her mother. In two different shots, the girl can be seen taking two bites and then jumping up to continue playing. The complainant objected to the product being presented as a healthy snack. The Chairman wonders whether there is a non-permitted health claim, namely: eating *Kinder Milk Slice* replenishes spent energy. The Chairman sees in images and text a suggested link between the consumption of the snack and the health effect. The Advertising Code Committee (*RCC*), however, thinks otherwise and rejects the complaint.

Why did the Chairman and the Advertising Code Committee (*RCC*) have a different view? It all comes down to timing. The Advertising Code Committee (*RCC*) seems to have watched the television commercial (more) closely and notes that the girl reacts enthusiastically (energetically) to receiving the Milk Slice while the packaging is still closed. At that point, she hasn't taken a bite, and therefore there is no link between eating *Kinder Milk Slice* and the effect on health. In case-law, the requirements for a health claim are further clarified. According to the definition, a health claim must meet four requirements:

i. a *claim*, which
ii. *states, suggests* or *implies*
iii. that there is a *relationship* between the food category, a food or a constituent thereof, and
iv. *health.*

The first two requirements are often regarded as relatively easy to satisfy. This is not entirely correct, because in order to be able to speak of a 'claim', two requirements must be met: it must be a) a *non-mandatory message* that b) relates to *certain properties.*

The fact that in practice the requirement under a) of 'certain properties' is not always carefully assessed also becomes clear when applying the third requirement: a relationship. The judgments of the CJEU in *Deutsches Weintor*[476] and *Green-Swan Pharmaceuticals*[477] are regularly cited – including in the context of enforcement activities by the Dutch Food and Consumer Product Safety Authority (NVWA) – to argue that the relationship can be interpreted broadly. This reasoning is too simplistic. What is often forgotten is that even for such a broad relationship to exist, there must first be a claim, i.e. that it concerns a *specific, qualified relationship* that relates to *certain properties* of the *specific food.* The German Federal Court of Justice expressed this well in the Energy & Vodka ruling.[478] General statements are by definition not a claim, because it must concern specific properties of the food in question that distinguish that food from similar foods.

The Advertising Code Committee (*RCC*) concludes that the girl's jumping up after eating the milk slice does not qualify as a health claim.

The relationship is missing. Although the ACC does not address the specific conditions for a claim and the suggested relationship with health as follows from

476 CJEU 6 September 2012, C-544/10, ECLI:EU:C:2012:526 (*Deutsches Weintor*). See Chapter 9 of this book.

477 CJEU 18 July 2013, C-299/12, ECLI:EU:C:2013:501 (*Green – Swan Pharmaceuticals*).

478 BGH 9 October 2014, ZR 167/12 (Energy and Vodka) https://dejure.org/dienste/vernetzung/rechtsprechung?Gericht=BGH&Datum=09.10.2014&Aktenzeichen=I%20ZR%20167/12.

case-law, the conclusion is still correct. Nowhere in the commercial are particular, distinctive properties of the *Kinder Milk Slice* emphasised in order to establish a specific, qualified relationship.

What about the voice-over? While the mother takes a Milk Slice for her daughter from the fridge, we can hear: '*And to quickly get back to playing, there is Kinder Milk Slice.*' When saying the last words of this sentence, you see that the girl accepts the snack from her mother. Does this suggest a relationship? Possibly, but as discussed above, a suggestion of a relationship is not sufficient to constitute a health claim. There must be a specific, qualified relationship to certain properties of the food. That is a high bar, and it is not met here.

Ebba Hoogenraad
Partner food law and advertising law at Hoogenraad & Haak, Amsterdam

CHAPTER 58

Arla climate neutral – link to label

Annotated by Ebba Hoogenraad

Board of Appeal 6 July 2022, 2022-00472 (*Arla climate neutral*)

Dairy, environmental claim, climate neutral, argumentation, Articles 2 and 3 Environmental Advertising Code (MRC)

58.1 Summary

Arla organic dairy meets the requirements of the Climate Neutral Group's (CNG) certification label. Its packaging bears the 'climate neutral' logo. The Stichting Even Geen Vlees (Take A Break From Meat Foundation) objected and the Advertising Code Committee (ACC) upheld their complaint. Climate neutral is an absolute environmental claim, for which the extremely heavy burden of proof rests on Arla. Arla does not meet this burden, according to the ACC.

In appeal, Arla argues that the claim 'climate neutral' as the company used and intended is not an absolute claim, but a relative claim, namely 'climate neutral' according to the Climate Neutral Group's certification requirements.

The Board of Appeal (BoA) recognises that Arla cannot be denied the right to invoke the climate neutral label in advertisements. If the claim in the advert is clearly linked to the label, the average consumer understands that Arla meets the requirements of the label. The BoA checks all advertisements from Arla (on the packaging and website). In the advertisements the link with the label is not (yet) sufficiently made. Therefore, the advertisements are in violation of Article 2 MRC (sustainability claim not made sufficiently specific or omission because essential information is omitted). But the advertisements do not violate Article 3 MRC (providing evidence of the sustainability claims) because Arla meets the licence requirements of the CNG label.

58.2 Complaint and main considerations

The complaint by *Stichting Even Geen Vlees* concerns the designation 'climate neutral' on the packaging of Arla organic dairy. The essence of the complaint is that the environmental claim 'climate neutral' used by Arla creates the impression that Arla organic has no effect on the climate. According to the complainant, this environmental claim cannot be proven.

 | HTTPS://DOI.ORG/10.1163/9789004751699-061

The ACC upheld the complaint in a very detailed decision: with 'climate neutral' a guaranteed result is claimed, namely full compensation of greenhouse gases emitted by Arla organic. The average consumer interprets this claim that without a doubt, on balance, no harmful effect on the climate occurs in the production of Arla organic dairy. In order to actually be able to speak of 'climate neutral', measures must be taken to ensure that the influence of these substances on the climate, as long as they are present in the atmosphere, is neutral on balance, according to the ACC.

Admittedly, it has not been disputed that Arla participates in certified compensation projects that meet certain recognised (theoretical) standards, namely the Climate Neutral Group label of the Climate Neutral Group. The ACC considers it sufficiently plausible that those projects involve a degree of greenhouse gas compensation.

But according to the ACC, Arla has not shown that it is absolutely certain that full and permanent compensation through its tree planting projects is guaranteed, also taking into account that it has not been shown that there is full scientific consensus on the extent of compensation through tree planting projects. Moreover, several projects in which Arla invests involve estimates (rather than factual data), while uncertain future events such as fire, expropriation or a changed political situation can also have an effect.

Therefore, the ACC concludes that the accuracy of the absolute claim 'climate neutral' has not been demonstrated to the extent required by Article 3 MRC. The claim is therefore misleading within the meaning of Article 2 MRC. Arla's defence that the designation 'climate neutral' simply appears in the Carbon Neutral Group's (CNG) 'Climate Neutral Product' label with which Arla organic is certified does change the ACC's decision. In the advertisements, independently of this label and without reservations, it is claimed that Arla organic is 'climate neutral'.

58.3 Appeal

The grievances

Among other things, the grievances elaborate on CNG's Climate Neutral Certification standard. Arla meets all these requirements. Arla is very transparent towards consumers, stating that it is climate neutral through compensation and reduction, with a reference to its website for further explanation.

During the hearing, Arla explicitly stated that this is not an absolute environmental claim, but a relative environmental claim, as the claim is specifically related to the CNG label and Arla is entitled to use this label for its organic dairy products. This makes this case different from previous BoA cases, where the BoA has ruled strictly on the use of climate-neutral or CO_2-neutral claims. Those cases did involve an absolute environmental claim.

The Board of Appeal

The core of the complaint is that the environmental claim 'climate neutral' used by Arla can be insufficiently justified by the available factual data. During the ACC proceedings, the discussion then focused on whether Arla could sufficiently substantiate this claim. The ACC assumed that this was a standalone absolute environmental claim, i.e. a claim that Arla uses independently of CNG's climate neutral label.

At the hearing, however, Arla explicitly states that the claim is not absolute, because it is related to a certification label and is also intended to be used in relation to the CNG label. Arla states that it only uses the claim 'climate neutral' in combination with the CNG label and with reference to the CNG standard. According to Arla, the existence of a standard means that the claim is not unlimited or absolute, but rather defined by the CNG standard.

Arla argues that it meets the requirements for the use of the certification label and thus for the use of the relative claim. It further argues that it clearly explains to consumers that the claim 'climate neutral' is related to the CNG standard and the associated compensation and reductions, namely (1) calculation of all greenhouse gases according to a recognised carbon footprint method, (2) reduction of emissions in line with the Paris Climate Agreement, (3) compensation of the remainder of emissions via carbon credits from FSC-certified tree planting projects and (4) transparency.

In a fundamental consideration the BoA says that it is not in dispute that Arla organic dairy has been certified according to the CNG standard and Arla organic is therefore authorised to bear the certification label. The label, specifically in relation to dairy, has been qualified by Dutch organisation *Milieu Centraal* as a 'top certification label for climate'. The complainant has not disputed that the label entails additional requirements. On the basis of the above, Arla cannot be denied the right to refer to that label in advertisements. If Arla makes it clear in the advertisements that the claim is specifically related to that certification label, the average consumer will understand that Arla is entitled to carry this label in the context of neutralising the CO_2 emissions from the organic dairy production in accordance with the requirements of the certification label.

This leads to the question whether from every advertisement sufficiently follows that Arla is invoking the CNG label.

The BoA then discusses all the individual advertisements and the listed texts and references to the website. The conclusion is that Arla's explanation of the environmental claim 'climate neutral' on the packaging and on the website is not directly recognisable to the average consumer as a reference to the CNG climate neutral label. This label is not clearly identified on the side of the packaging as the basis for the climate neutral environmental claim. The BoA finds the statement on the packaging 'Approved in accordance with the standard of Climate Neutral Group' insufficiently specific to assume that the average consumer, on that basis alone, will recognise that an environmental claim related to the certification label is made. The mere reference to CNG's website on the packaging does not make this clear either.

The same applies to the blue 'climate neutral product' logo on the front of the packaging. A factor here is that it has not been shown that this is a label that is now so well known to the public that the logo itself and its interpretation and meaning can be considered generally known and needs no further explanation.

On the Arla website, the information about the CNG label can only be found indirectly by actively searching for the information and clicking on the appropriate hyperlinks. The other advertisements do not contain any reference to the CNG label, which means that the average consumer is not informed that this is a CNG label-related claim.

This leads the BoA to conclude that Arla's specific interpretation of the claim 'climate neutral' is insufficiently clear to the average consumer. As a result, the consumer will regard this claim as a stand-alone environmental claim, i.e. separate from the CNG label. In appeal Arla notes that this was not its intention. Therefore the BoA merely notes that Arla has not made the specific interpretation of the climate-neutral claim sufficiently clear to the average consumer: Arla has not made it directly apparent by clear and eye-catching wording that the claim is specifically related to the CNG label.

The statements are misleading within the meaning of Article 2 MRC in conjunction with Article 8.3 and under c of the NRC. This is due to Arla's omission of essential information about environmental aspects that the average consumer requires in order to make an informed decision.

Thus, the BoA concludes – on different grounds than the ACC – that the statements are contrary to Article 2 MRC. However, the BoA rules that there is no violation of Article 3 MRC. After all, in connection with the burden of proof regarding the climate-neutral claim, Arla can rely on the fact that its organic dairy is certified according to the CNG label.

58.4 Discussion

A question of principle, and a principled and also pragmatic judgment.

According to the established line in BoA rulings, the term climate neutral is an absolute term. A very heavy burden of proof rests on the advertiser. This applies in particular if CO_2 compensation takes place via certified (tree planting) projects. There is no scientific consensus as to whether CO_2 neutrality can actually be achieved. Hence the line taken in the Shell rulings and applied to Arla by the ACC. This entails that Arla must demonstrate with solid, independent, verifiable and generally recognised evidence that in practice, i.e. factually, effectively guaranteed, there is full compensation of Arla's organic emissions (remaining after reduction). The ACC ruling discusses in detail the various requirements, studies and CO_2 compensation projects as raised at first instance. The conclusion: Arla cannot substantiate the claim 'CO_2 neutral'.

And still, something does not feel quite right. It is not in dispute that Arla organic dairy is certified to the standard of the CNG and can therefore bear the label with the claim 'climate neutral'.

Surely anyone who has a licence for this – rated as a top quality label by Milieu Centraal – should be able to use it?

During the hearing, Arla invoked this argument. And with effect. It led to the BoA's fundamental statement: Arla cannot be denied the right to invoke that CNG label in advertising. If the claim is clearly linked to the certification label, the average consumer will understand that Arla meets the requirements of the certification label.

Next, the BoA assesses whether, in this case, the statements themselves sufficiently show that Arla is invoking the CNG label. In other words, is there a clear link between 'climate neutral' and the label? The BoA concludes that this is not the case. The statements do not clearly refer to the certification label as the basis for the environmental claim 'climate neutral'. Because the link to the certification label is missing, the average consumer will consider the claim as a stand-alone absolute environmental claim. So – on different grounds than the ACC – the BoA comes to the conclusion there is a violation of Article 2 MRC.

The significance of this ruling for the industry is that i) absolute sustainability claims may be used as part of labels, while ii) there is a requirement that this can only be done by always clearly referring in the statement itself that the claim is about sustainable/climate-neutral, in the context of the relevant certification label.

The BoA's ruling also provides further guidance for this: a reference on the side of the pack is insufficient. The same applies to the text 'approved according to the Climate Neutral Group standard'. In those situations the link to the certification label is not specific enough. A single reference to the website with more information is also not enough. However, once a label has become widely known to the average consumer, the standard will be less strict.

Ebba Hoogenraad
partner in food law & advertising law, Hoogenraad & Haak, Amsterdam

CHAPTER 59

Sterke Start yoghurt – black-tailed godwit is environmental claim

Annotated by Ebba Hoogenraad

Advertising Code Committee 22 December 2022 (2022-00329 *FrieslandCampina grutto*)[479]

Environmental claims, sustainability claims, presentation of evidence, Article 2 and Article 3 Dutch Environmental Advertising Code (MRC)

59.1 Summary

Can an image of a meadow bird be an environmental claim? Yes. In this remarkable decision, the Advertising Code Committee (ACC) concluded that an image of a black-tailed godwit, in the context of the advertisement, is an environmental claim. At issue was an advertisement for 'Campina Sterke Start [strong start] yoghurt'. The advertisement depicted, among other things, a farmer on his land, a black-tailed godwit, a butterfly and daisies. Throughout the proceedings at the ACC, it was established that the intensification of agriculture was identified by organisations as Natuurmonumenten and Sovon as the cause of the decline in the population of the black-tailed godwit. The ACC believes that the average consumer is aware of this impact and rules that, the advertisement taking into consideration the whole context of the image (farmer's land with cow, milk churn, windmill and black-tailed godwit) suggests an environmental benefit of the yoghurt. The advertisement is not simply an image that reflects the landscape for the dairy product. The average consumer will think that FrieslandCampina is taking into account the interests of the endangered black-tailed godwit when producing this yoghurt.

Now that it is established that the black-tailed godwit is considered an environmental claim, it is up to FrieslandCampina to prove that measures that are beneficial to the black-tailed godwit are indeed being taken. FrieslandCampina is not able to provide this evidence. Therefore, the complaint is upheld and an ACC recommendation follows.

479 https://www.reclamecode.nl/uitspraken/grutto/voeding-en-drank-2022-00329/386094/

 | HTTPS://DOI.ORG/10.1163/9789004751699-062

59.2 Complaint and main considerations

The complaint concerns an advertisement in a Dutch newspaper for '*Campina Sterke Start [strong start] yoghurt*'. Roughly translated, the advert says, 'We are proud of our new strong start yoghurt. For the past 150 years we have been making the most delicious Campina products. That is why we are giving away treats all across the Netherlands today'. The advert depicts an image of a map of the Netherlands with amongst others grass, a windmill, a farmer next to a cow, a milk churn, a black-tailed godwit, daisies and a butterfly. See image 59.1 below.

The crux of the complaint is that the images used in the advertisement, more specifically a black-tailed godwit, a butterfly and daisies, refer to environmental aspects linked to the production of the '*Sterke Start yoghurt*' product. According to the complainant, this could mislead consumers. The complainant submitted various studies by '*Natuurmonumenten*' and *Sovon* (knowledge centre for distribution and trends of wild birds in the Netherlands) which show that the intensification of agriculture has been identified as the cause of the sharp decline in the number of breeding pairs of the black-tailed godwit (in 1975 there were 120,000 black-tailed godwits, but in 2023 only about 30,000).

According to the Advertising Code Committee (*Reclame Code Commissie: RCC*), the complainant made it sufficiently clear that the image of the black-tailed godwit, in the entire context of the image, could evoke associations with environmental

FIGURE 59.1 Campina Sterke Start advert, as seen in NRC, 24 June 2022.

aspects related to the production of the yoghurt. The Advertising Code Committee (*RCC*) assumes that the majority of readers of the advertisement are aware that agriculture has been identified as the cause of the decline in the population of the black-tailed godwit.

In its defence FrieslandCampina stated that the advertisement did not imply a specific claim regarding nature, animals and/or climate. According to FrieslandCampina, the images, including the black-tailed godwit, are characteristic of Dutch grassland. The images are meant to illustrate that the product originates from Dutch soil.

The Advertising Code Committee (*RCC*) does not go along with FrieslandCampina's defence. By depicting the black-tailed godwit in the foreground, alongside other elements of nature such as flowers and a butterfly, the impression is created that these nature elements are closely connected with the farmer's area of work. As a result, the advertisement falls within the scope of the Environmental Advertising Code (MRC). The circumstance that FrieslandCampina did not intend to make this environmental claim is not relevant. According to the Advertising Code Committee (*RCC*), in contrast to what the defence alleges, it is not sufficiently clear that the advertisement only intends to refer to the Dutch origin of the yoghurt.

In addition, the Advertising Code Committee (*RCC*) points out that FrieslandCampina repeatedly emphasised its extra efforts in the field of nature, animals and climate, and that the Sterke Start yoghurt complies with the On the Way to PlanetProof sustainability label. This cannot help FrieslandCampina according to the Advertising Code Committee (*RCC*), since that label appears on the packaging but not in the advertisement.

The Advertising Code Committee (*RCC*) believes that FrieslandCampina's efforts regarding nature, animals and climate can have positive effects to some extent, but on the other hand, any environmental claim must be demonstrably correct according to Article 2 MRC.

According to the Advertising Code Committee (*RCC*), there is no evidence that FrieslandCampina has taken measures that have a positive impact on the welfare of the black-tailed godwit. On the contrary, the complainant has substantiated that only a very few dairy cattle farms are actively stimulating biodiversity (temporary flooding of grasslands, adjusted mowing regime, no ryegrass). In conclusion, the images constitute a misleading environmental claim within the meaning of Article 2 MRC.

59.3 Discussion

An environmental claim is a claim that suggests that a product or service has a positive, less or no impact on the environment. This may include the environment in general or certain aspects of the environment, such as air, water, soil, ecosystems, biodiversity or climate.

This ACC decision was issued under the regime of the Environmental Advertising Code. On 1 February 2023, the MRC was replaced by the Sustainability Advertising Code (CDR). This new code incorporates the rationale of previous rulings on sustainability (the umbrella term for environmental or ethical claims).

This decision shows that the term 'environmental claim' needs to be interpreted very broadly. Even a picture of a black-tailed godwit can be an environmental claim in a certain context. The ACC substantiates in detail which circumstances can play a role. In other words, the context is relevant. So in this particular case, what matters is that the black-tailed godwit symbolises the negative consequences of intensive agriculture. After all, agriculture has been identified as the cause of the sharp decline in the population of the black-tailed godwit. It is also relevant that the black-tailed godwit is depicted in the farmer's 'working area'. This gives the impression that the population of the black-tailed godwit has been considered during the production of the yoghurt. However, FrieslandCampina has not been able to prove the latter.

FrieslandCampina argues that the image is meant to illustrate the Dutch grasslands. The product originates from Dutch soil and the black-tailed godwit is the national bird of the Netherlands. Therefore, the black-tailed godwit would not be an environmental claim. But the intention is irrelevant; what matters is the consumer interpretation.

An environmental claim is never far away. A nice picture of the national meadow bird of the Netherlands can already qualify as an environmental claim. This will probably not have occurred to the person who made the advertisement. Would it have helped if the Frisian symbol, a lapwing, had been depicted? Probably not, because that meadow bird is also under threat.

Does this mean that nothing is permitted anymore, that everything is an environmental claim? The ACC explicitly mentions FrieslandCampina's efforts because they are, among other things, authorised to use the On the Way to PlanetProof label on their dairy products. But relevant is, first of all, that the label must be included in the advertisement itself in order to make a general sustainability claim more specific. And secondly, the burden of proof remains: the advertiser must be able to substantiate the consumer interpretation of the sustainability claim with valid evidence.

Ebba Hoogenraad
partner in food law and advertising law at Hoogenraad & Haak, Amsterdam

CHAPTER 60

Warsteiner – Alkoholfrei ≠ alcohol free ≠ 0.0% alcohol

Annotated by Lisanne Steenbergen

Advertising Code Committee 9 July 2019, 2019/00152 (*Warsteiner Alkoholfrei*)

Article 2 Dutch Advertising Code (NRC), Article 7 Regulation (EC) 1169/2011 (FIC Regulation), Article 7c Warenwetbesluit Gereserveerde aanduidingen (Dutch Commodities Act Decree on Reserved Indications), Regulation 764/2008 (old), alcohol advertising, alcohol-free beer, labelling, mutual recognition.

60.1 Summary

The label of the German beer brand Warsteiner contains the word ‘Alkoholfrei’ (alcohol free). According to the complainant, this is misleading, as the beer gives the impression of being ‘alcohol free’ while it contains 0.5% alcohol by volume. Warsteiner claims to have done nothing wrong; the label is not misleading. The beer contains between 0.1% and 0.5% alcohol and can therefore be called ‘Alkoholfrei’ according to German law. Under Dutch law the designation ‘alcohol free’ also does not mean that the product is completely free of alcohol. Dutch legislation allows products with an ‘alcoholvrij’ (alcohol free) label to contain a maximum of 0.1% alcohol. Warsteiner invoked the principle of mutual recognition; the beer is lawfully on the market in Germany and should therefore also be allowed on the Dutch market. The chairman of the Advertising Code Committee (ACC) agreed; the beer is permitted on the Dutch market under the designation of ‘Alkoholfrei’ and even indicated as ‘alcoholvrij’ (even though this term has a different meaning in the Netherlands: 0.1% alcohol). Despite a successful invocation of the principle of mutual recognition, the chairman finds it necessary to assess whether the consumer is misled by the beer’s effective alcohol content by volume. The label reads ‘Alc. <0.5% vol’ to indicate that the beer contains less than 0.5% but more than 0.1% alcohol. According to the chairman, this is not clear enough. The average consumer might interpret this indication, in combination with the wording ‘Alkoholfrei’ and ‘alcoholvrij’ elsewhere on the label, as an indication that the product does not contain any alcohol at all. Warsteiner receives a recommendation to the extent it is necessary.

 | HTTPS://DOI.ORG/10.1163/9789004751699-063

60.2 Complaint and main considerations

'Alkoholfrei' is written in clearly legible text on the front of a bottle of beer from the German beer brand Warsteiner. The complainant notes that the beer contains 0.5% alcohol by volume and believes the label is 'hugely misleading' and even dangerous to his health.

Warsteiner responds to the complaint by saying that the label is not misleading. The beer contains less than 0.5% alcohol by volume. And according to German law a product is permitted to be called 'Alkoholfrei' if it contains up to 0.5% alcohol by volume. The rationale behind this rule would be that an alcohol content of up to 0.5% has no (for alcoholic beverages) typical effect on the (health of) persons consuming it. Dutch legislation allows the use of the designation 'alcoholvrij' (alcohol free) in the case of beer containing no more than 0.1% alcohol by volume. According to Warsteiner, that provision does not prevent the use of the designation 'Alkoholfrei' for beer containing less than 0.5% alcohol on the Dutch market. Warsteiner invokes the principles of mutual recognition and free movement of goods, on the basis of which a product lawfully marketed in another EU Member State (Germany) should in principle also be allowed on the market of another EU Member State (the Netherlands). In addition, Warsteiner has deliberately chosen the German-language designation 'Alkoholfrei', to alert (Dutch) consumers to the fact that the beer comes from Germany. The alcohol percentage (<0.5%) is also clearly mentioned on the label, which precludes the possibility of misleading the consumer.

The chairman asks Warsteiner to share the whole label and also receives a print set of the sales materials for Warsteiner Alkoholfrei. These materials, intended for the Dutch market, state in Dutch: 'alcoholvrij bier' (alcohol-free beer) . The chairman asks Warsteiner two questions: i) why was the decision made to place 'alcoholvrij' on the label instead of 'Alkoholfrei', and ii) why is the indication 'Alc. <0.5% vol' used instead of a more specific indication of the effective alcohol volume?

The first question is not really answered by Warsteiner. Warsteiner repeats that the beer drinker will understand that the product is German beer and that this beer is permitted to be called 'Alkoholfrei' as it contains less than 0.5% alcohol. Warsteiner also notes that in the Netherlands the indication 'less than 0.5%' will suffice; a more specific indication is not required.

The chairman notes that the complaint is directed against the label of Warsteiner Alkoholfrei Pilsener. The information requested shows that on the back of the label, in the Dutch-language section, the indication 'alcoholvrij bier' has been used. As Warsteiner claims the beer contains less than 0.5% alcohol, but more than 0.1%, the beer is not permitted to be called 'alcoholvrij' according to Dutch law. Article 7c of the Commodities Act Decree on Reserved Indications allows the use of the indication 'alcoholvrij' only for beer that contains a maximum of 0.1% (but no more than 0.1%) alcohol.

This case involves a special circumstance, which is that the beer is produced in Germany and marketed there as 'Alkoholfrei' in accordance with German law. In Germany, the term 'Alkoholfrei' may be used for beer containing up to 0.5% alcohol. Warsteiner's reliance on the principle of mutual recognition has proved successful. The sale of the German beer with the designation 'Alkoholfrei' in the Netherlands does not conflict with the Dutch rules on the use of the designation 'alcoholvrij' (alcohol free). The premise is that the use of 'Alkoholfrei' is lawful, therefore it must be assumed that the Dutch information including the indication 'alcoholvrij bier' is also in accordance with the law.

On the basis of mutual recognition, everything seems to be fine for Warsteiner, but the chairman is still in two minds as to whether the consumer is misled by the effective alcoholic strength by volume. Using the standard legal framework of the Court of Justice rulings in Gut Springenheide, Commission v Germany and Teekanne, the chairman concludes that the label is misleading.[480]

The chairman reasons that the average consumer, when faced with the indications 'Alkoholfrei' and 'alcoholvrij', is likely to assume that no alcohol is present in the drink at all. This corresponds with consumers expectations, given that the average consumer would expect the same from products labelled 'free of [ingredient X]'. Yet, according to the chairman, this is not true in all cases. The legislator has, for example, stipulated that products with nutrition claims such as 'fat free' and 'sugar free' are still permitted to contain a little fat and a little sugar, respectively. A similar situation applies to 'Alkoholfrei' and 'alcoholvrij'. The use of the indications is permitted if the alcohol content is below a certain limit. In other words, a little bit of alcohol is allowed. With this in mind, the chairman believes that the average consumer will not necessarily be misled when confronted with the words 'alkoholfrei' and 'alcoholvrij'. All the circumstances of the case need to be taken into account to determine whether the consumer is misled. In this context it is relevant that there is also beer on the Dutch market that contains no alcohol at all (so-called 0.0 beers). Producers should avoid a situation in which consumers confuse an alkoholfrei/alcoholvrij beer with a 0.0 beer.

To inform consumers about this, Warsteiner has included 'Alc. <0.5% vol' on the back of the bottle, above the nutrition table. According to the chairman, this single statement is insufficiently clear to communicate that the beer contains alcohol to some extent. The average consumer could also interpret 'Alc. <0.5% vol' in combination with 'Alkoholfrei' and 'alcoholvrij' as indications of a 0.0 beer. The chairman considers the effective alcohol content essential information that needs to be provided to the consumer. The chairman takes into account that some consumers will not want to consume alcohol at all for whatever reason (pregnancy, breastfeeding, religion, other). Warsteiner will therefore have to provide clearer information about the effective alcohol percentage.

---

480 CJEU 16 July 1998, C-210/96, ECLI:EU:C:1998:369 (Gut Springenheide), CJEU 26 October 1995, C-51/94, ECLI:EU:C:1995:352 (Commission v Germany), ECJ 4 June 2015, C-195/14, ECLI:EU:C:2015:361 (*Teekanne*). For Teekanne see Chapter 11 of this book.

To conclude: Warsteiner's label does not comply with the requirement of Article 7 FIC Regulation, which states that food information must not be misleading with regard to the nature of the foodstuff. Therefore, the advertisement violates Article 2 of the Dutch Advertising Code. The chairman issues Warsteiner with a recommendation to the extent necessary.

60.3 Discussion

This decision shows how mutual recognition works in practice. *Warsteiner Alkoholfrei* complies with German law and can therefore be sold in Germany. The principle of mutual recognition and the principle of free movement of goods allows that the same product should also be able to be marketed in other EU Member States. After all, the product is lawfully on the market in one EU Member State (Germany), so then the product name must also be allowed in another EU Member State (in this case, the Netherlands). As a result the special situation arises that on the basis of mutual recognition, a German beer with a German term Alkoholfrei on the label is lawfully on the Dutch market. The fact that the packaging material also contained the Dutch term 'alcoholvrij', with a different meaning than 'Alkoholfrei', complicates matters. How will the average consumer perceive the information? Due to the principle of mutual recognition, the chairman does not get to assess this question. Because lawful on the German market equals lawful in the Netherlands.

At the time of this decision, Regulation 764/2008 was still in force; at the moment we have Mutual Recognition Regulation 2019/515. The latter regulation aims to speed up the free movement of goods not subject to harmonised rules. To invoke mutual recognition, the producer of the goods can draw up a clear mutual recognition declaration that allows the producer to easily demonstrate that their product is lawfully on the market in another EU Member State. The annex to the Mutual Recognition Regulation 2019/515 contains a handy outline for such a declaration containing all the required information.

In this case, the chairman concluded that German beer was lawfully on the Dutch market on the basis of mutual recognition. Is that the end of the story? No, because the principle of mutual recognition does not say anything about the possible deception of the Dutch average consumer regarding the alcohol content by volume, therefore the CJEU Gut Springenheide consumer standard is applied.

The chairman notes that when confronted with the indications 'Alkoholfrei' and 'alcoholvrij', the average consumer will initially think that the beer is completely free of alcohol. This is in line with the average consumer's usual thoughts regarding 'free from' claims. This is the main rule, but exceptions are also possible, for example for nutrition claims such as 'fat free' and 'sugar free'. The same applies to 'alcoholvrij': the product contains a small amount of alcohol. In the Netherlands alcohol free means 0.1% and in Germany 0.5%. The chairman assumes that the average consumer, when seeing 'Alkoholfrei' and 'alcoholvrij', is aware that the product may contain a small amount of alcohol. The question is whether this knowledge should indeed be expected of the average consumer. Linguistically, 'free of' claims are very clear: the product should contain nothing, zero, 0.0%. That the legislator provides for exceptions by means of specialist food law is probably not common knowledge. But fortunately, there is CJEU Teekanne: from this ruling follows that we have to look at the entire label to determine whether the consumer is being misled. The meaning of 'Alkoholfrei' and 'alcoholvrij' (in different EU Member States) may or may not be known to the average consumer, and the other indications on the label also need to be taken into account. This also applies to the indication 'Alc. <0.5% vol'. According to Warsteiner it is crystal clear. The chairman holds a different view. 'Alc. 0.5% vol' could mean various things in the eyes of the average consumer: for example, that the beer contains 0.4%, or for example 0.2%, and it could also mean that it doesn't contain a drop of alcohol (0.0%).

Especially for a specific audience that is looking for a product without any alcohol, it is important to know for sure what the alcohol content of the product is. This qualifies the alcohol percentage on this label as essential information that should be communicated in clear language. Interestingly, the chairman considered that the specific combination of the elements 'Alc. <0.5% vol' and 'Alkoholfrei'/'alcoholvrij' is confusing for consumers. The decision could also have gone the other way for Warsteiner. Earlier, the chairman concluded that the average consumer is aware that 'free from' claims are not always completely free of nutrient X in practice. The average consumer is also assumed to be aware of the 0.0 variants on the market. Yet things go wrong for Warsteiner, and a recommendation follows to the extent necessary with regard to the alcoholic strength by volume. All in all, this decision could also be considered a victory for Warsteiner: a successful mutual recognition appeal! This is something we don't often see in ACC rulings.

Another encore: in a different case, where Warsteiner largely displayed the German word 'Alkoholfrei' in a television commercial for its Alkohol frei beer, the chairman also ruled that the principle of mutual recognition can be invoked, i.e. the beer can lawfully be sold on the Dutch market. However, on the grounds of deception, it still goes wrong for Warsteiner; the chairman considered that consumers should be prevented from confusing German Warsteiner beer with beer that is actually totally alcohol free (0.0). In this case Warsteiner also requested a so-called 'eyebrow decision' (conclusion that the advertisement is misleading without a recommendation). That request is denied because the chairman finds that there has been a case of repeated deception.[481]

Lisanne Steenbergen
food law and advertising lawyer at Hoogenraad & Haak, Amsterdam

481 Advertising Code Committee (ACC) 1 March 2021, file 2021/00043 (*Warsteiner Alkoholfrei*)

www.ingramcontent.com/pod-product-compliance
Ingram Content Group UK Ltd.
Pitfield, Milton Keynes, MK11 3LW, UK
UKHW051022091026
15624UKWH00012B/623